AF892231

Methods in Molecular Biology

Series Editor
John M. Walker
School of Life and Medical Sciences
University of Hertfordshire
Hatfield, Hertfordshire, UK

For further volumes:
http://www.springer.com/series/7651

For over 35 years, biological scientists have come to rely on the research protocols and methodologies in the critically acclaimed *Methods in Molecular Biology* series. The series was the first to introduce the step-by-step protocols approach that has become the standard in all biomedical protocol publishing. Each protocol is provided in readily-reproducible step-by-step fashion, opening with an introductory overview, a list of the materials and reagents needed to complete the experiment, and followed by a detailed procedure that is supported with a helpful notes section offering tips and tricks of the trade as well as troubleshooting advice. These hallmark features were introduced by series editor Dr. John Walker and constitute the key ingredient in each and every volume of the *Methods in Molecular Biology* series. Tested and trusted, comprehensive and reliable, all protocols from the series are indexed in PubMed.

Multienzymatic Assemblies

Methods and Protocols

Edited by

Haralambos Stamatis

Department of Biological Applications and Technologies, University of Ioannina, Ioannina, Greece

Humana Press

Editor
Haralambos Stamatis
Department of Biological Applications
and Technologies
University of Ioannina
Ioannina, Greece

ISSN 1064-3745 ISSN 1940-6029 (electronic)
Methods in Molecular Biology
ISBN 978-1-0716-2268-1 ISBN 978-1-0716-2269-8 (eBook)
https://doi.org/10.1007/978-1-0716-2269-8

© The Editor(s) (if applicable) and The Author(s), under exclusive license to Springer Science+Business Media, LLC, part of Springer Nature 2022
This work is subject to copyright. All rights are solely and exclusively licensed by the Publisher, whether the whole or part of the material is concerned, specifically the rights of translation, reprinting, reuse of illustrations, recitation, broadcasting, reproduction on microfilms or in any other physical way, and transmission or information storage and retrieval, electronic adaptation, computer software, or by similar or dissimilar methodology now known or hereafter developed.
The use of general descriptive names, registered names, trademarks, service marks, etc. in this publication does not imply, even in the absence of a specific statement, that such names are exempt from the relevant protective laws and regulations and therefore free for general use.
The publisher, the authors and the editors are safe to assume that the advice and information in this book are believed to be true and accurate at the date of publication. Neither the publisher nor the authors or the editors give a warranty, expressed or implied, with respect to the material contained herein or for any errors or omissions that may have been made. The publisher remains neutral with regard to jurisdictional claims in published maps and institutional affiliations.

This Humana imprint is published by the registered company Springer Science+Business Media, LLC, part of Springer Nature.
The registered company address is: 1 New York Plaza, New York, NY 10004, U.S.A.

Preface

This book provides comprehensive methods and protocols about multienzymatic assemblies.

The chapters are grouped by main topic, starting with methodologies describing expression and purification of multienzyme assemblies, state-of-the-art techniques for the structural and biochemical characterization, followed by examples of the immobilization of multienzymatic assemblies onto natural biopolymers and nano-scaffolds as well as cascade biocatalytic transformations mediated by free or immobilized multienzymatic systems.

Written in the format of the highly successful Methods in Molecular Biology series, each chapter includes an introduction to the topic, lists necessary materials and reagents, and includes tips on troubleshooting and known pitfalls and step-by-step, readily reproducible laboratory protocols.

Authoritative and cutting-edge, Multienzymatic Assemblies: Methods and Protocols aims to be a useful practical guide for researchers seeking technical methods to current aspects of multienzyme systems and related applications.

Ioannina, Greece *Haralambos Stamatis*

Contents

Contributors

APOSTOLOS ALISSANDRATOS • *Research School of Chemistry, Australian National University, Canberra, ACT, Australia; CSIRO Synthetic Biology Future Science Platform, Australian National University, Canberra, ACT, Australia*

SONGON AN • *Department of Chemistry and Biochemistry, University of Maryland Baltimore County (UMBC), Baltimore, MD, USA; Program in Oncology, Marlene and Stewart Greenebaum Comprehensive Cancer Center, University of Maryland, Baltimore, MD, USA*

IO ANTONOPOULOU • *Biochemical Process Engineering, Department of Civil, Environmental and Natural Resources Engineering, Luleå University of Technology, Luleå, Sweden*

ASKIN S. ASLAN-ÜZEL • *Institute of Biochemistry, University of Greifswald, Greifswald, Germany*

CHRISTOFFEL P. S. BADENHORST • *Institute of Biochemistry, University of Greifswald, Greifswald, Germany*

HADI BAHARIFAR • *Department of Medical Nanotechnology, Applied Biophotonics Research Center, Science and Research Branch, Islamic Azad University, Tehran, Iran*

NATALIA BARTZOKA • *Research School of Chemistry, Australian National University, Canberra, ACT, Australia*

SHIVAPRASAD BHUVANENDRAN • *Center for Genetic Medicine Research, Children's National Hospital, Washington, DC, USA*

UWE T. BORNSCHEUER • *Institute of Biochemistry, University of Greifswald, Greifswald, Germany*

JOYCE C. BREGER • *Center for Bio/Molecular Science and Engineering, Code 6900, Washington, DC, USA*

LJUBICA CALDOVIC • *Center for Genetic Medicine Research, Children's National Hospital, Washington, DC, USA; Department of Genomics and Precision Medicine, School of Medicine and Health Sciences, The George Washington University, Washington, DC, USA*

NIKOLAOS CHALMPES • *Department of Materials Science & Engineering, University of Ioannina, Ioannina, Greece*

ALEXANDRA V. CHATZIKONSTANTINOU • *Biotechnology Laboratory, Department of Biological Applications and Technologies, University of Ioannina, Ioannina, Greece*

WILFRED CHEN • *Department of Chemical and Biomolecular Engineering, University of Delaware, Newark, DE, USA*

XI CHEN • *Department of Chemistry, The Chinese University of Hong Kong, Hong Kong SAR, China*

CHRISTOS L. CHOCHOS • *Institute of Chemical Biology, National Hellenic Research Foundation, Athens, Greece*

PAUL CHRISTAKOPOULOS • *Biochemical Process Engineering, Department of Civil, Environmental and Natural Resources Engineering, Luleå University of Technology, Luleå, Sweden*

JAMES CHRISTENSON • *Bethel University, St. Paul, MN, USA*

PABLO DOMÍNGUEZ DE MARÍA • *Sustainable Momentum, SL, Las Palmas de Gran Canaria, Canary Islands, Spain*

GREGORY A. ELLIS • *Center for Bio/Molecular Science and Engineering, Code 6900, Washington, DC, USA*
JINGLIN FU • *Department of Chemistry, Rutgers University–Camden, Camden, NJ, USA; Center for Computational and Integrative Biology, Rutgers University–Camden, Camden, NJ, USA*
YUNZHENSHAN GAO • *State Key Laboratory of Chemical Resources Engineering, Beijing University of Chemical Technology, Beijing, China*
ARCHONTOULA GIANNAKOPOULOU • *Biotechnology Laboratory, Department of Biological Applications and Technologies, University of Ioannina, Ioannina, Greece*
ELENA GKANTZOU • *Laboratory of Biotechnology, Department of Biological Applications and Technologies, University of Ioannina, Ioannina, Greece*
DIMITRIOS GOURNIS • *Department of Materials Science & Engineering, University of Ioannina, Ioannina, Greece*
NADIA GUAJARDO • *Programa Institucional de Fomento a la Investigación, Desarrollo e Innovación, Universidad Tecnológica Metropolitana, Santiago, Chile*
DIMITRA HOUHOULA • *Department of Food Science and Technology, Faculty of Food Sciences, University of West Attica, Athens, Greece*
ELISAVET IOANNOU • *Laboratory of Enzyme Technology, Department of Biotechnology, School of Applied Biology and Biotechnology, Agricultural University of Athens, Athens, Greece*
JYOTI JAISWAL • *Center for Genetic Medicine Research, Children's National Hospital, Washington, DC, USA; Department of Genomics and Precision Medicine, School of Medicine and Health Sciences, The George Washington University,, Washington, DC, USA*
ANKARAO KALLURI • *Department of Material Science, University of Connecticut, Storrs, CT, USA*
ERIN L. KENNEDY • *Department of Chemistry and Biochemistry, University of Maryland Baltimore County (UMBC), Baltimore, MD, USA*
KAMYAR KHOSHNEVISAN • *Medical Nanotechnology and Tissue Engineering Research Center, Shahid Beheshti University of Medical Sciences, Tehran, Iran; Department of Tissue Engineering and Applied Cell Sciences, School of Advanced Technologies in Medicine, Shahid Beheshti University of Medical Sciences, Tehran, Iran*
SVYATOSLAV KONDRAT • *Institute of Physical Chemistry, Polish Academy of Sciences, Warsaw, Poland; Max-Planck-Institut für Intelligente Systeme, Stuttgart, Germany; IV. Institut für Theoretische Physik, Universität Stuttgart, Stuttgart, Germany*
CHALLA VIJAYA KUMAR • *Department of Chemistry, University of Connecticut, Storrs, CT, USA; Department of Material Science, University of Connecticut, Storrs, CT, USA; Department of Molecular and Cellular Biology, University of Connecticut, Storrs, CT, USA*
MINJOUNG KYOUNG • *Department of Chemistry and Biochemistry, University of Maryland Baltimore County (UMBC), Baltimore, MD, USA; Program in Oncology, Marlene and Stewart Greenebaum Comprehensive Cancer Center, University of Maryland, Baltimore, MD, USA*
NIKOLAOS E. LABROU • *Laboratory of Enzyme Technology, Department of Biotechnology, School of Applied Biology and Biotechnology, Agricultural University of Athens, Athens, Greece*
ERIC VON LIERES • *Forschungszentrum Jülich, IBG-1: Biotechnology, Jülich, Germany*
MIAO LIU • *Department of Chemistry, The Chinese University of Hong Kong, Hong Kong SAR, China*

THOMAS D. LOAN • *Research School of Chemistry, Australian National University, Canberra, ACT, Australia*

HASSAN MALEKI • *Nano Drug Delivery Research Center, Health Technology Institute, Kermanshah University of Medical Sciences, Kermanshah, Iran*

MANSI MALHOTRA • *Department of Chemistry, University of Connecticut, Storrs, CT, USA*

IGOR L. MEDINTZ • *Center for Bio/Molecular Science and Engineering, Code 6900, Washington, DC, USA*

SUNG WON OH • *Department of Chemistry, Rutgers University–Camden, Camden, NJ, USA; Center for Computational and Integrative Biology, Rutgers University–Camden, Camden, NJ, USA*

HIDEKI ONAGI • *Research School of Chemistry, Australian National University, Canberra, ACT, Australia*

ANASTASSIOS C. PAPAGEORGIOU • *Turku Bioscience Centre, University of Turku and Åbo Akademi University, Turku, Finland*

PRAKASH PARAJULI • *Department of Chemistry and Biochemistry, University of Maryland Baltimore County (UMBC), Baltimore, MD, USA*

MICHAELA PATILA • *Biotechnology Laboratory, Department of Biological Applications and Technologies, University of Ioannina, Ioannina, Greece*

IOANNIS V. PAVLIDIS • *Department of Chemistry, University of Crete, Heraklion, Greece*

ANGELIKI C. POLYDERA • *Biotechnology Laboratory, Department of Biological Applications and Technologies, University of Ioannina, Ioannina, Greece*

MARYAM RAEESZADEH-SARMAZDEH • *Department of Chemical and Materials Engineering, University of Nevada, Reno, NV, USA*

ULRIKA ROVA • *Biochemical Process Engineering, Department of Civil, Environmental and Natural Resources Engineering, Luleå University of Technology, Luleå, Sweden*

EVA D. SCHUITEN • *Institute of Biochemistry, University of Greifswald, Greifswald, Germany*

ANASTASIA SKONTA • *Laboratory of Biotechnology, Department of Biological Applications and Technologies, University of Ioannina, Ioannina, Greece*

HARALAMBOS STAMATIS • *Laboratory of Biotechnology, Department of Biological Applications and Technologies, University of Ioannina, Ioannina, Greece*

QING SUN • *Department of Chemical Engineering, Texas A&M University, College Station, TX, USA*

KIMIHIRO SUSUMU • *Optical Sciences Division, Code 5611, U.S. Naval Research Laboratory, Washington, DC, USA; Jacobs Corporation, Hanover, MD, USA*

QINGYUN TANG • *Institute of Biochemistry, University of Greifswald, Greifswald, Germany*

SHEN-LONG TSAI • *Department of Chemical Engineering, National Taiwan University of Science and Technology, Taipei City, Taiwan*

ALIKI TSAKNI • *Department of Food Science and Technology, Faculty of Food Sciences, University of West Attica, Athens, Greece*

ANDREAS-GEORGIOS VASIOS • *Laboratory of Biotechnology, Department of Biological Applications and Technologies, University of Ioannina, Ioannina, Greece*

SCOTT A. WALPER • *Center for Bio/Molecular Science and Engineering, Code 6900, Washington, DC, USA*

ZHICHENG WANG • *Department of Chemistry, Rutgers University–Camden, Camden, NJ, USA; Center for Computational and Integrative Biology, Rutgers University–Camden, Camden, NJ, USA*

QIXIN WEI • *Department of Chemistry, The Chinese University of Hong Kong, Hong Kong SAR, China*

JIANG XIA • *Department of Chemistry, The Chinese University of Hong Kong, Hong Kong SAR, China*

YOU YONG • *Department of Chemical Engineering, Tsinghua University, Beijing, China*

YIFEI ZHANG • *State Key Laboratory of Chemical Resources Engineering, Beijing University of Chemical Technology, Beijing, China; Beijing Advanced Innovation Center for Soft Matter Science and Engineering, Beijing University of Chemical Technology, Beijing, China*

Chapter 1

Recombinant Expression and Purification of Large Bacterial Multienzyme Assemblies for Biosynthetic Processes

James Christenson

Abstract

Many biosynthetic transformations have strict spatial and temporal requirements that necessitate the physical association of multiple enzymes for proper function. Here, we describe protocols for obtaining large multienzyme assemblies (>500 kDa) by recombinant expression in *Escherichia coli*. We focus on assemblies from stand-alone enzymes joined by intermolecular forces rather than multiple catalytic domains from a single polypeptide chain. Details are given for strategies to optimize protein expression and to design a multi-affinity tag purification scheme for large multienzyme assemblies. These insights are drawn from our study of bacterial hydrocarbon biosynthesis.

Key words Large multienzyme assemblies, Large enzyme complex, Recombinant expression, Protein purification, Hydrocarbon biosynthesis, OleBCD

1 Introduction

Living cells continually balance thousands of chemical transformations to sustain homeostasis. In many cases, the spatial and temporal regulation of these transformations is accomplished by physically grouping catalysts into multienzyme assemblies. The mammalian pyruvate dehydrogenase complex is a classic example of a high-molecular weight assembly [1]. In bacterial biosynthetic pathways, multienzyme assemblies allow greater metabolic control and are commonly utilized when pathway intermediates are unstable, hydrophobic, or toxic to the cell. Non-covalent association of stand-alone catalysts overcomes these biosynthetic challenges; however, such association often complicates research efforts, particularly when assemblies are large (>500 kDa). This chapter specifically discusses recombinant expression and purification techniques for large bacterial multienzyme assemblies beginning with transformed *E. coli* on an agar plate and ending with isolated multienzyme assemblies.

Haralambos Stamatis (ed.), *Multienzymatic Assemblies: Methods and Protocols*,
Methods in Molecular Biology, vol. 2487, https://doi.org/10.1007/978-1-0716-2269-8_1,
© The Author(s), under exclusive license to Springer Science+Business Media, LLC, part of Springer Nature 2022

There are two primary methods to access large multienzyme assemblies for study: purification from a native host and purification from a recombinant host. Utilizing the native bacterial host is generally preferable because it greatly increases the chance of generating physiologically relevant assemblies; however, this is not practical or not possible in many situations. Native bacteria may express assemblies in very low quantities making characterization difficult. Purification procedures are often more complex because assemblies lack recombinantly added purification aids (6× His-tag, Strep tag, etc.). Additionally, the vast majority of bacteria cannot be cultivated in a lab or require strict biosafety guidelines that hinder routine cultivation [2]. The opposing strategy—recombinant expression—overcomes these challenges by introducing the genes into an alternate bacterial host. However, this process requires prior knowledge of the enzymes involved in the assembly, and great precautions must be taken to ensure that assembly characteristics are not simply an artifact of production.

Our insights come from investigating bacterial hydrocarbon biosynthesis as a potential source of biofuels. Specifically, we studied long-chain hydrocarbon biosynthesis (Ole pathway) where bacterial cells assemble multiple copies of three enzymes (OleB, OleC, and OleD) into complexes larger than 1 MDa [3]. The assemblies couple two fatty acids head-to-head, ultimately forming a long-chain alkene (olefin) at the juncture [4]. The multienzyme assemblies are beneficial for sequestering a toxic beta-lactone intermediate and efficiently transferring hydrophobic intermediates that are insoluble in the aqueous cytosol [3]. Figure 1 shows an overview of the methods detailed in this chapter starting from transformed *E. coli*. The procedure is divided into four parts: (1) protein expression, (2) His-tag affinity column, (3) FLAG-tag affinity column, and (4) size exclusion column. The post-purification characterization of complexes by electron microscopy, native gel western blots, and activity assays are not discussed as these are somewhat specific to our system, but this information is published elsewhere [3]. These methods share significant overlap with recombinantly expressed affinity tag purifications of single proteins, but key differences and insights for working with large multienzyme assemblies are provided throughout.

2 Materials

Prepare all solutions with analytical grade chemicals and lab-grade deionized water (purity 18.2 MΩ). Specific vendors are noted for less common items.

Expression Combinations				
1	OleA	His-OleB	OleC-Flag	OleD
2	His-OleA	OleB	OleC-Flag	OleD
3	OleA	His-OleB	OleC	OleD
4		His-OleB	OleC-Flag	OleD
5		OleB	OleC-Flag	His-OleD
6		OleB	OleC-Flag	OleD-His

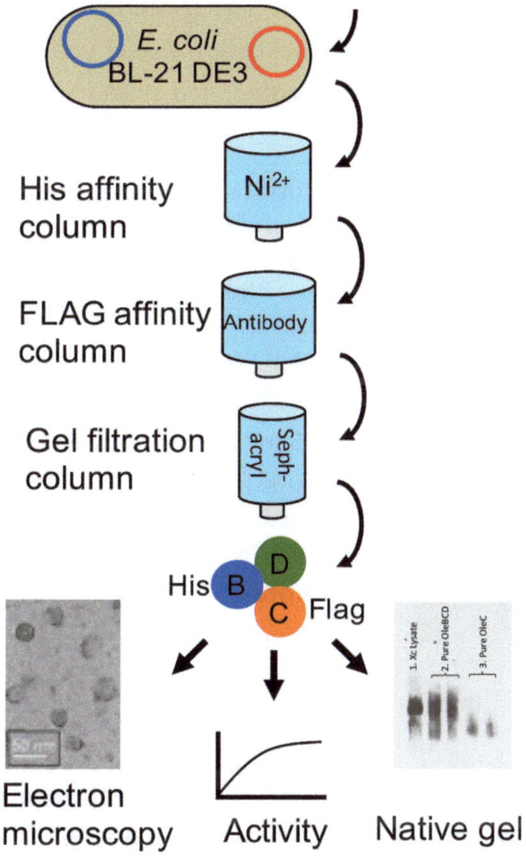

Fig. 1 Purification overview of multienzyme assemblies from recombinant organisms. Highlighted are the four main steps in this procedure: (1) protein expression, (2) His-tag affinity column, (3) FLAG-tag affinity column, (4) size exclusion column. Reproduced from Christenson, 2017 with permission from American Society for Microbiology [3]

2.1 Expression

1. Luria-Bertani agar plate containing *E. coli* colonies transformed with the plasmid(s) encoding the multienzyme assemblies (*see* **Note 1**).
2. 50 mg/mL kanamycin stock solution: Dissolve 500 mg of kanamycin sulfate in 10.0 mL of deionized water.
3. 30 mg/mL streptomycin stock solution: Dissolve 300 mg of streptomycin sulfate in 10.0 mL of deionized water.

4. 100 mM IPTG stock solution: Dissolve 0.238 g IPTG in 10.0 mL of deionized water (*see* **Note 2**).
5. Luria-Bertani (LB) media: Dissolve 25 g of LB powder in 1.0 L of deionized water in a 2-L Erlenmeyer flask. Take 3.0 mL of this and add it to a test tube for an inoculation culture. Cover the Erlenmeyer and the test tube with aluminum foil and autoclave to sterilize.
6. Incubation shaker with holders for 2-L flask and test tubes.
7. Refrigerated shaker with holders for flasks and test tubes (*see* **Note 3**).

2.2 Cell Lysis and His-tag Purification

1. Purification buffer: 200 mM NaCl, 10% v/v glycerol, 20 mM Na_2HPO_4 (pH 7.4). Dissolve 11.7 g of NaCl, 100 mL of glycerol, and 2.84 g of Na_2HPO_4 in 1.0 L of deionized water (*see* **Note 4**).
2. His-column elution buffer: purification buffer containing 500 mM imidazole (pH 7.4). Take 250 mL of purification buffer and add 8.51 g of imidazole (*see* **Note 5**).
3. 15 mL plastic conical tubes and 1.5 mL microcentrifuge tubes.
4. 10 mL glass or plastic pipette and bulb for resuspending pelleted cells.
5. Centrifuge and bottles capable of spinning at 35,000 × *g* (*see* **Note 6**).
6. French press or bacterial sonicator for cell lysis.
7. 10 mL plastic syringe with luer lock tip.
8. 0.45 μm polyether sulfone membrane filter with luer lock.
9. HisTrap HP column (1 mL) or any equivalent resin for binding of 6× His-tagged proteins.
10. Protein purification system: $\text{ÄKTA}_{\text{FPLC}}$ or equivalent system.
11. HisTrap HP storage buffer: 20% v/v ethanol in deionized water. Mix 100 mL of ethanol with 400 mL of deionized water.

2.3 FLAG-tag Purification

1. 15 mL plastic conical tubes and 1.5 mL microcentrifuge tubes.
2. Anti-FLAG M2 Resin (Sigma) or any equivalent resin containing and antibody for DYKDDDK tag binding.
3. Empty glass chromatography column and stopper valve for anti-FLAG resin approximately 10 mL in size.
4. 6 in. of 1/8″ inner diameter tigon tubing to speed column flow (optional).
5. FLAG elution buffer: purification buffer containing 100 μg/mL FLAG peptide. Dissolve 1.0 mg of FLAG peptide in 10 mL of purification buffer from **step 1** of Subheading 2.2 (*see* **Note 7**).

6. FLAG regeneration buffer: 0.10 M glycine adjusted to pH 3.5 using 6.0 M HCl.
7. FLAG storage buffer: 50% v/v glycerol, 10 mM HNa_2PO_4 (pH 7.4), 150 mM NaCl, 0.02% w/v sodium azide. Sodium azide prevents microbial growth, but it is toxic. Adjust the pH to 7.4 using 6.0 M HCl.

2.4 Gel Filtration

1. Gel filtration column: HiPrep 16/60 Sephacryl S-400 HP.
2. Blue Dextran: 50 mg/vial.
3. Gel filtration high molecular weight calibration kit (*see* **Note 8**).
4. 15 mL conical tubes and 1.5 mL microcentrifuge tubes.
5. Amicon 30,000 MWCO spin filter (50 mL conical) for concentrating sample.
6. Protein purification system: $\text{ÄKTA}_{\text{FPLC}}$ or equivalent system.
7. Gel filtration storage buffer: 20% v/v ethanol. Mix 100 mL of ethanol with 400 mL of deionized water.

3 Methods

Complete all procedural steps at room temperature, but keep protein samples on ice or at 4 °C when possible. Exceptions are noted.

SDS-PAGE is key for tracking the success of protein purification. By collecting samples throughout the procedure and visualizing them by SDS-PAGE, we optimized our methods for our specific proteins to create the protocols detailed below. Figure 2 shows a representative gel of a successful purification of hydrocarbon biosynthesis proteins.

3.1 Protein Expression

1. Prepare a starter culture in the test tube with 3.0 mL of LB broth and add 3.0 μL of both the kanamycin and streptomycin antibiotic stock solutions. Final concentrations in the test tube are 0.050 mg/mL and to 0.030 mg/mL, respectively.
2. Pick a single colony from your plate and inoculate the 3.0 mL test tube. Incubate this at 37 °C overnight shaking at 225 rpm. (*see* **Note 1** for discussion on bacterial plate).
3. The next day (~18 h), transfer 1.0 mL of the overnight starter culture into 1.0 L of fresh LB media. Add 1.0 mL of kanamycin and streptomycin stock solutions to maintain selective pressure, and continue growth shaking at 37 °C.
4. When the media reaches an optical density (OD 600 nm) of 0.4–0.6, transfer the flask to an incubating shaker at 16 °C for 30 min (*see* **Note 9**).

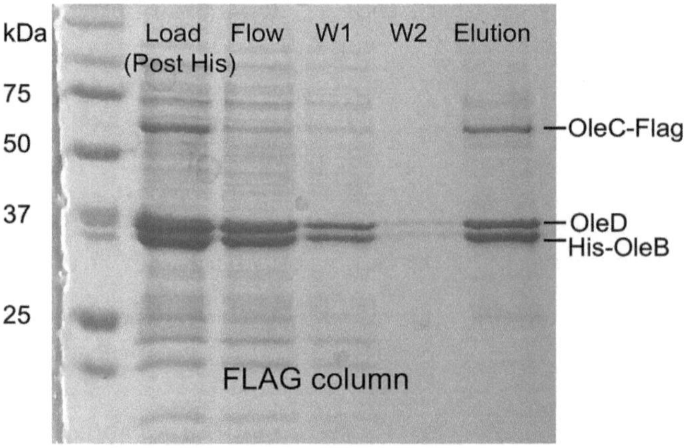

Fig. 2 SDS-PAGE showing the successful purification of the OleBCD multienzyme assembly after His-tag and FLAG-tag columns. Reproduced from Christenson, 2017 with permission from American Society for Microbiology [3]

5. Following 30 min, induce the cells by adding 1.0 mL of 100 mM IPTG (0.10 mM final concentration) to initiate protein production. Continue growth at 16 °C (*see* **Note 10**).
6. After overnight growth at 16 °C (~18 h), pour bacterial cells into a pre-massed centrifuge bottle before pelleting at 5000 × *g* for 10 min in an appropriately balanced centrifuge.
7. Pour off spent media for appropriate biohazard disposal before inverting the container with cells onto a paper towel to drain residual spent media. After a few minutes, mass the container with cells for future yield calculations (*see* **Note 11**).

3.2 His-tag Affinity Column

1. Resuspend the cell pellet in 6.0 mL of purification buffer pre-chilled in a refrigerator (*see* **Note 12**).
2. For lysis, perform three passes in French press according to the manufacturer's recommendation for your equipment (*see* **Note 13**). Specifically, load resuspended cells into chilled metal cylinder before loading the cylinder and plunger onto the press (*see* **Note 14**). Set the pressure to 1500 psi and extrude the cells at a rate of approximately 1.0 mL/min into a tube chilled in an ice bucket. Repeat this step two additional times.
3. Transfer the lysate to a pre-chilled centrifuge tube and spin balanced at 35,000×*g* for 1 h at 4 °C. This separates soluble proteins from large cellular debris.
4. After centrifugation, pour the supernatant, which should contain your expressed assemblies, into a separate conical tube and store on ice, being careful not to disturb the pellet. Save a small sample (<0.1 g) of the pellet on ice for future analysis by SDS-PAGE (*see* **Note 15**).

5. Pass the supernatant through a 0.45 μm pore filter to remove any residual large cellular debris (*see* **Note 16**).
6. Pass the filtered supernatant over the clean nickel affinity resin according to the manufacturer's instructions to achieve binding of 6× His-tagged assemblies to the resin. When using the 1.0 mL HisTrap system, flow supernatant at a rate of 1.0 mL/min or less (*see* **Note 17**).
7. Once the sample is bound, wash out contaminant proteins. Flow 5 column volumes (CV) of a solution containing 20% elution buffer and 80% purification buffer (1.0 mL/min) to dislodge any weakly bound material (*see* **Note 18**).
8. To elute the proteins, flow 10 CV of elution buffer (100%) over the column at 0.50 mL/min. Collect the elution in 1.0 mL fractions (10 tubes total) (*see* **Note 19**).
9. Equilibrate the His-column in 20% ethanol in deionized water for long-term storage.
10. Determine which fractions contain protein by inspecting the ultraviolet trace (280 nm) from your instrument or by performing a Bradford assay.
11. Proceed directly to FLAG affinity column or flash freeze the fractions with highest protein content in liquid nitrogen (*see* **Note 20**).

3.3 FLAG-tag Affinity Column (See Note 21)

1. Fix the empty column with the stopper closed to a ring stand in a refrigerator at 4 °C.
2. Create a slurry of the FLAG affinity resin in the stock bottle and pore this into the column. Use a few mL of purification buffer to rinse any remaining resin from the stock bottle into the column. A 1.0 mL resin bed is sufficient for our experiments (*see* **Note 22**).
3. Allow the buffer to drain before rinsing the column with 5 column volumes (CV) of purification buffer. Repeat this step two additional times to ensure the resin is clean (*see* **Note 23**).
4. Pour the fractions from the His-affinity column containing the multienzyme assemblies onto the column and allow the sample to flow through the column.
5. Wash the column with 5 CV of purification buffer a total of three times to remove non-binding materials.
6. Elute the multienzyme assemblies by adding 5 CV FLAG elution buffer. Repeat this a second time to remove all bound material. Collect the multienzyme assemblies in 1.0 mL fractions (*see* **Note 24**).

7. Add 5 CV of regeneration buffer to the column and allow this to drain (*see* **Note 25**).
8. Wash the column with 10 CV of storage buffer to remove the low pH regeneration buffer. Seal the column ends well and store at 4 °C for future use.

3.4 Gel Filtration Column

1. Attach the gel filtration column to your protein purification system and wash the column with 2 CV of purification buffer at a rate of 1.0 mL/min.
2. Pass the purified assemblies over the gel filtration column at a rate of 1.0 mL/min (*see* **Note 26**).
3. Collect 1 mL fractions after the void volume of the column is exceeded. This is approximately 45 mL for the HiPrep 16/60 Sephacryl S-400 HP (*see* **Note 27**).
4. Determine which fractions contain protein by inspecting the ultraviolet trace (280 nm) from your instrument or by performing SDS-PAGE.
5. Proteins are often dilute coming off the gel filtration column. Use spin concentrators with a 30,000 molecular weight cut off (MWCO) to concentrate multiple desired fractions (*see* **Note 28**).
6. Proceed directly to characterization work or flash freeze the content in liquid nitrogen.

4 Notes

1. In our experience, selecting the appropriate vectors for protein expression is the primary determinant of success when purifying multienzyme assemblies from recombinant sources. However, every multienzyme assembly is different and will likely require the optimization of any standard protocol. Below is a description of our system with principles that guided our decisions. Our long-chain hydrocarbon biosynthesis pathway contained four genes: *oleA*, *oleB*, *oleC*, and *oleD*. All four co-localized genes are found in diverse clades of the bacterial kingdom [5]. We chose to work with the genes from *Xanthomonas campestris* for two reasons. First, it is a gram-negative bacterium and therefore likely compatible for recombinant expression in *E. coli*. Second, we had prior experience in the lab with the organism. All genes were codon optimized for *E. coli*. We utilized two pET duet vectors, each capable of expressing two genes. One 6× His-tag and one FLAG tag were always added to facilitate purification. Prior experience with individually purifying the Ole proteins guided tag location selection, but we tried multiple permutations and used SDS-

PAGE to determine the presence or absence of the four proteins (Fig. 2). After multiple attempts, it became clear that the assemblies did not include the first pathway enzyme, OleA, and we dropped it from subsequent trials. We tried more tag permutations on the three remaining proteins and were most successful with an N-terminal 6× His-tag on *oleB* expressed alone in pET28b+ vector and a FLAG tag on the C-terminus of *oleC* expressed in tandem with *oleD* from a pCDF duet vector. Three overall takeaways are listed below.

(a) Multiple affinity tags are extremely beneficial to ensuring a physiologically relevant complex. If one of the proteins in an assembly spuriously associates with host cell proteins, the second tag purification step can eliminate these erroneous assemblies. The non-physiological association can easily be tracked by SDS-PAGE by attempting two purifications and reversing the order of the affinity column steps. For example, starting with the His-affinity column reveals proteins associated with the His-tagged protein and spurious associations are likely lost in the FLAG affinity column.

(b) Trial and error are needed to identify correct tag and expression combinations. We are not sure why some of our combinations failed, but it is likely that the complex could not form with tags in certain locations.

(c) If we were to start the project again today, we would attempt to insert the whole four-gene cluster directly into a plasmid for recombinant expression. The greatest challenge in recombinant expression of multienzyme assemblies is ensuring that the resulting assembly is physiologically relevant. By utilizing native promoters and ribosome binding sites in the cluster, the levels of each individual protein have a greater chance of being correctly balanced in the host system. Multiple tags can still be inserted either during DNA synthesis or by standard molecular biology techniques.

2. IPTG is light sensitive. Wrap in aluminum foil and store frozen when not in use. The stock solution can be kept frozen and covered for months.
3. A refrigerated shaker is not required, but it may be helpful. Protein expression is a function of temperature and some multienzyme assemblies may be expressed better below ambient temperature.
4. The pH will be too high (basic) with Na_2HPO_4. Use a pH meter and 6.0 M HCl to achieve the correct pH by dropwise addition of the acid to the stirring buffer.

5. Imidazole is a base. The solution's pH must be adjusted to 7.4 again as in **Note 4**.
6. A lower speed centrifuge will work. However, protein purifications will be less pure as larger cellular debris will not be pelleted and removed prior to column purifications.
7. The sequence of the flag peptide is DYKDDDK, and this will outcompete the tagged protein to allow elution. This oligopeptide can be synthesized by many university core facilities or ordered from a manufacturer. We avoided other elution procedures that required different pH or chelating metals as they could disrupt our multienzyme assemblies.
8. The highest commercially available molecular weight calibration kit only goes up to 669 kDa. For many assemblies this is sufficient, but extrapolation is used for higher assemblies. Calibration of gel filtration assumes proteins are generally spherical in shape. Alternatively, proteins like bacterial ribosomes (2.7 MDa) could be used as higher molecular weight markers.
9. Inoculating the 1.0 L of media to an optical density of 0.5 takes approximately 3 h. Bacterial growth is typically limited by oxygen diffusion into media, so at least 2 L flasks should be used to give sufficient surface area to the 1.0 L of media. Growing up 1.0 L of cells in multiple smaller flasks is also an option.
10. This step can be optimized for your system. In our opinion, this optimization is less important than optimizing the initial vector(s) that will be used for expression (*see* **Note 1**). In our pET vector system, induction is temperature dependent, and induction at 37 °C results in rapid production of enzymes. However, this can lead to protein aggregation and ultimately loss of soluble multienzyme assemblies. By cooling the media prior to induction, our protein expression is attenuated to avoid this issue. The amount of IPTG can also be optimized. Our concentration of IPTG saturates LacI binding and therefore expression is maximized.
11. After this step, cell pellets can be frozen at −80 °C for many months for purification at a more convenient time. While some proteins may be affected by this, we have never detected any altered activity or characteristics because of a single freeze-thaw cycle while assemblies remained within cells.
12. Pellets are best resuspended by physically breaking the pellet with a metal scapula before repeatedly drawing up and dispensing the suspension using a 10 mL pipette. Resuspension of cells is critical for efficient lysis. Clumps of cells will not lyse well, and protein yield will be decreased. Keep cells on ice. Adding a protease inhibitor like PMSF may be helpful to avoid

proteolytic digestion during the purification, but our purification did not appear to be affected by leaving it out.

13. French press is the best lysis technique for our system. Sonication works but is less efficient. Critically, avoid chemical lysis additives, such as detergents, as they may disrupt the noncovalent interactions of the multienzyme assemblies.
14. Heat generated by the passage through the French press can warm the sample and denature proteins. Chill the French press metal cylinder (jacket) and plunger by storing it in a refrigerator or immerse it in an ice water bath or before being dried and loaded with cells. We do not re-chill the cylinder and plunger between the three passes.
15. This whole procedure assumes that the multienzyme assemblies under investigation are soluble rather than membrane bound. As a reference, a bacterial ribosome is 2.7 MDa and is typically isolated from soluble fractions from spins at 100,000 × *g* [6]. Soluble complexes should remain in solution at 35,000 × *g*. After this spin, the pellet and supernatant can be run on SDS-PAGE to determine if proteins are ending up in the insoluble material during this spin. Unless there is reason to believe your enzyme assembly is membrane bound, the presence of excessive desired protein in this pellet likely means your expression procedures are causing insoluble aggregation of proteins. In our system, multiple expression conditions lead to the presence of hydrocarbon biosynthesis proteins in this insoluble pellet. We altered factors such as lowering the growth temperature, attenuating the amount of inducer (IPTG), using weaker promoters in our vectors for protein production, and selecting different affinity tag combinations (C-terminal 6× His vs N-terminal 6× His). Much of this was trial and error.
16. For reference, an *E. coli* ribosome is 2.7 MDa and approximately 20 nm in diameter.
17. We use a ÄKTA$_{\text{FPLC}}$ fast protein liquid chromatography instrument to pump our purification and elution buffers over our column. Either a syringe or gravity flow will work fine.
18. It is critical to know the approximate concentration of imidazole required to elute your protein. For our system, ~200 mM is required, so a wash with 100 mM (20% elution buffer) is appropriate. A gradient can be used initially to discover the elution point of your sample.
19. There is a tradeoff between purity and concentration of your final material. Very slowly increasing the concentration of imidazole using a linear gradient over a long period of time will likely result in a pure sample, resolved from contaminants. However, the sample will be more dilute since elution will not be instantaneous. A stepwise gradient rapidly elutes your

material, but contaminants with similar binding affinities will be included in your fractions.

20. Flash freezing did not appear to make an appreciable difference in characteristics of our protein assemblies.
21. The Sigma Aldrich technical bulletin for the anti-FLAG M1 resin is an excellent resource for this procedure.
22. The resin is never allowed to run dry after this point in the procedure. A small layer of liquid (a few mm) is always left on the resin.
23. Affixing a small piece of tubing to the bottom of the column stopper increases the rate of flow. The mass of liquid in the tube effectively pulls more liquid through the column.
24. The FLAG peptide can mask absorbance at 280 nm or Bradford assay readings that are frequently used to determine the presence of protein. When collecting fractions, consider the column volume and the volume of tubing after the stopper to estimate when your sample will be eluted. With a 1.0 mL column volume and approximately 1 mL of tubing below the stopper, our protein does not elute immediately upon the addition of FLAG elution buffer.
25. As described in the Sigma anti-FLAG M1 resin technical bulletin, the resin should not be in the regeneration buffer for more than 20 min.
26. Injecting a smaller volume of a more concentrated sample gives better resolution of multienzyme assemblies.
27. Use blue dextran to determine the exact void volume of your column. The void volume is the volume of liquid your column contains and represents the fastest possible volume a protein could elute. Wash the column with 2 CV of purification buffer before injecting 0.50 mL of blue dextran. These have very large components that will be unable to interreact with any of the pores within the gel and therefore will elute rapidly to define the exact void volume for your column.
28. When spin concentrating, we experienced very little precipitation at concentrations below 10 mg/mL. Spin the concentrators for 10 min and mix before repeating this cycle to achieve the desired protein concentration. Mixing of samples is needed because the bottom of the upper chamber becomes more concentrated than the bulk solution.

Acknowledgement

This work was conducted in the lab of Dr. Lawrence Wackett at the University of Minnesota. I thank him for his support and mentorship. Financial support came from the NIH's Biotechnology Training Grant and internal University of Minnesota funds. Many others at the University of Minnesota made this research possible. Additionally, I thank Dr. Matthew Jensen, Troy Biernath, and Jordan Holthe for helpful comments in the revision process.

References

1. Vijayakrishnan S, Kelly SM, Gilbert RJC, Callow P, Bhella D, Forsyth T, Lindsay JG, Byron O (2010) Solution structure and characterization of the human pyruvate dehydrogenase complex core assembly. J Mol Biol 399(1): 71–93. https://doi.org/10.1016/j.jmb.2010.03.043
2. Pham VHT, Kim J (2012) Cultivation of unculturable soil bacteria. Trends Biotechnol 30(9): 475–484. https://doi.org/10.1016/j.tibtech.2012.05.007
3. Christenson JK, Jensen MR, Goblirsch BR, Mohamed F, Zhang W, Wilmot CM, Wackett LP (2017) Active multienzyme assemblies for long-chain olefinic hydrocarbon biosynthesis. J Bacteriol 199(9):e00890–e00816. https://doi.org/10.1128/JB.00890-16
4. Christenson JK, Richman JE, Jensen MR, Neufeld JY, Wilmot CM, Wackett LP (2017) β-lactone synthetase found in the olefin biosynthesis pathway. Biochemistry 56(2):348–351. https://doi.org/10.1021/acs.biochem.6b01199
5. Sukovich DJ, Seffernick JL, Richman JE, Gralnick JA, Wackett LP (2010) Widespread head-to-head hydrocarbon biosynthesis in bacteria and role of OleA. Appl Environ Microbiol 76(12):3850–3862. https://doi.org/10.1128/AEM.00436-10
6. Mehta P, Woo P, Venkataraman K, Karzai AW (2012) Ribosome purification approaches for studying interactions of regulatory proteins and RNAs with the ribosome. Methods Mol Biol 905:273–289. https://doi.org/10.1007/978-1-61779-949-5_18

Chapter 2

Multi-dimensional Fluorescence Live-Cell Imaging for Glucosome Dynamics in Living Human Cells

Songon An, Prakash Parajuli, Erin L. Kennedy, and Minjoung Kyoung

Abstract

Fluorescence live-cell imaging that has contributed to our understanding of cell biology is now at the frontline of studying quantitative biochemistry in a cell. Particularly, technological advancements of fluorescence live-cell imaging and associated strategies in recent years have allowed us to discover various subcellular macromolecular assemblies in living human cells. Here we describe how real-time dynamics of a multienzyme metabolic assembly, the "glucosome," that is responsible for regulating glucose flux at subcellular levels, has been investigated in both 2- and 3-dimensional space of single human cells. We envision that such multi-dimensional fluorescence live-cell imaging will continue to revolutionize our understanding of how intracellular metabolic pathways and their network are functionally orchestrated at single-cell levels.

Key words Live-cell Imaging, Fluorescence Microscopy, Lattice Light-sheet Microscopy, Glucosome, Metabolic Condensate, Metabolic Complex, Macromolecular Assembly, Glycolysis, Glucose Metabolism, Cancer Metabolism

1 Introduction

Fluorescence microscopy has been one of the popular techniques in the field of cell biology and has contributed significantly to our current understanding of cell anatomy [1, 2]. However, its application had been technically limited when studying the dynamic nature of intracellular biomolecules in living cells due to the lack of subcellular specificity necessary to visualize specific biomolecules at single-cell levels [3, 4]. Recent scientific and technical developments in biophysics and biochemistry have allured us to explore previously untouchable territories of life sciences [3–6]. Multi-dimensional fluorescence live-cell imaging has now opened a foreseeable avenue of cellular biochemistry by revealing the existence of multienzyme metabolic complexes in living cells and thus their real-time functional contributions to the cell [7–9]. Here we demonstrate how we have investigated subcellular dynamics of a

Haralambos Stamatis (ed.), *Multienzymatic Assemblies: Methods and Protocols*,
Methods in Molecular Biology, vol. 2487, https://doi.org/10.1007/978-1-0716-2269-8_2,
© The Author(s), under exclusive license to Springer Science+Business Media, LLC, part of Springer Nature 2022

multienzyme metabolic assembly, namely the "glucosome," that regulates glucose flux in single human cells under fluorescence live-cell imaging [10–12]. Our strategy has allowed us to advance our knowledgebase of not only how glucose metabolism is regulated in space and time as part of a functional metabolic network in human cells, but also its functional association with other cellular processes at the level of single human cells.

2 Materials

2.1 Human Cancer Cell Lines

Glucosomes have been successfully visualized in various human cancer cell lines, including HeLa (cervical adenocarcinoma) [12], Hs578T (breast carcinoma) [12], BT-549 (breast ductal carcinoma), MCF-7 (breast adenocarcinoma), and MDA-MB-436 (breast adenocarcinoma).

2.2 Cell Culture and Imaging

1. General disposable plasticware and small equipment for mammalian cell culture.
2. Glass-bottomed 35-mm cell culture petri dishes.
3. DFG medium: 450 mL of Dulbecco's modified eagle medium with high glucose (DMEM) is supplemented with 50 mL of fetal bovine serum (FBS), 55 mg of sodium pyruvate, 25 μL of 1 g/mL gentamycin sulfate.
4. RFG medium: 450 mL of Roswell Park Memorial Institute 1640 (RPMI 1640) is supplemented with 50 mL of dialyzed FBS (*see* **Note 1**) and 25 μL of 1 g/mL gentamycin sulfate.
5. Gibco Opti-MEM-I reduced serum medium (Opti-MEM-I).
6. Lipofectamine 2000 transfection reagent.
7. A plasmid encoding human phosphofructokinase liver type that is fused with monomeric enhanced green fluorescent protein (PFKL-mEGFP) [12].
8. Buffered-saline solution (BSS) [7, 12]: Add about 700 mL autoclaved water to a 1 L graduate cylinder. Add 20 mL of 1 M HEPES, pH 7.4, 33.8 mL of 4 M NaCl, 5 mL of 1 M KCl, 2 mL of 0.5 M $MgCl_2$, 3.6 mL of 0.5 M $CaCl_2$, and 1.9 mL of 3 M D-glucose. Mix and make up to 1 L with autoclaved water. Perform sterile filtration and store at 4 °C.
9. Nikon Eclipse Ti inverted C2 confocal microscope equipped with a 60× objective lens (Nikon CFI Plan Apo TIRF, 1.45 NA), a monochrome CCD camera (Photometrics CoolSnap EZ), and a filter set appropriate for mEGFP detection.
10. Imaging software for wide-field and confocal microscopy: Nikon NIS-Elements.

11. Lattice light-sheet microscope (LLSM) and imaging software [10, 13].
12. Image analysis software: ImageJ [14, 15] and MATLAB (version 9.7.0 (R2019b), The MathWorks Inc)).

3 Methods

All media and reagents described here were warmed up to room temperature for a few hours or 37 °C for an hour before use. Please note that all procedures were carried out at room temperature unless otherwise specified.

3.1 Human Cell Culture

1. A frozen stock vial of each cell line (HeLa, Hs578T, BT-549, MCF-7, and MDA-MB-436) that contains ~ 2–3 × 10^6 cells is thawed and cultured in a T25 tissue culture flask containing the DFG medium.
2. After the cells are maintained in the DFG medium for about 2 weeks by sub-culturing them in every 3–4 days, they are passed into the DFG medium or the RFG medium following 3–4 days sub-culture for another 2 weeks before preparing samples for live-cell imaging (*see* **Note 2**).
3. The cells are now ready for transient transfection and imaging in every passage cycle for another 2 months.

3.2 Transient Transfection

1. Cells are seeded into a glass bottom 35 mm petri dish in a total of 2 mL volume of the desired growth medium without antibiotics (i.e., DF or RF media) for wide-field and confocal microscopy. Alternatively, cells are seeded into 5 mm round coverslips that are placed on a glass bottom 35 mm petri dish for lattice light-sheet microscopy (LLSM) (*see* **Note 3**). Cells should be uniformly distributed on the glass-bottom imaging area of the dish or on the 5 mm round coverslips by rocking it slowly and incubating overnight at 37 °C under 5% CO_2 and 95% humidity.
2. The next day, transient transfection is performed using Lipofectamine 2000 when cells are approximately 70–80% confluent at the time of transfection (*see* **Note 4**).
3. A plasmid expressing PFKL-mEGFP as a glucosome marker (0.8 μg) [11, 12] is added into 50 μL of Opti-MEM-I. In a separate tube, 1.5 μL of Lipofectamine 2000 is added into 50 μL of Opti-MEM-I. Each mixture is gently mixed and incubated individually for 5 min at room temperature (*see* **Note 5**).
4. Combine both mixtures, mix gently, and incubate for 30 min at room temperature.

5. While incubating the plasmid-transfection reagent mixture, take out cell dishes from a CO_2 incubator and rinse the cells twice with 2 mL of Opti-MEM-I. Then, add 1 mL of Opti-MEM-I to each dish (*see* **Note 6**).
6. After 30 min incubation of **step 4**, add 900 μL of Opti-MEM-I into the plasmid-transfection reagent mixture.
7. Add the entire plasmid-transfection reagent mixture (~ 1 mL) dropwise to the dish containing cells (*see* **Note** 7).
8. Incubate the cells at 37 °C in a CO_2 incubator (95% humidity) for 5–6 h.
9. Exchange the medium with fresh growth medium without antibiotics and continue incubation for 18–24 h at 37 °C in a CO_2 incubator (95% humidity).

3.3 2-Dimentional (2D) Visualization of Glucosome Using Wide-Field or Confocal Microscopy

1. Remove the cell dishes from the CO_2 incubator and visualize for adherence under a cell culture microscope (*see* **Note 8**).
2. Rinse the cells with 2 mL of 1× BSS solution at room temperature.
3. Add 2 mL of 1× BSS solution and incubate for 10 min at room temperature. Repeat this step twice.
4. Add 2 mL 1× BSS solution again to the cells and incubate for 1–2 h at room temperature prior to imaging (*see* **Note 9**).
5. Imaging is now performed at ambient temperature (~25 °C) with the Nikon Eclipse Ti inverted C2 confocal microscope. PFKL-mEGFP is shown to form spatially resolved glucosomes at various sizes in living human cells (Fig. 1). We have then categorized them into 3 sub-classes [12, 16]; namely, small (<0.1 μm^2) (*see* **Note 10**), medium (<3 μm^2) (*see* **Note 11**), and large sizes (<8 μm^2) of glucosome assemblies (*see* **Note 12**).
6. Colocalization Analysis: If cells are transfected with two plasmids, dually transfected cells are imaged in each color channel. Digitally merged images are analyzed using the ImageJ software for colocalization between PFKL-mEGFP and another protein that is fused with either mCherry or mOrange fluorescent protein [12].
7. Addition of temporal resolution: Since glucosomes are being visualized in live cells, real-time dynamics of glucosomes can be monitored as a function of time in the presence and/or absence of additives to investigate the importance of glucosome dynamics in biological processes of interest [11, 12].

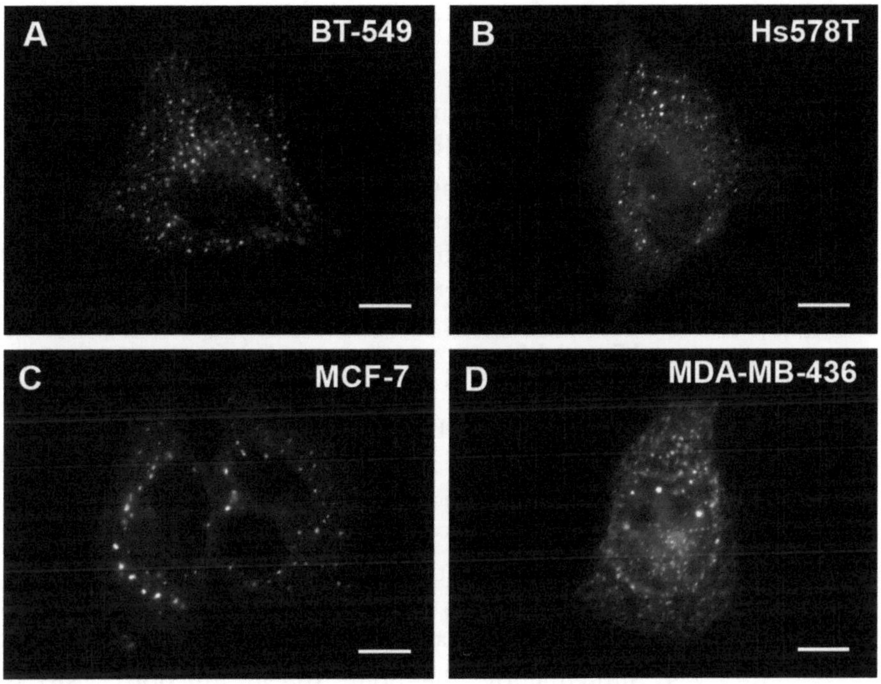

Fig. 1 Glucosomes in Human Breast Cancer Cells. Representative images of PFKL-mEGFP as a glucosome marker are from BT-549 (**A**), Hs578T (**B**), MCF-7 (**C**), and MDA-MB-436 (**D**) cells. Scale bars, 10 μm

3.4 3D Visualization of Glucosome Using Lattice Light-Sheet Microscopy (LLSM)

1. After rinsing the cells as described in **steps 1–3** in Subheading 3.3 above, add 2 mL 1× BSS solution again to the cells along with a diluted solution of TetraSpeck™ fluorescent beads and allow the beads to settle on the glass surface for at least 10 min (*see* **Note 13**).
2. Prior to LLSM imaging, it is important to align illumination pathways daily as described below in **steps 3–8** in Subheading 3.4.
3. Using a 30 mm cage alignment plate with Ø0.9 mm hole prior to the spatial light modulator (SLM) while projecting a single beam pattern in the "Continuous" illumination mode, ensure that all laser beams are centered on the pattern. This can be done by carefully adjusting the knobs on the mounted filters that are close to the individual lasers.
4. Align the single beams through the annular mask by matching the illuminated single beam pattern to the position of the annular mask. This can be observed in the back focal plane guppy camera and adjusted with the knobs on the second mirror before the annular mask [13].
5. Use the flip mirror to direct the lasers into the illumination objective and examine that the exiting beam of illumination is

symmetrical. Add diluted fluorescein and sulforhodamine B solutions in the sample chamber to visualize Bessel beams. Using the Orca Flash 4.0 camera, adjust the focus on the detection objective using the travel stage knob until the beam is in focus (*see* **Note 14**).

6. Using the "Continuous Z-stack" mode without moving the detection objective piezo stage, ensure that the plane of illumination is orthogonal to the plane of detection. Adjust the travel stage on the flip mirror behind the illumination objective until both ends of the Bessel beam focus at the same time (*see* **Notes 15** and **16**).
7. Ensure that the entire plane of Bessel beams is in focus by switching the SLM to a multibeam pattern. Knobs on the mirrors prior to the flip mirror can be adjusted until perfect foci of all Bessel beams in the multibeam patterns are in the same plane.
8. If any of the later adjustments have changed, revisit earlier adjustment steps (i.e., **steps 2–7** in Subheading 3.4) to ensure that no additional alignment is needed (*see* **Note 17**).
9. Subsequently, the point spread function (PSF) should be obtained daily as follows (i.e., **steps 10–14** in Subheading 3.4).
10. Dilutions of TetraSpeck™ fluorescent microspheres are sonicated at room temperature for 5–10 min. Afterwards, a single droplet is added to clean a 5 mm coverslip. The fluorescent beads adhere to the coverslip surface after 10 min of undisturbed incubation at room temperature.
11. Securely load the coverslip into the LLSM sample holder and rinse the sample chamber with fresh 1× BSS. Bring the beads into the detection region using the "Motion Console" mode that controls the sample piezo stages.
12. Using a single isolated bead with the "Autofocus Bead" mode, determine the z-galvo offsets that are associated with each laser to ensure optimal illumination for all laser channels. This step allows to overlap the different laser illumination in the same sample plane to each other (*see* **Notes 18** and **19**).
13. Switch acquisition to the "single Z-stack" mode and obtain 3D images of a bead for each color channel to determine the PSF daily (*see* **Note 20**).
14. Change acquisition to the "XZ PSF" mode and capture a cross-section of the pattern. If any variance in illumination is observed, repeat from **steps 2–7** in Subheading 3.4 to ensure that all regions of the experimental data are valid for image analysis (*see* **Note 21**).
15. LLSM Imaging with live cells is now performed at ambient temperature (~25 °C).

16. Clean the sample chamber thoroughly with fresh 1× BSS.
17. Load the 5 mm coverslip that contains transfected cells into the sample holder and position in the sample chamber (*see* **Note 22**).
18. 3D sample scans are taken of each sample for each illumination channel by moving the sample stage in the "SI Scan" mode. The step size of the sample piezo is approximately half of the z-dimension of the PSF. All sample specific factors including the PSF are used in the image deconvolution process using the 3D deconvolution algorithm in MATLAB [13, 17] (*see* **Notes 23** and **24**).
19. TetraSpeck™ fluorescent beads are isolated from the captured deconvolved images in all color channels. To register two color channels, the centroids of the beads obtained from the "3D Objects Counter" plugin [18] in the ImageJ software are calibrated using the imtranslate syntax of MATLAB. After alignment of the color channels, beads and non-target cells are cleared from the processed images for analysis (*see* **Note 25**).
20. PFKL-mEGFP as a glucosome marker [11, 12] is now visualized at various sizes in living human cells. The "3D Objects Counter" plugin [18] is used to identify glucosome assemblies and to create an 8-bit binary 3D objects mask of the identified glucosome assemblies (*see* **Note 26**).
21. Colocalization Analysis (Fig. 2): If cells are transfected with two plasmids, dually transfected cells are imaged in each color channel. The "Coloc 2" plugin [14] is used to assign mCherry-fusion protein to Channel 1, mEGFP-fusion protein to Channel 2, and the object mask to the Region-of-Interest (ROI) Mask [10]. Note that the colocalized TetraSpeck™ beads are served as an internal colocalization control.
22. Addition of temporal resolution: Since glucosomes are being visualized in live cells, real-time dynamics of glucosomes in 3D space can be monitored as a function of time in the presence and/or absence of additives to investigate the importance of glucosome dynamics in biological processes of interest under LLSM imaging [10].

4 Notes

1. We have dialyzed FBS against 0.9% NaCl solution at 4 °C for ~2 days using 12–14 kDa MWCO dialysis membrane [12, 16].
2. This means that it will take about 1 month for cells to be ready for glucosome imaging from a frozen stock. This protocol

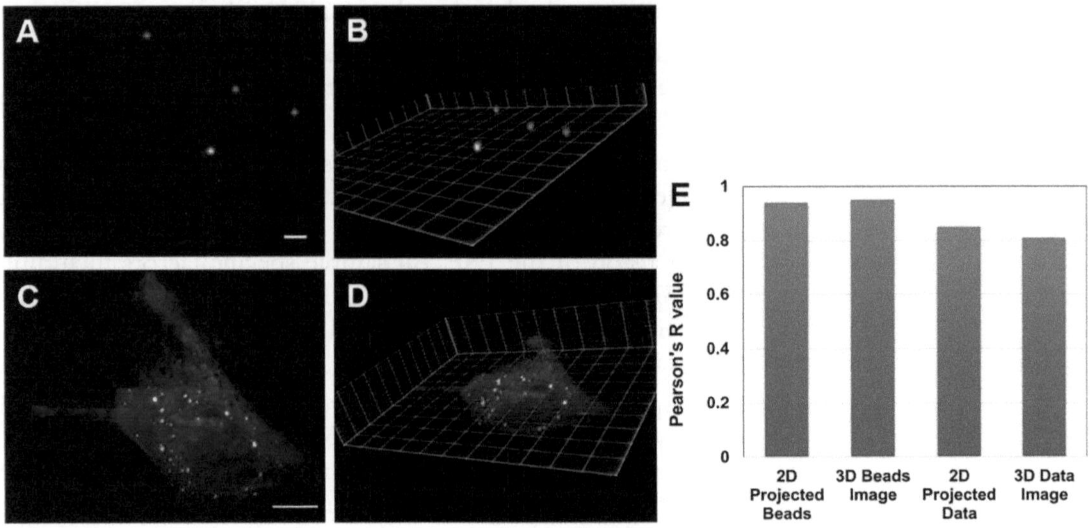

Fig. 2 3D Colocalization Analysis. Colocalized TetraSpeck™ beads overlaid in the red and green channels are shown as a 2D projection (**A**) or as a 3D image (**B**). mCherry-fusion proteins (in red) colocalized with mEGFP-fusion proteins (in green) are shown as a 2D projection (**C**) or as a 3D image (**D**). The Pearson's coefficients are calculated by the "Coloc 2" plugin in the ImageJ software (**E**). Note that the representative cell image here (**C** and **D**) shows PFKL-mCherry in red and mEGFP-PKM2 in green. Yellow color indicates their colocalization. Scale bars, 1 μm (**A**) or 10 μm (**C**)

appears to be one of the indispensable factors that influence on the reproducibility of our imaging data.

3. The 5 mm round coverslips are cleaned in 1 M KOH solution, rinsed with autoclaved Millipore water, and subsequently sterilized under UV light prior to placing them on glass bottom 35 mm petri dishes.
4. For HeLa, Hs578T, and MCF-7 cells, we seed ~2×10^5 cells per petri dish in 2 mL of a growth medium without antibiotics [10–12]. However, for BT-549 and MDA-MB-436 cells, ~3×10^5 cells per petri dish appear to work better.
5. The amount of the plasmid for glucosome imaging should be determined for each cell line for the best result by evaluating various experimental factors, specifically considering the ratio (μg/μL) of the plasmid to Lipofectamine 2000, and the dilution ratio of the plasmid-transfection reagent mixture to Opti-MEM-I. At the same time, transfection efficiency of the protein also varies with the type of cell lines. It is necessary to optimize at least the described parameters above to ensure the optimal transfection efficiency for reproducible results. Note that transfection efficiency should be determined for each cell line by counting fluorescent cells over the total number of cells. If dual-color imaging and colocalization experiment are desired, a second plasmid expressing a protein of interest with mCherry

or mOrange fluorescence protein is added into the plasmid mixture together with PFKL-mEGFP in a total of 50 μL of Opti-MEM-I. In this case, the amount of the second plasmid and the ratio of two plasmids should be optimized for best co-transfection efficiency in addition to the parameters mentioned for single transfection above.

6. We usually take out cell dishes and perform this step ~20–25 min after **step 4** in Subeading 3.2.
7. This step usually takes ~30–60 s per plasmid-transfection reagent mixture per dish.
8. Reduction in cell population is anticipated because of the cytotoxicity of the transfection reagent. In addition, degree of such reduction in cell population depends on the cell line of interest.
9. HeLa cells have been experimentally shown to require at least 1 or 2 h to be equilibrated at a desired temperature prior to imaging [19]. This step is particularly important if one would like to evaluate the effect of additives on glucosome assemblies from live cells.
10. Due to the optical diffraction limit of light, the sizes of fluorescent assemblies in single cells cannot be determined if they are smaller than 0.1 μm^2. We have thus established 0.1 μm^2 as a first threshold to distinguish small sizes of glucosomes from others.
11. 3 μm^2 is then set as a second threshold because, unlike cancer cells, in human non-cancerous breast tissue cells (i.e., Hs578Bst) we have barely observed glucosome assemblies that are larger than 3 μm^2 [12]. To emphasize glucosomes that are larger than 3 μm^2 as biologically independent entities, the medium size of glucosomes is defined to range from 0.1 to 3 μm^2.
12. We also observe apparent fluorescent aggregates, which are larger than 8 μm^2, inside ~5% of transfected cells. The aggregate formation is determined to be independent of the identity of transfected proteins [7, 12, 16]. We also confirmed using a biophysical technique, so-called fluorescent recovery after photobleaching, that these aggregates contain only immobile fractions of transfected proteins [12, 16]. Therefore, ~5% of transfected cells in our cell culture conditions appear to suffer from ectopic overexpression.
13. Although TetraSpeck™ fluorescent microspheres are shown to be not cell-permeable nor cause distress inside cells, we still recommend diluting the fluorescent beads so that a few beads can be observed within a field of view.
14. Multiple modes of acquisition are used throughout the alignment and experimental process in the LLSM. The

"Continuous" mode is used to quickly observe the plane of illumination and detection and the z-galvo and objective piezo stage are not moving in this mode. When bringing the single Bessel beam into focus or observing the total plane focus of the multibeam pattern it is important not to dither the x-galvo.

15. When using the "Continuous Z-stack" mode to make the illumination and detection planes orthogonal, the z-galvo mirror is mobile but the detection objective piezo and the x-galvo are stationary.
16. When using the "Continuous Z-stack" mode to capture the PSF daily, the detection objective piezo and the z-galvo mirror are moving in unison while the x-galvo dithers.
17. It is recommended that after the thorough alignment, permanent adjustable iris diaphragms are added throughout the laser pathways to reduce time needed for total realignment later. The iris diaphragms are located after all lasers are combined, after the acousto-optic tunable filter, prior to the cylindrical lenses, and after the annular mask.
18. Additional adjustment using the "Autofocus Bead" mode from the LLSM imaging software is desired throughout the experiment to keep the illumination plane aligned with the detection plane.
19. For the "Autofocus Bead" and the "XZ PSF" modes, the z-galvo and objective piezo are moving independently and the x-galvo is dithering. The purpose of "Autofocus Bead" is to determine the best offset position for the z-galvo to ensure that the illumination and detection planes are perfectly matched for each other.
20. The PSF observed at any given day of experiment should be symmetrical. The size of the PSF varies depending on wavelengths, choices of SLM patterns, media conditions, and specific alignments, which are why it must be imaged every day at the time of experiment.
21. The purpose of the "XZ PSF" mode is to calculate the cross-section of illumination at the sample plane which provides insight into alignment before imaging.
22. We usually allow cells to equilibrate with 1 x BSS imaging solution for ~ 60 min prior to LLSM imaging.
23. In order to avoid cross-channel bleed-through or crosstalk, LLSM is equipped with a filter wheel station containing laser-specific emission filters prior to the Orca Flash 4.0 camera, and the laser channels were manually selected for illumination.
24. The "SI Scan" mode holds the z-galvo and objective piezo stationary, dithers the x-galvo, and moves the sample piezo stage through the illumination plane.

25. Using the beads as a colocalization standard is important because spectrally distinct PSFs present different sizes due to the diffraction limit and chromatic aberration. Such PSF differences result in imperfect colocalization factors, including the Pearson's R values obtained from the "Coloc 2" plugin analysis [14].
26. Glucosomes vary in size and intensity, so intensity thresholds and size filters in the "3D Objects Counter" plugin [18] of the ImageJ software are used to determine the sizes of glucosome assemblies in LLSM imaging.

Acknowledgement

We wish to thank all former and current members who have contributed to the described protocol. We would also like to thank Dr. Gerald M. Wilson and Dr. Jiayuh Lin for sharing MCF-7 and MDA-MB-436 cell lines, respectively. This work is financially supported by the National Institutes of Health: R01GM134086 (M.K.), R01GM125981 (S.A.), R03CA219609 (S.A.), T32GM066706 (E.L.K.), and R25GM55036 (E.L.K.).

References

1. North AJ (2006) Seeing is believing? A beginners' guide to practical pitfalls in image acquisition. J Cell Biol 172:9–18
2. Pearson H (2007) The good, the bad and the ugly. Nature 447:138–140
3. Giepmans BN, Adams SR, Ellisman MH et al (2006) The fluorescent toolbox for assessing protein location and function. Science 312: 217–224
4. Prescher JA, Bertozzi CR (2005) Chemistry in living systems. Nat Chem Biol 1:13–21
5. Marks KM, Nolan GP (2006) Chemical labeling strategies for cell biology. Nat Methods 3: 591–596
6. Wouters FS, Verveer PJ, Bastiaens PI (2001) Imaging biochemistry inside cells. Trends Cell Biol 11:203–211
7. An S, Kumar R, Sheets ED et al (2008) Reversible compartmentalization of de novo purine biosynthetic complexes in living cells. Science 320:103–106
8. Kohnhorst CL, Schmitt DL, Sundaram A et al (2016) Subcellular functions of proteins under fluorescence single-cell microscopy. Biochim Biophys Acta 1864:77–84
9. Schmitt DL, An S (2017) Spatial organization of metabolic enzyme complexes in cells. Biochemistry 56:3184–3196
10. An S, Jeon M, Kennedy EL et al (2019) Phase-separated condensates of metabolic complexes in living cells: purinosome and glucosome. Methods Enzymol 628:1–17
11. Jeon M, Kang HW, An S (2018) A mathematical model for enzyme clustering in glucose metabolism. Sci Rep 8:2696
12. Kohnhorst CL, Kyoung M, Jeon M et al (2017) Identification of a multienzyme complex for glucose metabolism in living cells. J Biol Chem 292:9191–9203
13. Chen BC, Legant WR, Wang K et al (2014) Lattice light-sheet microscopy: imaging molecules to embryos at high spatiotemporal resolution. Science 346:1257998
14. Schindelin J, Arganda-Carreras I, Frise E et al (2012) Fiji: an open-source platform for biological-image analysis. Nat Methods 9: 676–682
15. Schneider CA, Rasband WS, Eliceiri KW (2012) NIH Image to ImageJ: 25 years of image analysis. Nat Methods 9:671–675
16. Kyoung M, Russell SJ, Kohnhorst CL et al (2015) Dynamic architecture of the

purinosome involved in human de novo purine biosynthesis. Biochemistry 54:870–880

17. Gao L, Shao L, Chen BC et al (2014) 3D live fluorescence imaging of cellular dynamics using Bessel beam plane illumination microscopy. Nat Protoc 9:1083–1101
18. Bolte S, Cordelieres FP (2006) A guided tour into subcellular colocalization analysis in light microscopy. J Microsc 224:213–232
19. Verrier F, An S, Ferrie AM et al (2011) GPCRs regulate the assembly of a multienzyme complex for purine biosynthesis. Nat Chem Biol 7: 909–915

Chapter 3

Mechanisms and Effects of Substrate Channelling in Enzymatic Cascades

Svyatoslav Kondrat and Eric von Lieres

Abstract

Substrate or metabolite channelling is a transfer of intermediates produced by one enzyme to the sequential enzyme of a reaction cascade or metabolic pathway, without releasing them entirely into bulk. Despite an enormous effort and more than three decades of research, substrate channelling remains the subject of continuing debates and active investigation. Herein, we review the benefits and mechanisms of substrate channelling in vivo and in vitro. We discuss critically the effects that substrate channelling can have on enzymatic cascades, including speeding up or slowing down reaction cascades and protecting intermediates from sequestration and enzymes' surroundings from toxic or otherwise detrimental intermediates. We also discuss how macromolecular crowding affects substrate channelling and point out the galore of open questions.

Key words Enzymatic cascade, Direct channelling, Channelling by proximity, Metabolon, Enzyme complex, Multifunctional enzyme, Macromolecular crowding

1 Introduction

Enzyme-catalyzed reactions play a vital role in living systems and have numerous biotechnological applications. Typically, an enzymatic reaction does not occur alone but goes along with other reactions as part of a cascade or metabolic pathway. Substrate or metabolite channelling of such enzymatic cascades is the passing of intermediates produced by one enzyme to the sequential enzyme of a cascade, avoiding the release and complete mixing of the intermediates with bulk phase. Substrate channelling has been suggested to play an essential role in the organization and regulation of metabolism [1–3] and enhancing metabolic fluxes [4–9]. However, the benefits and significance of channelling for in vivo systems and in biotechnology have been the subject of continuing debates [10–33].

Haralambos Stamatis (ed.), *Multienzymatic Assemblies: Methods and Protocols*,
Methods in Molecular Biology, vol. 2487, https://doi.org/10.1007/978-1-0716-2269-8_3,
© The Author(s), under exclusive license to Springer Science+Business Media, LLC, part of Springer Nature 2022

Evidence had emerged in early 1960 that enzymes are not randomly distributed inside living cells but assemble into functionally significant aggregates. In their 1966 review, Reed and Cox described α-keto acid dehydrogenase complexes, as well as fatty acid synthetases and tryptophan synthetase [34]. They suggested that enzyme assemblies might alter the kinetic properties of enzymes and enhance the overall efficiency due to enzyme proximity. Although kinetic experiments indicated a direct transfer of intermediates between sequential enzymes as early as in 1950 [35], the concept of substrate channelling had developed only in the mid-1980. Srivastava and Bernhard reported on direct transfer of reduced nicotinamide adenine dinucleotide (NADH) between glyceraldehyde-3-phosphate dehydrogenase (GPDH) and alcohol dehydrogenase (ADH) [36]. The same authors reported on the channelling of NADH between GDPH and L-lactate dehydrogenase (LDH) in the glycolytic pathway [37], causing heated debates in the literature [10–13]. At the same time, work on the tricarboxylic acid (TCA) cycle [38, 39] suggested the dynamic formation of multienzyme complexes, termed by Srere as metabolons [40].

Since that early work, numerous stable and transient enzyme complexes have been identified and characterized, helping reveal the physiological significance and potential benefits of substrate channelling [4, 41]. We divide these benefits into two groups. **(A)** Enhancement:

1. Increasing global reaction rates. It is mentioned as advantage of channelling frequently [2, 4–9]. However, as several articles point out [27–29, 31, 33, 42], and as we discuss below, channelling per se can enhance reaction velocities only transiently and may even slow them down in a steady state.
2. Enhancing reactions by avoiding metabolite's solvation–desolvation steps [7, 43].

(B) Channelling can also play a protective role, likely facilitating regulatory mechanisms in vivo [1–3]. These protective functions are:

3. Protecting unstable intermediates from degradation. For instance, carboxy phosphate, channelled by carbamoyl phosphate synthetase [44], has a half-life of about 70 ms [45, 46], which makes bulk reactions with this substrate inefficient or impossible at low enzyme concentrations.
4. Preventing substrate competition between different enzymes or metabolic pathways. Many metabolites are substrates for more than one enzyme. For instance, oxaloacetate is a substrate for malate dehydrogenase and citrate synthase [47], hydrogen peroxide is a substrate for horseradish peroxidase and catalase, and there are many more examples.

5. Protecting cells from toxic or otherwise detrimental intermediates [7, 48], for instance, by reducing the metabolite pool size below the poisonous level [49, 50].
6. Circumventing unfavorable equilibria, e.g., preventing hydrophobic intermediates from diffusing out of the aqueous cell environment. An example is an indole channelled in tryptophan synthase [2, 5, 51], which we discuss below (Subheading 2.1.1).

In this chapter, we first review the types and the mechanisms of channelling (Subheading 2). We provide only a general overview of these mechanisms and refer the interested readers to more detailed biochemical reviews (refs. [2, 5, 52, 53] for multifunctional enzymes; refs. [8, 29, 32, 48, 54–56] for metabolons and enzyme clusters; refs. [7, 9, 29, 57, 58] for synthetic complexes and biotechnological applications). In Subheading 3, we discuss critically some of the benefits of channelling listed above and show how simple kinetic and reaction–diffusion models can lead to new physical insights clarifying the effects of substrate channelling. In Subheading 4, we comment on a relation between substrate channelling and macromolecular crowding, relevant to in vivo systems. We finish with a few concluding remarks in Subheading 5.

2 Types and Mechanisms of Channelling

We divide channelling into three types. In direct channelling, two sequential enzymes of a reaction cascade form a permanent, stable complex, sometimes called a multifunctional enzyme. A molecular tunnel connects the active sites of this complex, preventing the intermediates from diffusing into bulk (Fig. 1a). In proximity channelling, two or more enzymes are located near each other; for example, they are synthetically conjugated with a molecular linker (Fig. 1c). In this case, the intermediates are only partly channelled, i.e., they can escape from an enzyme complex and diffuse into bulk.

Non-covalent interactions (e.g., electrostatic or hydrophilic/phobic interactions) between an intermediate and an enzyme complex may guide the intermediate to the second active site of the complex, reducing the leakage of proximity channelling; an example is an electrostatic "highway" suitable for charged intermediates (Fig. 1b). We refer to this type of channelling as guided channelling to distinguish it from non-leaky direct channelling and more leaky proximity channelling. Enzyme complexes providing guided channelling can be permanent (multifunctional enzymes), transient, or synthetic and may have varying degrees of transfer efficiency.

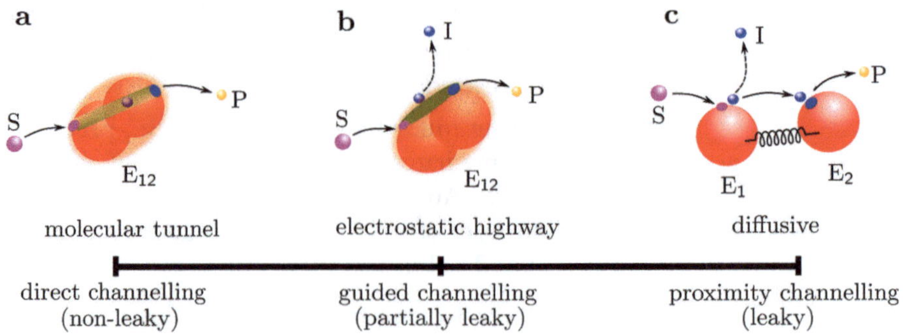

Fig. 1 Types of substrate channelling. Substrate S is transformed by enzyme E_1 into intermediate I, which is transformed by enzyme E_2 into product P. E_{12} is an enzyme complex or a multifunctional enzyme consisting of enzymes E_1 and E_2. (**a**) Direct channelling of intermediates via a molecular tunnel running inside E_{12} (*cf.* Fig. 3.2a). (**b**) Guided channelling with an "electrostatic highway" running on the surface of E_{12} (*cf.* Fig. 3.2b and Fig. 3.3 for examples). (**c**) Proximity channelling via conjugation or immobilization of enzymes in close vicinity of each other (*cf.* Fig. 3.4 for an example). Note that in (**b**) and (**c**), substrate channelling is not perfect, i.e., such enzyme complexes lose intermediates (dash arrows)

2.1 Multifunctional Enzymes

X-ray crystallographic studies have revealed two distinct molecular mechanisms of transferring intermediates in multifunctional enzymes [5]. The most widespread mechanism involves molecular channels (or tunnels), connecting active sites of an enzyme, while the other one relies on electrostatic interactions between intermediates and the enzyme surface. We only briefly describe these mechanisms and refer the interested readers to [2, 5, 52, 53] for more detailed biochemical reviews.

2.1.1 Molecular Channels

Molecular channels are ubiquitously present in enzymes. Typically, these channels connect the active sites, hidden inside enzymes, with the exterior and likely play a significant role in enzyme specificity. An analysis of 4306 enzymes showed that over 64 % of them possess channels longer than 1.5 nm and over 86 % of the enzymes contain channels at least 0.5 nm long, with a typical length of about 2 nm [59]. Thus, it is not very surprising that most multifunctional enzymes providing substrate channelling developed physical tunnels connecting their active sites.

A classic example is tryptophan synthase, which catalyzes the final two reactions of tryptophan biosynthesis [2, 5, 51, 60]. Tryptophan synthase consists of two subunits α and β and is usually found as an $\alpha\beta\beta\alpha$ tetramer (Fig. 2a). The α subunit catalyzes the formation of indole and glyceraldehyde-3-phosphate (G3P) from indole-3-glycerol phosphate (IGP). The β subunit catalyzes tryptophan from indole and serine in a pyridoxal phosphate (PLP)-dependent reaction. The two active sites are connected by a 2.5 nm long hydrophobic tunnel, allowing for diffusion of indole. Since indole is hydrophobic, it would quickly diffuse out of a cell if released into bulk. Thus, the indole channelling is essential for tryptophan biosynthesis.

Other examples of bifunctional enzymes include glutamine phosphoribosylpyrophosphate amidotransferase (GPAtase), featuring a 2 nm long tunnel, and carbamoyl phosphate synthetase (CPSase), which has a nearly 10 nm long tunnel [44, 61]. 4-Hydroxy-2-oxovalerate aldolase-aldehyde dehydrogenase (DmpFG) has a 2.9 nm long, water-filled tunnel [62]. Brownian dynamics simulations show that it takes about 25 ns for acetaldehyde (an intermediate) to travel inside this tunnel from the DmpG to DmpF active site [63].

2.1.2 Electrostatic Channelling

Electrostatic channelling is diffusional transfer of a charged intermediate across the oppositely charged enzyme surface controlled by electrostatic interactions. This channelling mechanism has been first discovered in bifunctional thymidylate synthase-dihydrofolate reductase (TS-DHFR). In bacteria, DHFR and TS are separate enzymes that catalyze sequential reactions in the thymidylate pathway. In protozoa and some plants, these two enzymes are coded in the same polypeptide chain, resulting in a bifunctional TS-DHFR enzyme [64, 65]. The crystal structure analysis showed no tunnel connecting the active sites of DHFR and TS. Instead, the study of charge distribution of the TS-DHFR surface revealed a positively charged electrostatic "highway" linking the TS active site to the 4 nm-distant DHFR site, which allows diffusive transport of the negatively charged oxidized dihydrofolate (an intermediate) on the TS-DHFR surface [66] (Fig. 2b). This

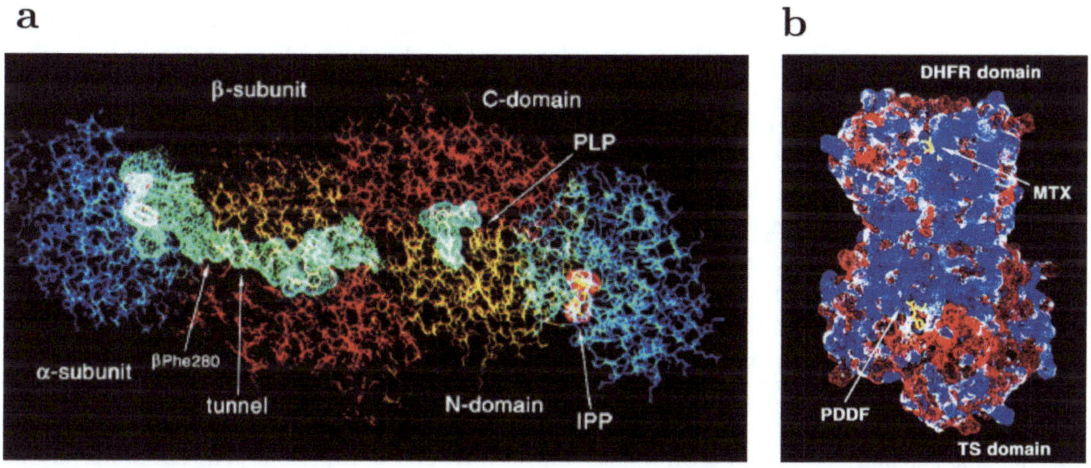

Fig. 2 Examples of bifunctional enzymes. (**a**) Tryptophan synthase. A molecular tunnel (green) connects the active sites of α subunit (blue) and β subunit (red and orange). Indole-3-propanol phosphate (IPP) locates the active site of α subunit, and pyridoxal phosphate (PLP) locates the active site of β subunit. (**b**) Thymidylate synthase–dihydrofolate reductase (TS-DHFR). Charge on the TS-DHFR surface is shown by red (negative) and blue (positive). Positively charged electrostatic highway (blue) connects the DHFR and TS active sites, indicated by white arrows and yellow spots (PDDF is 10-propargyl-5.8-dideazafolate and MTX is methotrexate). Reproduced from [5] under Creative Commons (CC) license

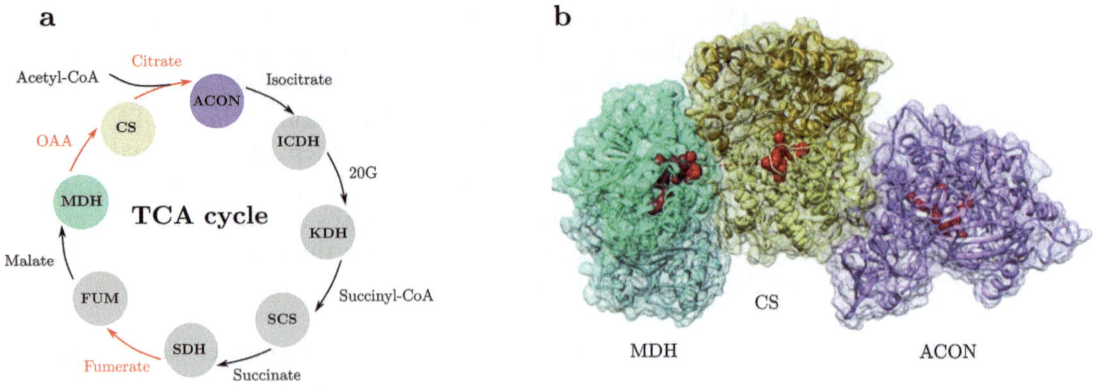

Fig. 3 Channelling and metabolons of the TCA cycle. (**a**) Schematic of tricarboxylic acid (TCA) cycle. The arrows show the transformations of metabolites between sequential enzymes, indicated in filled circles. The colored circles form a metabolon shown in (**b**). The red color denotes intermediates channelled between the corresponding enzymes. Apart from oxaloacetate (OAA) and citrate [73], channelling has also been reported for fumerate [82]. (**b**) Protein structure of a transient malate dehydrogenase–citrate synthase–aconitase (MDH-CS–ACON) complex [73]. The red spheres show the locations of the active sites. Panel (**b**) is reproduced from [55] under Creative Commons (CC) license

channelling mechanism has been confirmed in a number of Brownian dynamics simulation and kinetic studies [67–70].

Apart from TS-DHFR, electrostatic channelling has been reported for mitochondrial malate dehydrogenase–citrate synthase (MDH-CS) and CS–aconitase (ACON) [38, 39, 71–73] (which are, however, metabolons, *cf.* Subheading 2.2 and Fig. 3) and, more recently, for aldehyde dehydrogenase-hydratase (PaaZ enzyme) [74]. Liu et al. have introduced electrostatic channelling in a synthetic complex composed of hexokinase (HK) and glucose-6-phosphate dehydrogenase (G6PDH), covalently conjugated by a cationic oligopeptide bridge [75].

As we have pointed out at the beginning of this section, unlike molecular tunnels, electrostatic highways are not 100 % efficient, i.e., not all metabolites produced at the first active site reach the second active site of such a complex. For instance, kinetic studies show the efficiency of about 80 % in TS-DHFR from *L. major* [66], while Brownian dynamics simulations indicate over 95 % efficiency at zero salt concentration but much lower efficiencies (about 50 % [67] or even lower [70]) at physiological salt concentrations.

2.2 Metabolons and Enzyme Clusters

A metabolon is a "supramolecular complex of sequential metabolic enzymes" [40]. Such complexes are distinct from multifunctional enzymes (Subheading 2.1) in that they are transient, i.e., they can assemble and disassemble over time. Due to the dynamic nature of metabolons, their detection in vivo is more challenging than detecting stable enzyme complexes. The detection methods are often indirect, for instance, an analysis of lag time (time before a reaction reaches equilibrium), which differs for channelled and non-channelled systems, or perturbing a cascade reaction in bulk and observing changes in the reaction kinetics. A useful method is isotope dilution, which relies on the addition of known amounts of isotopically enriched (labelled) precursor metabolites and unlabelled (or differently labelled) intermediates [76]. Similarly, in a ^{13}C metabolic flux analysis, a ^{13}C-labelled precursor is supplied to a cell, and the distribution of labels in steady state, which depends on channelling at branching points, is analyzed by comparing experimental and simulated data [77–79].

The first metabolon has been reported for mitochondrial tricarboxylic acid (TCA) cycle, also known as Krebs or citric acid cycle [38, 39, 71–73]. The TCA cycle plays a crucial role in maintaining respiratory cell functionality by oxidizing the acetyl group of acetyl coenzyme A (Acetyl-CoA) for energy production [80]. Robinson and Srere reported that transient complexes might regulate metabolic fluxes by dynamically associating/disassociating and suggested that eight enzymes of the TCA cycle form a metabolon through weak non-covalent interactions [38, 39, 71, 72]. (These enzymes are: mitochondrial malate dehydrogenase (MDH), citrate synthase (CS), aconitase (ACON), iso-citrate dehydrogenase (ICDH), α-ketoglutarate dehydrogenase (KDH), succinyl-CoA synthetase (SCS), succinate dehydrogenase (SDH), and fumarase (FUM); see Fig. 3a.) Vélot et al. proposed a quinary structure model for an MDH-CS–ACON complex as a part of the TCA metabolon and identified electrostatic channels connecting the MDH, CS, and ACON active sites [72]. Brownian dynamics simulations showed that the transfer efficiency of oxaloacetate (OAA, an intermediate) between the MDH and CS active sites ranges from 50 % to 90 %, depending on the culture [81]. A low-resolution structure of the MDH–CS–ACON complex was resolved in a recent study with mass spectrometry and in vivo chemical cross-linking (Fig. 3b) [73]. In addition to citrate channelled by the CS–ACON complex, isotope dilution experiments also revealed channelling of fumarate between the SDH and FUM enzymes [82].

The existence of metabolons has been reported for several other metabolic pathways, including glycolysis [76, 83], oxidative pentose phosphate pathway [84], various primary and secondary plant metabolisms [56], *etc.* Apart from metabolons, cells tend to organize enzymes into clusters or membrane-bound microdomains. The enzymes within these clusters are not linked and diffuse freely, providing benefits by the locally enhanced concentration of sequential enzymes [85–87]. Two notable examples are glycosomes and purinosomes. Glycosomes are natural glycolytic clusters found in trypanosomatids, which contain enzymes of the first several steps of glycolysis [88]. Purinosomes contain all six de novo purine biosynthesis enzymes and appear under conditions of high purine demand [89, 90].

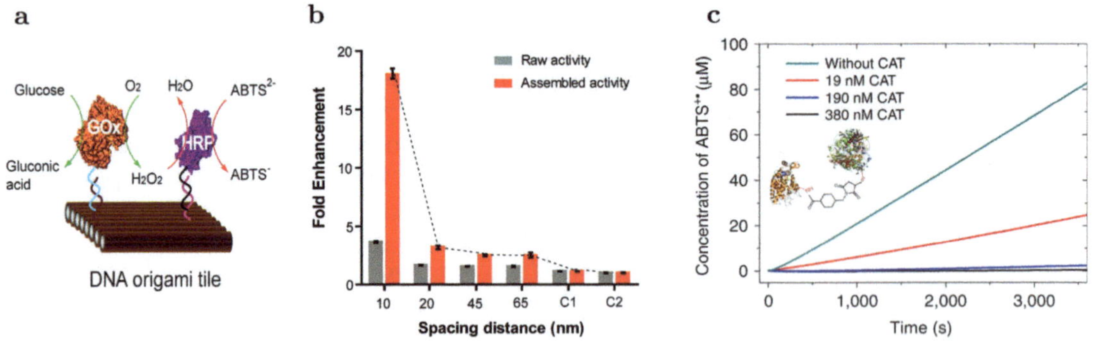

Fig. 4 Synthetic enzyme complexes. (**a**) Glucose oxidase (GOx) and horseradish peroxidase (HRP) attached to a DNA origami with well-controlled enzyme separation [95]. (**b**) Reaction velocity of the GOx–HRP cascade as a function of separation between the enzymes. The enhancement at 10 nm was attributed to enzyme proximity [95]. (**c**) For GOx–HPR covalently conjugated with a molecular linker [30], the velocity of the product (ABTS) formation depends on the concentration of catalase (competing enzyme) similarly as for a free, non-conjugated GOx–HRP cascade, despite the GOx–HRP proximity. For a channelling diagram rationalizing this behavior, see Fig. 3.7b. Panels (**a,b**) are adapted with permission from [95]. Copyright (2012) American Chemical Society. Panel (**c**) is reproduced from [30] under Creative Commons (CC) license

2.3 Synthetic Enzyme Complexes

Inspired by natural enzyme complexes in vivo, several approaches have been developed to conjugate, or co-localize sequential enzymes on synthetic scaffolds, aimed at generating artificial substrate channelling with the hope to accelerate the overall reaction rate [7, 91–97]. Artificial enzyme complexes can be constructed in vivo by gene fusion [98] or linking enzymes with scaffoldings [50], as well as in vitro via positionally dependent co-immobilizations, e.g., using DNA origami [92, 95, 99, 100] or natural scaffolding protein of cellulosomes [94, 101, 102].

Numerous studies have reported enhanced reaction kinetics, arguably due to enzyme proximity [7, 50, 92–95, 103–105]. For instance, Fu et al. [95] organized glucose oxidase (GOx) and horseradish peroxidase (HRP) on a DNA origami with well-controlled inter-enzyme separations (Fig. 4a). GOx catalyzes the oxidation of glucose and produces hydrogen peroxide (an intermediate), which is used by HRP to oxidize 2,2-azino-bis (3-ethylbenzothiazoline-6-sulphonic acid) (ABTS). The authors of [95] reported a 15-fold enhancement of reaction velocity for a GOx–HRP separation of 10 nm (Fig. 4b), which they attributed to GOx–HRP proximity. However, Zhang et al. challenged this view [30]. These authors demonstrated no enhancement for GOx–HRP covalently conjugated with a molecular linker, also in the presence of catalase, an enzyme competing for hydrogen peroxide (Fig. 4c). They suggested that the enhancement reported in [95] could be due to a locally reduced pH near the DNA nanostructure, which creates a favorable reaction environment for HRP [30]. They estimated that the pH lowers by two points in the vicinity of a DNA origami, leading to an enhancement similar to that observed in

[95]. However, how pH depends on the enzyme separation is not clear.

3 Effects of Channelling

3.1 Acceleration/Deceleration of Reaction Cascades

3.1.1 Direct Channelling

To understand how direct channelling influences reaction velocity, we consider a tandem reaction catalyzed by two sequential enzymes E_1 and E_2, described by the kinetic equations:

$$S + E_1 \underset{d_1}{\overset{a_1}{\rightleftharpoons}} SE_1 \overset{k_1}{\rightarrow} E_1 + I, \tag{3.1a}$$

and

$$I + E_2 \underset{d_2}{\overset{a_2}{\rightleftharpoons}} IE_2 \overset{k_2}{\rightarrow} E_2 + P. \tag{3.1b}$$

Here S is a substrate of the first enzyme, intermediate I is the product of the first reaction and a substrate for the second enzyme, and SE_1 and IE_2 are the standard Michaelis–Menten complexes; a_i and d_i are macroscopic association and dissociation rate constants, and k_i is the turnover number of i'th enzyme.

In an enzyme–enzyme complex, $E_{12} = E_1E_2$, the intermediates are channelled, and one can assume the following reaction kinetics [31]:

$$S + E_{12} \underset{d_{ch}}{\overset{a_{ch}}{\rightleftharpoons}} SE_{12} \overset{k_1}{\rightarrow} IE_{12} \overset{k_{ch}}{\longrightarrow} E_{12}I \overset{k_2}{\rightarrow} E_{12} + P, \tag{3.2}$$

where SE_{12} denotes a complex in which S is attached to the first active site of E_{12}. The substrate S in the SE_{12} complex is first transformed into I (complex IE_{12}) and then transferred along the tunnel to the second active site of E_{12} (complex $E_{12}I$), where it is transformed into the product P and released. Equation (3.2) assumes that the turnover numbers k_i are not affected by channelling, but the association/dissociation rates are, as they depend on diffusion of enzymes and enzyme complexes.

High Substrate Concentrations. For substrate concentrations, [S], larger than the Michaelis–Menten constants (*cf.* Eq. (3.5) below), the steady-state reaction rates are (if the first reaction is rate-limiting) [31]

$$\begin{aligned} v_{\text{non}} &= k_1[E_1] \text{and} \\ v_{\text{ch}} &= \left(\frac{1}{k_1} + \frac{1}{k_{ch}} + \frac{1}{k_2}\right)^{-1} [E_{12}], \end{aligned} \tag{3.3}$$

for the non-channelled and channelled cascades, respectively. Thus, for $[E_1] = [E_2] = [E_{12}]$, which we assume for a fair comparison, channelling cannot speed up the reaction velocity but slows it down. For a non-channelled cascade in a steady state, the intermediates are readily available for the E_2 enzyme in bulk. In contrast, in a E_{12} complex, an intermediate has to be converted from S, transported along the tunnel to the second active site of E_{12}, and only then it is converted to the product. In other words, in a E_{12} complex, the two reactions occur in series while they proceed in parallel if the cascade is non-channelled.

Low Substrate Concentrations. For substrate concentrations lower than the Michaelis–Menten constants, the steady-state reaction velocities are [31]

$$\begin{aligned} v_{\text{non}} &= \frac{k_1[\text{E}_1][\text{S}]}{K_M^{(1)}} \text{ and} \\ v_{\text{ch}} &= \frac{k_1[\text{E}_{12}][\text{S}]}{K_M^{\text{ch}}}, \end{aligned} \tag{3.4}$$

where

$$\begin{aligned} K_M^{\text{ch}} &= \frac{k_1 + d_{\text{ch}}}{a_{\text{ch}}} \text{ and} \\ K_M^{(1)} &= \frac{k_1 + d_1}{a_1} \end{aligned} \tag{3.5}$$

are the Michaelis–Menten constants. Taking again $[\text{S}_1] = [\text{S}_{12}]$, we see that for channelling to speed up a reaction cascade, K_M^{ch} must be smaller than $K_M^{(1)}$. The Michaelis–Menten constants depend on association and dissociation rates, which can be split into intrinsic (microscopic) and diffusion (k_D) rate constants [106, 107]; here, $k_D \approx 4\pi DR$, where R is the reaction radius and D is the mutual diffusion coefficient of a substrate and enzyme E_1 or complex E_{12}. Assuming that intrinsic association/dissociation rates are not affected by channelling and using Brownian dynamics simulations to determine k_D for the channelled and non-channelled model systems, Poshyvailo et al. showed that $K_M^{\text{ch}} \approx K_M^{(1)}$ [31]. This result means that in this case, too, channelling does not enhance the overall reaction velocity in steady state.

Similar conclusions hold for more complex systems, such as various bi-substrate reactions and enzyme complexes allowing for more than one intermediate in their channels [31]. Although channelling is unlikely to speed up the steady-state reaction velocity in all these cases, a channelled cascade may, however, proceed faster at short times, before the concentration of intermediates in the non-channelled system has built up [31]. One can speculate that this transient enhancement is beneficial for cell metabolism when exposed to fluctuating environmental conditions. Still, the role of channelling in the biophysical processes of living cells remains unclear.

3.1.2 Enzyme Proximity

Several articles have indicated that in steady state, enzyme proximity cannot enhance reaction velocity if the intermediates are not sequestrated [27–29, 31, 33, 42]. The conservation law dictates that the rate of forming intermediary metabolites and their conversion into the product (or next metabolite in a cascade) must be equal in steady state; otherwise, the concentration of intermediates would change over time. This argument is mathematically rigorous and independent of the position of enzymes, reaction rates, diffusion constants, and other factors. It means that similarly to direct channelling (Subheading 3.1.1), proximity channelling cannot enhance reaction velocity in a steady-state.

We consider the effect of proximity before the steady state is achieved and assume that initially, when a cascade reaction commences, the concentration of intermediates, [I], is zero. As the reaction progresses, [I] increases to its steady-state value, but this increase is non-uniform in space and time. During this stage, the highest concentration of I is in the vicinity of the first enzyme, meaning that the velocity of a cascade reaction can indeed benefit from enzyme proximity. Figure 5 shows how the distribution of intermediates changes over time, as obtained from the solution of the reaction–diffusion model for the GOx–HRP cascade [28]. By 1 s, the concentration of hydrogen peroxide at the second enzyme (HRP) is practically indistinguishable from the bulk value, which means that the GOx–HRP cascade can benefit from enzyme proximity at most within the first second.

By comparing the influx of intermediates to the second enzyme coming directly from the first enzyme and from the bulk, Idan and Hess [28] have estimated that these fluxes become comparable at time $t_{ch} = (k_{D\ell}[E_{12}])^{-1}$, where $[E_{12}]$ is the concentration of enzyme complexes and $k_{D\ell} = 4\pi D\ell$, with D being the mutual diffusion coefficient and ℓ the separation between the active sites. Note that t_{ch} does not depend on enzyme activities, i.e., it is determined

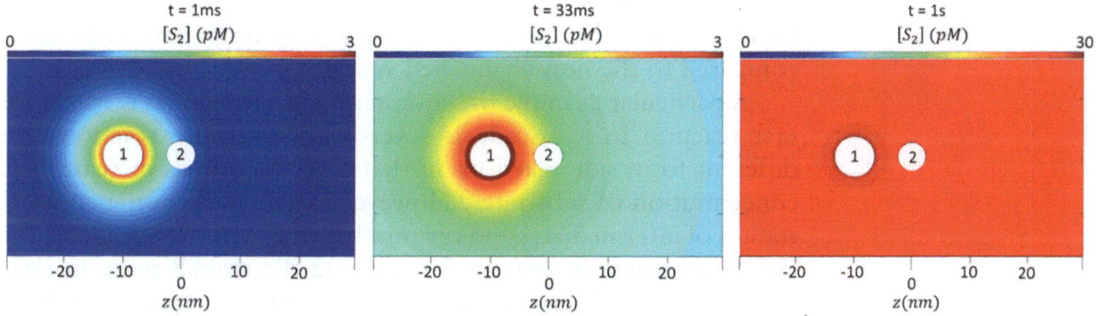

Fig. 5 Evolution of hydrogen peroxide around GOx–HRP complex. Concentration of hydrogen peroxide (here $[S_2]$) as obtained from the solution of the reaction–diffusion model for GOx–HRP cascade at three different times. GOx and HRP are denoted 1 and 2. Reproduced with permission from [28]. Copyright (2013) American Chemical Society

solely by diffusive (D) and geometric ($[E_{12}]$ and ℓ) properties. The time t_{ch} decreases with increasing the enzyme concentration and increases for slower intermediates and larger separations. Taking the diffusion coefficient of hydrogen peroxide [95] $D = 10^3 \mu m^2 s^{-1}$ and GOx–HRP separation $\ell = 10$ nm, we find $t_{ch} \approx 10$ ms for enzyme concentration $[E_{12}] = 1$ nM and just 10 μs for $[E_{12}] = 1$ μM. For slower metabolites, this time is slightly larger, for instance, $t_{ch} \approx 23$ ms for $D = 600\ \mu m^2 s^{-1}$ corresponding to glucose in water at room temperature.

3.2 Protection of Intermediates

Channelling may play a central role in protecting intermediary metabolites from sequestration due to degradation or competing or reversible reactions. The effect of sequestration is the reduced bulk concentration of intermediates, which slows down a cascade reaction. Protecting intermediates from competing reactions, by channelling them to the second active site of an enzyme complex, may enhance the overall reaction velocity. We now quantify this enhancement for direct and proximity channelling.

3.2.1 Direct Channelling

We consider again a tandem reaction (3.1a, b) with the direct channelling given by Eq. (3.2). Sequestration of intermediates can be described by adding an equation:

$$\frac{d[I]}{dt} = -k_{seq}[I] \tag{3.6}$$

to Eqs. (3.1a, 3.1b). For a side reaction, the "sequestration" rate $k_{seq} = [E_3]k_3/K_M^{(3)}$, where E_3 is an enzyme competing for intermediates I and k_3 and $K_M^{(3)}$ are its turnover number and Michaelis–Menten constant, respectively. We assume that this reaction is absent in the channelled cascade.

The steady-state solution to these equations is presented in Fig. 6a, which shows that channelling is not always beneficial to a reaction cascade. There is a threshold value of k_{seq} (or $[E_3]$), above which channelling accelerates the cascade (shaded areas in Fig. 6a). In this region, however, the reaction can be a few folds faster compared to the non-channelled system.

A particular example of sequestering intermediates is when the first reaction in a cascade is reversible. Reversible reactions are different from side reactions in that reverse catalysis increases the concentration of substrates. However, it also reduces the concentration of intermediates and the number of E_1 enzymes available for the direct reaction, potentially reducing the overall reaction rate. Instead of Eqs. (3.1a, b), the following kinetic equation describes a reversible reaction:

$$S + E_1 \underset{d_1}{\overset{a_1}{\rightleftharpoons}} SE_1 \underset{\bar{k}_1}{\overset{k_1}{\rightleftharpoons}} E_1 + I, \tag{3.7}$$

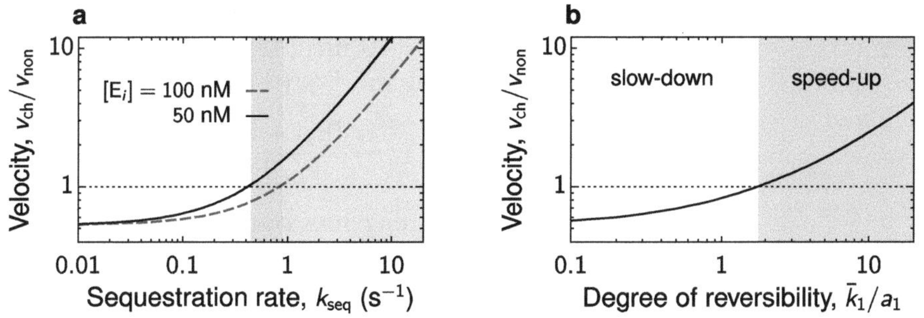

Fig. 6 Effect of direct, non-leaky channelling. (**a**) Channelling can speed up a cascade reaction if the intermediates degrade or participate in side reactions; in the latter case, $k_{seq} \approx [E_3]k_3/K_M^{(3)}$, where E_3 is a competing enzyme and k_3 and $K_M^{(3)}$ are its turnover number and Michaelis–Menten constant, respectfully. (**b**) Channelling can enhance reaction velocity if the first reaction in a cascade is reversible (Eq. (3.7) in lieu of Eqs. (3.1a, b)). The ratio v_{non}/v_{ch} is independent of [E$_i$]. In both panels, the concentrations of enzymes and enzyme–enzyme complexes are equal (i.e., $[E_i] = [E_1] = [E_2] = [E_{12}]$). The shadowed areas show the regions where channelling speeds up the reaction. Note that in panel (**a**), there are two overlapping shaded areas. Adapted from [31] under Creative Commons (CC) license

where $\bar{k}_1$ is the rate of the reverse I→S reaction catalyzed by E_1. We assume that the enzyme complex does not catalyze the reverse reaction. Figure 6b shows that similarly to competing reactions, there is a threshold value of the degree of reversibility ($\bar{k}_1/a_1$), above which channelling enhances the overall reaction velocity. Unlike competing reactions, this threshold value depends only on the rate constants but not on the enzyme concentration (provided $[E_1]=[E_2]=[E_{12}]$).

3.2.2 Enzyme Proximity

To estimate to what extent enzyme proximity protects metabolites from sequestering, Kuzmak et al. considered a model system shown schematically in Fig. 1c and solved the appropriate reaction–diffusion system analytically. They arrived at an approximate expression for reaction enhancement [33]:

$$\delta v = \frac{v_{ch} - v_{non}}{v_{non}} = \frac{k_{E_3}[E_3] + k_{E_2}[E]}{k_{D\ell}[E]}\left[1 - \left(\frac{3\ell^3}{4\pi[E]}\right)^{1/3}\right], \tag{3.8a}$$

where $k_{E_2} = k_2/K_M^{(2)}$ is the rate constant of the second enzyme of a cascade, [E] is the concentration of enzymes and enzyme complexes, assumed the same (i.e., $[E]=[E_1]=[E_2]=[E_{12}]$), and $[E_3]$ is the concentration and $k_{E_3} = k_3/K_M^{(3)}$ the rate constant of an enzyme competing for the intermediates (here, as before, k_i and $K_M^{(i)}$ ($i = 1, 2, 3$) are the turnover number and the Michaelis–Menten constant of i'th enzyme, respectfully). For small separations or low enzyme concentrations, satisfying $[E] \ll 3/(4\pi\ell^3)$ ($\approx$0.4 mM for $\ell = 10$ nm), the second term in the square brackets can be neglected and Eq. (3.8a, b) simplifies to

$$\delta v = \frac{k_{E_3}[E_3] + k_{E_2}[E]}{k_{D\ell}[E]}. \tag{3.8b}$$

Suppose that competing enzymes outperform the second enzyme of a cascade, i.e., $k_{E_3}[E_3] \gg k_{E_2}[E]$. In this case, Eqs. (3.8a, b) reduces to $\delta v = k_{E_3}[E_3]/(k_{D\ell}[E])$ and the enhancement is determined by how fast the intermediates are sequestrated compared to their diffusion rate. This limit has been considered by Idan and Hess [28], who arrived at a similar equation for δv.

In the opposite limit $k_{E_3}[E_3] \ll k_{E_2}[E]$, δv depends only weakly on k_{E_3} and $[E_3]$, which might seem surprising and suggest that enzyme proximity can speed up reaction velocity also in the absence of competing enzymes. However, competing enzymes are essential as they remove the intermediates from the system, making the conservation-law argument (beginning of Subheading 3.1.2) not any more valid.

Equations (3.8a, b) implies a linear relation between the concentrations of enzyme complexes ($[E]$) and competing enzymes ($[E_3]$):

$$\begin{aligned} &(k_{D\ell}\delta v - k_{E_2})[E] \\ &\quad = k_{E_3}[E_3]. \end{aligned} \tag{3.9}$$

This equation gives the concentration of conjugates, $[E]$, below which proximity channelling provides an enhancement δv or higher, or the concentration of competing enzymes, $[E_3]$, above which the reaction enhancement is higher than δv.

We consider, as an example, a popular GOx–HRP conjugate and catalase as competing enzymes (Subheading 2.3, Fig. 4). Figure 7a compares reaction velocities of the channelled and non-channelled systems as a function of catalase concentration, [CAT]. Channelling is beneficial to the cascade only at high (millimolar) concentrations of catalase. The threshold value of [CAT], above which channelling speeds up the reaction, decreases with decreasing [GOx–HRP] but reaches reasonably low (micromolar) values only for subnanomolar concentrations of GOx–HRP conjugates. A diagram in Fig. 7b summarizes this behavior, showing, for comparison, the catalase-densest system considered in [30] (380 nM [CAT] in Fig. 4c). It is also worth noting that approximate expression (3.9) agrees well with full numerical calculations.

Using Eq. (3.9), Kuzmak et al. [33] have estimated that for comparable concentrations of enzyme conjugates and competing enzymes, proximity can be advantageous to a reaction cascade only for "perfect" enzymes, catalyzing reactions at rates about

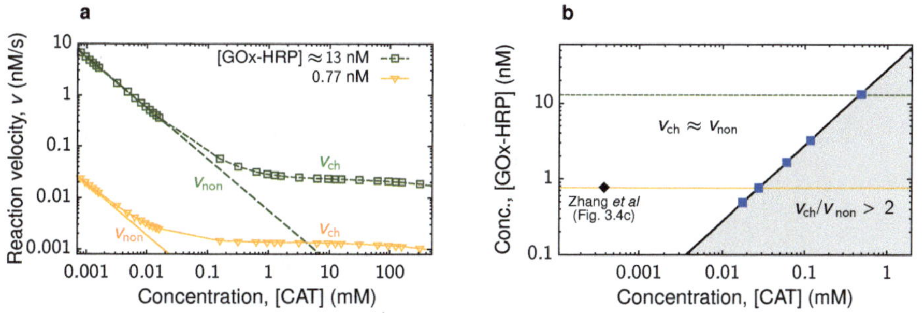

Fig. 7 Enzyme proximity for GOx–HRP cascade. (**a**) Proximity channelling can speed up a GOx–HRP reaction cascade if hydrogen peroxide (an intermediate) is sequestered by catalase (CAT). For high catalase concentrations, velocity of the channelled system can be orders of magnitude higher than of the non-channelled one. (**b**) Diagram showing the region where the enzyme proximity enhances reaction velocity at least twice compared to the non-channelled system. The thin horizontal lines show the values of [GOx–HRP] used in panel (**a**). The diamond shows the values of [GOx–HRP] and [CAT] from [30], closest to the enhancement region (380 nM CAT in Fig. 3.4c). The squares denote the results of full numerical calculations and the thick black line is given by Eq. (3.9). Adapted from [33] under Creative Commons (CC) license

$10^9 M^{-1}s^{-1}$ or higher. However, higher concentrations of competing enzymes may make channelling beneficial for slower enzymes, too.

3.3 Protection of Surroundings

Decreased concentration of intermediates, which might be toxic to a cell, has been frequently mentioned as a potential benefit of channelling [4, 7, 25, 41]. Channelling may also prevent high metabolite concentrations in bulk and consequently high cytoplasm viscosity, which would slow down or hamper biological processes [108]. Elevated metabolite concentration in the cytoplasm can also interfere with the regulation of other metabolic pathways.

For multifunctional enzymes providing direct, non-leaky channelling, the intermediates are not released into bulk, and hence the bulk concentration of intermediates is zero. However, suppose that both channelled and non-channelled cascades are present in a system. In that case, the concentration of intermediates can decrease or increase for an increasing fraction of enzyme complexes [17, 22–25, 31] (Fig. 8). The increase may look surprising, but it occurs, e.g., if the first enzyme in a cascade is more abundant than the second, with their concentrations satisfying $k_1[E_1] = k_2[E_2]$ (this relation maximizes the overall reaction velocity of a non-channelled cascade in steady state). When E_1 and E_2 enzymes form a E_{12} complex, the reduced number of E_2 enzymes cannot efficiently convert the metabolites produced by the first enzymes, which are in excess, leading to an increase in the concentration of these metabolites [31].

We note that enzyme proximity is not effective in protecting the enzymes' environment from intermediates, as Fig. 5 shows.

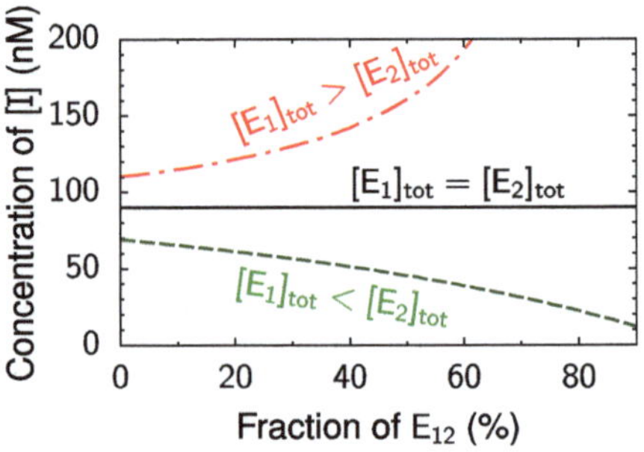

Fig. 8 Effect of channelling on metabolite pool size. Concentration of intermediates as a function of fraction of enzyme complexes, $[E_{12}]/[E_{12}]_{max}$, in a system. $[E_{12}]_{max}$ is the maximum possible concentration of E_{12} equal to the minimum of $[E_1]_{tot}$ and $[E_2]_{tot}$, where $[E_i]_{tot}$ is the total concentration of enzymes E_i (*i.e.*, the sum of free enzymes and enzymes assembled into E_{12}). The concentration of intermediates can increase or decrease with the formation of enzyme complexes. Adapted from [31] under Creative Commons (CC) license

Brownian dynamics simulations and kinetic studies of multifunctional enzymes [67, 67, 68, 68, 69, 69, 70, 70] and metabolons [81, 109] with electrostatic highways suggest that guided channelling may protect the surrounding to a certain degree. However, we have not found studies quantifying this effect on the level of metabolite pool size.

4 Macromolecular Crowding

In living cells, 20 % to 40 % of the volume is occupied by macromolecules [110, 111]. Such macromolecular crowding plays a crucial role in cellular life [112–114], affecting protein stability [115], chemical equilibria [112], gene regulation [116], enzymatic reactions [117–119], and diffusion of metabolites and macromolecules [120–125]. While macromolecular diffusion and single enzymatic reactions have been extensively investigated [117–125], the studies of substrate channelling under crowding are scarce [33, 126].

Equations (3.8a, 3.8b) and (3.9) suggest that the reduced metabolite mobility due to crowding [121] (implying the reduced $k_{D\ell}$) enhances reaction velocity more significantly than in a dilute system. Figure 9 illustrates this for transketolase–transaldolase (TK–TL) complex and triose-phosphate isomerase (TPI) as competing enzymes, showing that in crowded systems, enzyme proximity may indeed be more beneficial for reaction velocity. We note, however, that this result follows from an effective mean-field model, which does not take into account molecular size-specific effects, relevant to crowded environments [117–125]. More work

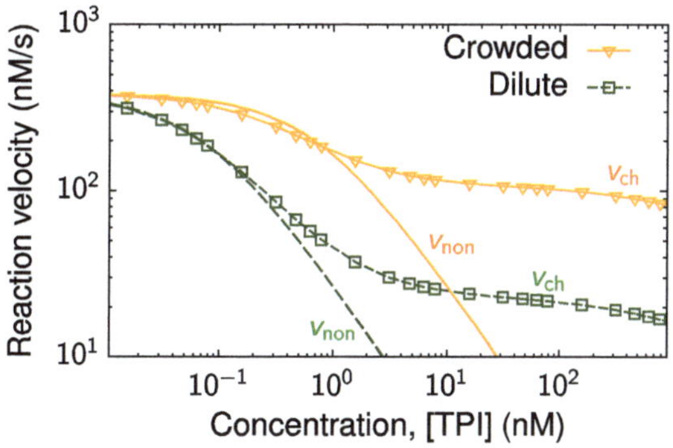

Fig. 9 Enzyme proximity for dilute and crowded systems. Proximity channelling of glyceraldehyde 3-phosphate (g3p) by a hypothetical transketolase–transaldolase (TK–TAL) complex with triose-phosphate isomerase (TPI) as competing enzymes. The diffusion coefficients of g3p are $D = 600\ \mu m^2 ns^{-1}$ in the dilute and $D = 60\ \mu m^2 ns^{-1}$ in the crowded system. Under crowding, the effect of enzyme proximity on reaction velocity can be stronger than in a dilute system. Adapted from [33] under Creative Commons (CC) license

is required to understand how such effects influence substrate channelling and clarify its role in cellular life.

5 Concluding Remarks

Decades of research have led to a wealth of information on substrate channelling in living cells. Numerous stable and dynamic enzyme complexes have been revealed and confirmed in various metabolic pathways with different methods (Subheadings 2.1 and 2.2). However, our understanding of these complexes and their physiological significance remains obscure. It has been challenging to demonstrate substrate channelling among more than three enzymes [8], while direct evidence of channelling even between two enzymes is rare. Many more questions remain unanswered. What are the mechanisms of metabolic regulation with substrate channelling? What is the role of macromolecular crowding? Why and how enzyme clusters and metabolons form, and what are their control mechanisms? How metabolic channelling affects the response to external stimuli?

In biotechnology, a significant effort has been channelled into bringing enzymes together (Subheading 2.3). Multiple studies reported several-fold enhancement of reaction velocities, presumably due to enzyme proximity. However, as we have seen in this review (Subheading 3.1), theoretical and modelling studies show that enzyme proximity cannot induce the enhancements observed experimentally, necessitating the revision of the experimental data. Such enhancements may originate from changes in the local enzymes' environments [30, 127], challenging substrate channelling as a tool for facilitated metabolic throughput. Can enzyme proximity nevertheless provide technological advantages? What are other aspects of channelling the biotechnology can exploit?

Answering these questions requires a collaborative work of physics, biology, and chemistry and a better understanding of raison d'être of substrate channelling. For now one can conclude that despite several decades of history and the enormous research effort, substrate channelling and its role in in vivo processes and biotechnology still remain elusive. Yet, it continues to be an attractive subject for further investigation.

Acknowledgements

This work was supported by National Science Centre grant No. 2017/25/B/ST3/02456 to S.K.

References

1. Kholodenko BN, Demin OV, Westerhoff HV (1993) Channelled' pathways can be more sensitive to specific regulatory signals. FEBS Lett 320:75–78. https://doi.org/10.1016/0014-5793(93)81661-I
2. Yon-Kahn J, Hervé G (2009) Mol Cell Enzymol; Springer, Berlin, Heidelberg, pp 679–722. https://doi.org/10.1007/978-3-642-01228-0_16
3. Seelig B (2017) Multifunctional enzymes from reduced genomes model proteins for simple primordial metabolism? Mol Microbiol 105:505–507. https://doi.org/10.1111/mmi.13742
4. Ovádi J (1991) Physiological significance of metabolic channelling. J Theor Biol 152: 1–22. https://doi.org/10.1016/S0022-5193(05)80500-4
5. Miles EW, Rhee S, Davies DR (1999) The molecular basis of substrate channeling. J Biol Chem 274:12193–12196. https://doi.org/10.1074/jbc.274.18.12193
6. Ovádi J, Saks V (2004) On the origin of intracellular compartmentation and organized metabolic systems. Mol Cell Biochem 256: 5–12. https://doi.org/10.1023/B:MCBI.0000009855.14648.2c
7. Zhang Y-HP (2011) Substrate channeling and enzyme complexes for biotechnological applications. Biotechnol Adv 29:715–725. https://doi.org/10.1016/j.biotechadv.2011.05.020
8. Schmitt DL, An S (2017) Spatial organization of metabolic enzyme complexes in cells. Biochemistry 56:3184–3196. https://doi.org/10.1021/acs.biochem.7b00249
9. Wang T, Qin X, Liang C, Yuan H (2018) Engineering substrate channeling in biosystems for improved efficiency. J Chem Technol Biotechnol 93:3364–3373. https://doi.org/10.1002/jctb.5731
10. Srivastava DK, Bernhard SA (1987) Mechanism of transfer of reduced nicotinamide adenine dinucleotide among dehydrogenases. Transfer rates and equilibria with enzyme-enzyme complexes. Biochemistry 26: 1240–1246. https://doi.org/10.1021/bi00379a006
11. Chock PB, Gutfreund H (1988) Reexamination of the kinetics of the transfer of NADH between its complexes with glycerol-3-phosphate dehydrogenase and with lactate dehydrogenase. Proc Natl Acad Sci 85: 8870–8874. https://doi.org/10.1073/pnas.85.23.8870
12. Srivastava DK, Smolen P, Betts GF, Fukushima T, Spivey HO, Bernhard SA (1989) Direct transfer of NADH between alpha-glycerol phosphate dehydrogenase and lactate dehydrogenase: fact or

misinterpretation? Proc Natl Acad Sci 86: 6464–6468. https://doi.org/10.1073/pnas.86.17.6464

13. Wu XM, Gutfreund H, Lakatos S, Chock PB (1991) Substrate channeling in glycolysis: a phantom phenomenon. Proc Natl Acad Sci 88:497–501. https://doi.org/10.1073/pnas.88.2.497
14. Knowles JR (1991) Calmer waters in the channel? J Theor Biol 152:53–55. https://doi.org/10.1016/S0022-5193(05)80509-0
15. Cárdenas ML (1991) Are the transitory enzyme-enzyme complexes found in vitro also transitory in vivo? If so, are they physiologically important? J Theor Biol 152: 111–113. https://doi.org/10.1016/S0022-5193(05)80522-3
16. Clegg JS (1991) The physiological significance of metabolite channeling: an idea whose time has come. J Theor Biol 152: 63–64. https://doi.org/10.1016/S0022-5193(05)80511-9
17. Cornish-Bowden A (1991) How much effect on free metabolite concentrations does channelling have? J Theor Biol 152:39–40. https://doi.org/10.1016/S0022-5193(05)80505-3
18. Easterby J (1991) Homeostasis, flexibility and conflict in the kinetic advantage of channelling. J Theor Biol 152:47–48. https://doi.org/10.1016/S0022-5193(05)80507-7
19. Gutfreund H, Chock P (1991) Substrate channeling among glycolytic enzymes: Fact or fiction. J Theor Biol 152:117–121. https://doi.org/10.1016/S0022-5193(05)80524-7
20. Srere PA (1991) Channeling: the pathway that cannot be beaten. J Theor Biol 152:23. https://doi.org/10.1016/S0022-5193(05)80501-6
21. Ovádi J (1991) Physiological significance of metabolite channelling: Author's response to commentaries. J Theor Biol 152:135–141. https://doi.org/10.1016/S0022-5193(05)80528-4
22. Cornish-Bowden A (1991) Failure of channelling to maintain low concentrations of metabolic intermediates. Eur J Biochem 195: 103–108. https://doi.org/10.1111/j.1432-1033.1991.tb15681.x
23. Mendes P Kell DB, Westerhoff HV (1992) Channelling can decrease pool size. Eur J Biochem 204:257–266. https://doi.org/10.1111/j.1432-1033.1992.tb16632.x
24. Cornish-Bowden A, Cardenas ML (1993) Channelling can affect concentrations of metabolic intermediates at constant net flux: artefact or reality? Eur J Biochem 213: 87–92. https://doi.org/10.1111/j.1432-1033.1993.tb17737.x
25. Menedes P, Kell DB, Westerhoff HV, Mendes P (1996) Why and when channelling can decrease pool size at constant net flux in a simple dynamic channel. Biochim Biophys Acta - Gen Subj 1289:175–186. https://doi.org/10.1016/0304-4165(95)00152-2
26. Cornish-Bowden A, Cárdenas ML, Letelier J-C, Soto-Andrade J, Abarzúa FG (2004) Understanding the parts in terms of the whole. Biol Cell 96:713–717. https://doi.org/10.1016/j.biolcel.2004.06.006
27. Idan O, Hess H (2012) Diffusive transport phenomena in artificial enzyme cascades on scaffolds. Nat Nanotechnol 7:769–770. https://doi.org/10.1038/nnano.2012.222
28. Idan O, Hess H (2013) Origins of activity enhancement in enzyme cascades on scaffolds. ACS Nano 7:8658–8665. https://doi.org/10.1021/nn402823k
29. Wheeldon I, Minteer SD, Banta S, Barton SC, Atanassov P, Sigman M (2016) Substrate channelling as an approach to cascade reactions. Nat Chem 8:299–309. https://doi.org/10.1038/nchem.2459
30. Zhang Y, Tsitkov S, Hess H (2016) Proximity does not contribute to activity enhancement in the glucose oxidasehorseradish peroxidase cascade. Nat Commun 7:13982. https://doi.org/10.1038/ncomms13982
31. Poshyvailo L, von Lieres E, Kondrat S (2017) Does metabolite channeling accelerate enzyme-catalyzed cascade reactions? PLoS One 12:e0172673. https://doi.org/10.1371/journal.pone.0172673
32. Sweetlove LJ, Fernie AR (2018) The role of dynamic enzyme assemblies and substrate channelling in metabolic regulation. Nat Commun 9:2136. https://doi.org/10.1038/s41467-018-04543-8
33. Kuzmak A, Carmali S, von Lieres E, Russell AJ, Kondrat S (2019) Can enzyme proximity accelerate cascade reactions? Sci Rep 9:455. https://doi.org/10.1038/s41598-018-37034-3
34. Reed LJ, Cox DJ (1966) Macromolecular Organization of Enzyme Systems. Annu Rev Biochem 35:57–84. https://doi.org/10.1146/annurev.bi.35.070166.000421
35. Cori CF, Velick SF, Cori GT (1950) The combination of diphosphopyridine nucleotide with glyceraldehyde phosphate dehydrogenase. Biochim Biophys Acta 4:160–169. https://doi.org/10.1016/0006-3002(50)90020-5

36. Srivastava DK, Bernhard, SA (1984) Direct transfer of reduced nicotinamide adenine dinucleotide from glyceraldehyde 3-phosphate dehydrogenase to liver alcohol dehydrogenase. Biochemistry 23: 4538–4545. https://doi.org/10.1021/bi00315a006
37. Srivastava D, Bernhard S (1986) Metabolite transfer via enzyme-enzyme complexes. Science (80-) 234:1081–1086. https://doi.org/10.1126/science.3775377
38. Robinson JB, Srere PA (1985) Organization of Krebs tricarboxylic acid cycle enzymes in mitochondria. J Biol Chem 260: 10800–10805. https://doi.org/10.1016/S0021-9258(19)85153-0
39. Robinson JB, Inman L, Sumegi B, Srere PA (1987) Further characterization of the Krebs tricarboxylic acid cycle metabolon. J Biol Chem 262:1786–1790. https://doi.org/10.1016/S0021-9258(19)75707-X
40. Srere PA (1985) The metabolon. Trends Biochem Sci 10:109–110. https://doi.org/10.1016/0968-0004(85)90266-X
41. Mathews CK (1993) The cell-bag of enzymes or network of channels? J Bacteriol 175: 6377–6381. https://doi.org/10.1128/JB.175.20.6377-6381.1993
42. Idan O, Hess H (2013) Engineering enzymatic cascades on nanoscale scaffolds. Curr Opin Biotechnol 24:606–611. https://doi.org/10.1016/j.copbio.2013.01.003
43. Dewar MJ, Storch DM (1985) Alternative view of enzyme reactions. Proc Natl Acad Sci 82:2225–2229. https://doi.org/10.1073/pnas.82.8.2225
44. Thoden JB, Holden HM, Wesenberg G, Raushel FM, Rayment I (1997) Structure of carbamoyl phosphate synthetase: A journey of 96 Å from substrate to product. Biochemistry 36:6305–6316. https://doi.org/10.1021/bi970503q
45. Sauers CK, Jencks WP, Groh S (1975) Alcohol-bicarbonate-water system. Structure-reactivity studies on the equilibriums for formation of alkyl monocarbonates and on the rates of their decomposition in aqueous alkali. J Am Chem Soc 97: 5546–5553. https://doi.org/10.1021/ja00852a038
46. Broussard TC, Pakhomova S, Neau DB, Bonnot R, Waldrop GL (2015) Structural analysis of substrate, reaction intermediate, and product binding in haemophilus influenzae biotin carboxylase. Biochemistry 54: 3860–3870. https://doi.org/10.1021/acs.biochem.5b00340
47. Geck MK, Kirsch JF (1999) A novel, definitive test for substrate channeling illustrated with the aspartate aminotransferase/malate dehydrogenase system. Biochemistry 38: 8032–8037. https://doi.org/10.1021/bi983029c
48. Jørgensen K, Rasmussen AV, Morant M, Nielsen AH, Bjarnholt N, Zagrobelny M, Bak S, Møller BL (2005) Metabolon formation and metabolic channeling in the biosynthesis of plant natural products. Curr Opin Plant Biol 8:280–291. https://doi.org/10.1016/j.pbi.2005.03.014
49. Kizer L, Pflera DJ, Pfleger BF, Keasling JD (2008) Application of functional genomics to pathway optimization for increased isoprenoid production. Appl Environ Microbiol 74: 3229–3241. https://doi.org/10.1128/AEM.02750-07
50. Dueber JE, Wu GC, Malmirchegini GR, Moon TS, Petzold CJ, Ullal AV, Prather KLJ, Keasling JD (2009) Synthetic protein scaffolds provide modular control over metabolic flux. Nat Biotechnol 27:753–759. https://doi.org/10.1038/nbt.1557
51. Hyde CC, Ahmed SA, Padlan EA, Miles EW, Davies DR (1988) Three-dimensional structure of the tryptophan synthase alpha 2 beta 2 multienzyme complex from Salmonella typhimurium. J Biol Chem 263: 17857–17871. https://doi.org/10.1016/S0021-9258(19)77913-7
52. Perham RN (2000) Swinging arms and swinging domains in multifunctional enzymes: Catalytic machines for multistep reactions. Annu Rev Biochem 69:961–1004. https://doi.org/10.1146/annurev.biochem.69.1.961
53. Huang X, Holden HM, Raushel FM (2001) Channeling of substrates and intermediates in enzyme-catalyzed reactions. Annu Rev Biochem 70:149–180. https://doi.org/10.1146/annurev.biochem.70.1.149
54. Haanstra JR, González-Marcano EB, Gualdrón-López M, Michels PA (2016) Biogenesis, maintenance and dynamics of glycosomes in trypanosomatid parasites. Biochim Biophys Acta Mol Cell Res 1863:1038–1048. https://doi.org/10.1016/j.bbamcr.2015.09.015
55. Fernie AR, Zhang Y, Sweetlove LJ (2018) Passing the Baton: Substrate channelling in respiratory metabolism. Research 2018: 1–16. https://doi.org/10.1155/2018/1539325
56. Zhang Y, Fernie AR (2021) Metabolons, enzyme-enzyme assemblies that mediate substrate channeling, and their roles in plant metabolism. Plant Commun 2:100081.

https://doi.org/10.1016/j.xplc.2020.100081
57. Jandt U, You C, Zhang YH, Zeng A-P (2013) Advances in biochemical engineering/biotechnology, vol 137, pp 41–65. Springer, Berlin, Heidelberg. https://doi.org/10.1007/10_2013_221
58. Pröschel M, Detsch R, Boccaccini AR, Sonnewald U (2015) Engineering of metabolic pathways by artificial enzyme channels. Front Bioeng Biotechnol 3:168. https://doi.org/10.3389/fbioe.2015.00168
59. Pravda L, Berka K, Svobodová Vaeková R, Sehnal D, Banáš P, Laskowski RA, Koča J, Otyepka M (2014) Anatomy of enzyme channels. BMC Bioinformatics 15:379. https://doi.org/10.1186/s12859-014-0379-x
60. Miles EW (1979) Adv. Enzymol. Relat. Areas Mol. Biol, vol 49, pp 127–186. Wiley Blackwell. https://doi.org/10.1002/9780470122945.ch4
61. Holden HM, Thoden JB, Raushel FM (1998) Carbamoyl phosphate synthetase: a tunnel runs through it. Curr Opin Struct Biol 8: 679–685. https://doi.org/10.1016/S0959-440X(98)80086-9
62. Manjasetty BA, Powlowski J, Vrielink A (2003) Crystal structure of a bifunctional aldolase-dehydrogenase: Sequestering a reactive and volatile intermediate. Proc Natl Acad Sci 100:6992–6997. https://doi.org/10.1073/pnas.1236794100
63. Smith NE, Vrielink A, Attwood PV, Corry B (2012) Biological channeling of a reactive intermediate in the bifunctional enzyme DmpFG. Biophys J 102:868–877. https://doi.org/10.1016/j.bpj.2012.01.029
64. Ferone R, Roland S (1980) Dihydrofolate reductase: thymidylate synthase, a bifunctional polypeptide from Crithidia fasciculata. Proc Natl Acad Sci 77:5802–5806. https://doi.org/10.1073/pnas.77.10.5802
65. Stroud RM (1994) An electrostatic highway. Nat Struct Biol 1:131–134. https://doi.org/10.1038/nsb0394-131
66. Knighton DR, Kan C-C, Howland E, Janson CA, Hostomska Z, Welsh KM, Matthews DA (1994) Structure of and kinetic channelling in bifunctional dihydrofolate reductase-thymidylate synthase. Nat Struct Biol 1: 186–194. https://doi.org/10.1038/nsb0394-186
67. Elcock AH, Potter MJ, Matthews DA, Knighton DR, Mc-Cammon J (1996) Electrostatic channeling in the bifunctional enzyme dihydrofolate reductase-thymidylate synthase. J Mol Biol 262:370–374. https://doi.org/10.1006/jmbi.1996.0520
68. Trujillo M, Donald RGK, Roos DS, Greene PJ, Santi DV (1996) Heterologous expression and characterization of the bifunctional dihydrofolate reductase-thymidylate synthase enzyme of toxoplasma gondii . Biochemistry 35:6366–6374. https://doi.org/10.1021/bi952923q
69. Elcock AH, Huber GA, McCammon JA (1997) Electrostatic channeling of substrates between enzyme active sites: comparison of simulation and experiment. Biochemistry 36: 16049–16058. https://doi.org/10.1021/bi971709u
70. Metzger VT, Eun C, Kekenes-Huskey PM, Huber G, McCammon JA (2014) Electrostatic channeling in P. falciparum DHFR-TS: Brownian dynamics and smoluchowski modeling. Biophys J 107:2394–2402. https://doi.org/10.1016/j.bpj.2014.09.039
71. Srere PA (1987) Complexes of sequential metabolic enzymes. Annu Rev Biochem 56: 89–124. https://doi.org/10.1146/annurev.bi.56.070187.000513
72. Vélot C, Mixon MB, Teige M, Srere PA (1997) Model of a quinary structure between Krebs TCA cycle enzymes: A model for the metabolon. Biochemistry 36:14271–14276. https://doi.org/10.1021/bi972011j
73. Wu F, Minteer S (2015) Krebs cycle metabolon: Structural evidence of substrate channeling revealed by cross-linking and mass spectrometry. Angew Chemie Int Ed 54: 1851–1854. https://doi.org/10.1002/anie.201409336
74. Sathyanarayanan N, Cannone G, Gakhar L, Katagihallimath N, Sowdhamini R, Ramaswamy S, Vinothkumar KR (2019) Molecular basis for metabolite channeling in a ring opening enzyme of the phenylacetate degradation pathway. Nat Commun 10:4127. https://doi.org/10.1038/s41467-019-11931-1
75. Liu Y, Hickey DP, Guo J-Y, Earl E, Abdellaoui S, Milton RD, Sigman MS, Minteer SD, Calabrese Barton S (2017) Substrate channeling in an artificial metabolon: A molecular dynamics blueprint for an experimental peptide bridge. ACS Catal 7: 2486–2493. https://doi.org/10.1021/acscatal.6b03440

76. Shearer G, Lee JC, Koo J-a, Kohl DH (2005) Quantitative estimation of channeling from early glycolytic intermediates to CO2 in intact Escherichia coli. FEBS J 272:3260–3269. https://doi.org/10.1111/j.1742-4658.2005.04712.x
77. Wiechert W (2001) 13C metabolic flux analysis. Metab Eng 3:195–206. https://doi.org/10.1006/mben.2001.0187
78. Williams TC, Sweetlove LJ, George Ratcliffe R (2011) Capturing metabolite channeling in metabolic flux phenotypes. Plant Physiol 157:981–984. https://doi.org/10.1104/pp.111.184887
79. Wiechert W, Niedenführ S, Nöh K (2015) Fundam. Bioeng, pp 97–142. Wiley-VCH Verlag GmbH & Co. KGaA: Weinheim, Germany. https://doi.org/10.1002/9783527697441.ch05
80. Fernie AR, Carrari F, Sweetlove LJ (2004) Respiratory metabolism: glycolysis, the TCA cycle and mitochondrial electron transport. Curr Opin Plant Biol 7:254–261. https://doi.org/10.1016/j.pbi.2004.03.007
81. Huang Y-mmM, Huber GA, Wang N, Minteer SD, McCammon JA (2018) Brownian dynamic study of an enzyme metabolon in the TCA cycle: Substrate kinetics and channeling. Protein Sci 27:463–471. https://doi.org/10.1002/pro.3338
82. Zhang Y, Beard KFM, Swart C, Bergmann S, Krahnert I, Nikoloski Z, Graf A, Ratcliffe RG, Sweetlove LJ, Fernie AR, Obata T (2017) Protein-protein interactions and metabolite channelling in the plant tricarboxylic acid cycle. Nat Commun 8:15212. https://doi.org/10.1038/ncomms15212
83. Graham JW, Williams TC, Morgan M, Fernie AR, Ratcliffe RG, Sweetlove LJ (2007) Glycolytic enzymes associate dynamically with mitochondria in response to respiratory demand and support substrate channeling. Plant Cell 19:3723–3738. https://doi.org/10.1105/tpc.107.053371
84. Debnam PM, Shearer G, Blackwood L, Kohl DH (1997) Evidence for channeling of intermediates in the oxidative pentose phosphate pathway by soybean and pea nodule extracts, yeast extracts, and purified yeast enzymes. Eur J Biochem 246:283–290. https://doi.org/10.1111/j.1432-1033.1997.00283.x
85. Buchner A, Tostevin F, Gerland U (2013) Clustering and optimal arrangement of enzymes in reaction-diffusion systems. Phys Rev Lett 110:208104. https://doi.org/10.1103/PhysRevLett.110.208104
86. Castellana M, Wilson MZ, Xu Y, Joshi P, Cristea IM, Rabinowitz JD, Gitai Z, Wingreen NS (2014) Enzyme clustering accelerates processing of intermediates through metabolic channeling. Nat Biotechnol 32:1011–1018. https://doi.org/10.1038/nbt.3018
87. Gopich IV (2021) Cluster channeling in cascade reactions. J Phys Chem B 125:2061–2073. https://doi.org/10.1021/acs.jpcb.0c11155
88. Parsons M, Furuya T, Pal S, Kessler P (2001) Biogenesis and function of peroxisomes and glycosomes. Mol Biochem Parasitol 115:19–28. https://doi.org/10.1016/S0166-6851(01)00261-4
89. An S, Kumar R, Sheets ED, Benkovic SJ (2008) Reversible compartmentalization of de Novo Purine biosynthetic complexes in living cells. Science (80-) 320:103–106. https://doi.org/10.1126/science.1152241
90. Pedley AM, Benkovic SJ (2017) A new view into the regulation of purine metabolism: The purinosome. Trends Biochem Sci 42:141–154. https://doi.org/10.1016/j.tibs.2016.09.009
91. Wilner OI, Shimron S, Weizmann Y, Wang Z-G, Willner I (2009) Self-assembly of enzymes on DNA scaffolds: En route to biocatalytic cascades and the synthesis of metallic nanowires. Nano Lett 9:2040–2043. https://doi.org/10.1021/nl900302z
92. Wilner OI, Weizmann Y, Gill R, Lioubashevski O, Freeman R, Willner I (2009) Enzyme cascades activated on topologically programmed DNA scaffolds. Nat Nanotechnol 4:249–254. https://doi.org/10.1038/nnano.2009.50
93. Delebecque CJ, Lindner AB, Silver PA, Aldaye FA (2011) Organization of intracellular reactions with rationally designed RNA assemblies. Science (80-) 333:470–474. https://doi.org/10.1126/science.1206938
94. You C, Myung S, Zhang Y-HP (2012) Facilitated substrate channeling in a self-assembled trifunctional enzyme complex. Angew Chemie Int Ed 51:8787–8790. https://doi.org/10.1002/anie.201202441
95. Fu J, Liu M, Liu Y, Woodbury NW, Yan H (2012) Interenzyme substrate diffusion for an enzyme cascade organized on spatially addressable DNA nanostructures. J Am Chem Soc 134:5516–5519. https://doi.org/10.1021/ja300897h
96. Fu J, Yang YR, Johnson-Buck A, Liu M, Liu Y, Walter NG, Woodbury NW, Yan H (2014) Multi-enzyme complexes on DNA scaffolds capable of substrate channelling with an artificial swinging arm. Nat Nanotechnol 9:531–536. https://doi.org/10.1038/nnano.2014.100

97. Yang YR, Liu Y, Yan H (2015) DNA nanostructures as programmable biomolecular scaffolds. Bioconjug Chem 26:1381–1395. https://doi.org/10.1021/acs.bioconjchem.5b00194
98. Bülow L (1991) Multienzyme systems obtained by gene fusion. Trends Biotechnol 9:226–231. https://doi.org/10.1016/0167-7799(91)90075-S
99. Rothemund PWK (2006) Folding DNA to create nanoscale shapes and patterns. Nature 440:297–302. https://doi.org/10.1038/nature04586
100. Dey S, Fan C, Gothelf KV, Li J, Lin C, Liu L, Liu N, Nijenhuis MAD, Saccà B, Simmel FC, Yan H, Zhan P (2021) DNA origami. Nat Rev Methods Prim 1:13. https://doi.org/10.1038/s43586-020-00009-8
101. Bayer EA, Morag E, Lamed R (1994) The cellulosome A treasure-trove for biotechnology. Trends Biotechnol 12:379–386. https://doi.org/10.1016/0167-7799(94)90039-6
102. Mingardon F, Chanal A, Tardif C, Bayer EA, Fierobe H-P (2007) Exploration of new geometries in cellulosome-like chimeras. Appl Environ Microbiol 73:7138–7149. https://doi.org/10.1128/AEM.01306-07
103. Moehlenbrock MJ, Toby TK, Waheed A, Minteer SD (2010) Metabolon catalyzed pyruvate/air biofuel cell. J Am Chem Soc 132:6288–6289. https://doi.org/10.1021/ja101326b
104. Xin L, Zhou C, Yang Z, Liu D (2013) Regulation of an enzyme cascade reaction by a DNA machine. Small 9:3088–3091. https://doi.org/10.1002/smll.201300019
105. Zhao Z, Fu J, Dhakal S, Johnson-Buck A, Liu M, Zhang T, Woodbury NW, Liu Y, Walter NG, Yan H (2016) Nanocaged enzymes with enhanced catalytic activity and increased stability against protease digestion. Nat Commun 7:10619. https://doi.org/10.1038/ncomms10619
106. Collins FC, Kimball GE (1949) Diffusion-controlled reaction rates. J Colloid Sci 4:425–437. https://doi.org/10.1016/0095-8522(49)90023-9
107. Berg OG (1978) On diffusion-controlled dissociation. Chem Phys 31:47–57. https://doi.org/10.1016/0301-0104(78)87025-6
108. Oliveira SM, Neeli-Venkata R, Goncalves NS, Santinha JA, Martins L, Tran H, Mäkelä J, Gupta A, Barandas M, Häkkinen A, Lloyd-Price J, Fonseca JM, Ribeiro AS (2016) Increased cytoplasm viscosity hampers aggregate polar segregation in Escherichia coli. Mol Microbiol 99:686–699. https://doi.org/10.1111/mmi.13257
109. Elcock AH, McCammon JA (1996) Evidence for electrostatic channeling in a fusion protein of malate dehydrogenase and citrate synthase. Biochemistry 35:12652–12658. https://doi.org/10.1021/bi9614747
110. Fulton AB (1982) How crowded is the cytoplasm? Cell 30:345–347. https://doi.org/10.1016/0092-8674(82)90231-8
111. Zimmerman SB, Trach SO (1991) Estimation of macromolecule concentrations and excluded volume effects for the cytoplasm of Escherichia coli. J Mol Biol 222:599–620. https://doi.org/10.1016/0022-2836(91)90499-V
112. Ellis R (2001) Macromolecular crowding: obvious but underappreciated. Trends Biochem Sci 26:597–604. https://doi.org/10.1016/S0968-0004(01)01938-7
113. Herzfeld J (2004) Crowding-induced organization in cells: spontaneous alignment and sorting of filaments with physiological control points. J Mol Recognit 17:376–381. https://doi.org/10.1002/jmr.703
114. Weiss M (2014) Int. Rev. Cell Mol. Biol, 1st edn, vol 307, pp 383–417. Elsevier Inc. https://doi.org/10.1016/B978-0-12-800046-5.00011-4
115. Gomez D, Huber K, Klumpp S (2019) On protein folding in crowded conditions. J Phys Chem Lett 10:7650–7656. https://doi.org/10.1021/acs.jpclett.9b02642
116. Tabaka M, Kalwarczyk T, Holyst R (2014) Quantitative influence of macromolecular crowding on gene regulation kinetics. Nucleic Acids Res 42:727–738. https://doi.org/10.1093/nar/gkt907
117. Pastor I, Pitulice L, Balcells C, Vilaseca E, Madurga S, Isvoran A, Cascante M, Mas F (2014) Effect of crowding by Dextrans in enzymatic reactions. Biophys Chem 185:8–13. https://doi.org/10.1016/j.bpc.2013.10.006
118. Maximova K, Wojtczak J, Trylska J (2019) Enzymatic activity of human immunodeficiency virus type 1 protease in crowded solutions. Eur Biophys J 48:685–689. https://doi.org/10.1007/s00249-019-01392-1
119. Skóra T, Popescu MN, Kondrat S (2021) Conformation-changing enzymes and macromolecular crowding. Phys Chem Chem Phys 23:9065–9069. https://doi.org/10.1039/D0CP06631A
120. Han J, Herzfeld J (1993) Macromolecular diffusion in crowded solutions. Biophys J 65:1155–1161. https://doi.org/10.1016/S0006-3495(93)81145-7
121. Dauty E, Verkman AS (2004) Molecular crowding reduces to a similar extent the

diffusion of small solutes and macromolecules: measurement by fluorescence correlation spectroscopy. J Mol Recognit 17: 441–447. https://doi.org/10.1002/jmr.709

122. Höfling F, Franosch T (2013) Anomalous transport in the crowded world of biological cells. Reports Prog Phys 76:046602. https://doi.org/10.1088/0034-4885/76/4/046602

123. Kondrat S, Zimmermann O, Wiechert W, Lieres EV (2015) The effect of composition on diffusion of macromolecules in a crowded environment. Phys Biol 12:046003. https://doi.org/10.1088/1478-3975/12/4/046003

124. Feig M, Yu I, Wang PH, Nawrocki G, Sugita Y (2017) Crowding in cellular environments at an atomistic level from computer simulations. J Phys Chem B 121: 8009–8025. https://doi.org/10.1021/acs.jpcb.7b03570

125. Skóora T, Vaghefikia F, Fitter J, Kondrat S (2020) Macromolecular crowding: How shape and interactions affect diffusion. J Phys Chem B 124:7537–7543. https://doi.org/10.1021/acs.jpcb.0c04846

126. Rohwer JM, Postma PW, Kholodenko BN, Westerhoff HV (1998) Implications of macromolecular crowding for signal transduction and metabolite channeling. Proc Natl Acad Sci 95:10547–10552. https://doi.org/10.1073/pnas.95.18.10547

127. Zhang Y, Hess H (2017) Toward rational design of high-efficiency enzyme cascades. ACS Catal 7:6018–6027. https://doi.org/10.1021/acscatal.7b01766

Chapter 4

Structural Characterization of Multienzyme Assemblies: An Overview

Anastassios C. Papageorgiou

Abstract

Multienzyme assemblies have attracted significant attention in recent years for use in industrial applications instead of single enzymes. Owing to their ability to catalyze cascade reactions, multienzyme assemblies have become inspirational tools for the in vitro construction of multienzyme molecular machines. The use of such molecular machines could offer several advantages such as fewer side reactions, a high product yield, a fast reaction speed, easy product separation, a tolerable toxic environment, and robust system operability compared to current microbial cell catalytic systems. Besides, they can provide all the benefits found in the use of enzymes, including reusability, catalytic efficiency, and specificity. Similar to single enzymes, multienzyme assemblies could offer economical and environmentally friendly alternatives to conventional catalysts and play a central role as biocatalysts in green chemistry applications. However, detailed characterization of multienzyme assemblies and a full understanding of their mechanistic details are required for their efficient use in industrial biotransformations. Since the determination of the first enzyme structure in 1965, structural information has played a pivotal role in the characterization of enzymes and elucidation of their structure–function relationship. Among the structural biology techniques, X-ray crystallography has provided key mechanistic details into multienzyme assemblies. Here, the structural characterization of multienzyme assemblies is reviewed and several examples are provided.

Key words Enzyme structure, Enzyme immobilization, Catalytic mechanism, Structure determination, X-ray, Biocatalysis, Channeling

1 Introduction

Enzymes are essential biological macromolecules that support life by catalyzing many biological processes. They are characterized by unique properties, such as specificity, high affinity, and catalytic efficiency, which make them different from other biological macromolecules. Owing to their properties, enzymes have attracted considerable interest for use in many biotechnological applications.

Enzymes require mild conditions to function, thus they can save key resources such as energy or water [1]. Furthermore, enzymes are attractive alternatives as biocatalysts in chemical

Haralambos Stamatis (ed.), *Multienzymatic Assemblies: Methods and Protocols*,
Methods in Molecular Biology, vol. 2487, https://doi.org/10.1007/978-1-0716-2269-8_4,
© The Author(s), under exclusive license to Springer Science+Business Media, LLC, part of Springer Nature 2022

reactions owing to their minimal impact on the environment and low costs. Taking into account the benefits of green chemistry, enzyme biocatalysis is now found in various traditional chemical processes in several fields [2]. Interestingly, the number of processes has increased, for example, from 60 in 1990 to several hundred in 2019 [3]. This trend is expected to continue and expand to even more areas owing to new emerging technologies in enzyme engineering, such as directed evolution [4].

Enzymes do not act alone inside the cells and they often participate in pathways that require a cascade of enzymatic reactions [5]. Recent progress in multienzyme reactions has experienced a rapid growth for scientific and industrial applications, particularly biotransformations, biosensors, and biomedical engineering [6]. Consequently, multienzyme complexes have drawn attention in synthetic biology for the in vitro construction of multienzyme molecular machines that could offer several advantages compared to current microbial cell catalytic systems, including fewer side reactions, a high product yield, a fast reaction speed, easy product separation, a tolerable toxic environment, and robust system operability [7]. An understanding of the structure and function of multienzyme assemblies is, however, important for better exploitation of their properties and to improve reaction rates and increase reaction efficiency. Structural information has played over the years a key role in the better characterization of multienzyme assemblies and an overview of our current understanding is provided here.

2 From Single Enzymes to Complex Multienzyme Structures

Structural studies on single enzymes have helped in understanding better more complicated enzyme systems, such as multienzyme assemblies. The number of enzyme structures determined so far accounts for ~70% of all entries in the Protein Data Bank (PDB) (Table 1). The first enzyme to be crystallized was jack bean urease in 1926 by James Sumner. In 1936, Desmond Bernal and Dorothy Crowfoot found that protein crystals, including crystals of chymotrypsin and pepsin, could give a good diffraction pattern after exposure to X-rays providing they were kept moist [8]. Linus Pauling proposed in 1948 the idea that enzymes bind with high specificity the transition state of the catalytic reaction [9]. A breakthrough in enzyme research came in 1965 when the first enzyme structure was determined by X-ray crystallography after the group of David Phillips succeeded in determining the crystal structure of lysozyme at 2 Å resolution [10]. The structure revealed a cleft with the right shape to accept the substrate and a mode of substrate binding that favored distortion (half-chair conformation) of the substrate towards a transition configuration. Further details of the

Table 1
Enzyme structures in the PDB (as of April 25th, 2021)[a]

Enzyme classification	No. of entries	Percentage of total PDB entries
Oxidoreductases (1)	19,021	10.7
Transferases (2)	37,551	21.2
Hydrolases (3)	47,583	26.9
Lyases (4)	8208	4.6
Isomerases (5)	4031	2.3
Ligases (6)	3486	2.0
Translocases (7)	2492	1.4

[a]Total PDB entries 177,009 (www.rcsb.org)

binding and interactions were provided in the co-crystal structures of lysozyme with various inhibitors even though the resolution was only at 6 Å [11].

The success with lysozyme led to the structure determination of other enzymes. In 1967, structures of carboxypeptidase A with substrates and inhibitors were determined at 6 Å resolution by William Lipscomb's group at Harvard [12]. The binding of substrates revealed conformational changes in the enzyme, especially in regions adjacent to the substrate. At the same time, the structure of ribonuclease A was determined [13]. Substrates were found to bind in a groove with high specificity. The structure of chymotrypsin in 1971 uncovered the classic mechanism in proteases (as well as esterases and other hydrolytic enzymes) based on a catalytic triad and the formation of an oxyanion hole [14].

More complicated enzyme structures became available owing to advances in X-ray intensity with the use of synchrotron radiation sources. The structure of RubisCO, for instance, a multimeric enzyme involved in atmospheric CO_2 fixation enzyme became available in the non-activated and activated forms [15, 16]. The role of co-factors in catalysis was made clear with the crystal structures of enzymes such as aspartate aminotransferase [17] that uses pyridoxal 5′-phosphate during catalysis.

The understanding of allosteric regulation was facilitated by the crystal structures of various allosteric enzymes, such as glycogen phosphorylase (GP) [18], aspartate transcarbamoylase (ATCase) [19, 20], and phosphofructokinase (PFK) [21]. Notably, GP crystals were also used as a model system in first attempts to study the catalytic reaction in the crystal by taking advantage of the fast data collection offered by synchrotron radiation [22].

Table 2
Examples of multienzyme assemblies

Multienzyme assembly	Biosynthetic pathway	Reference
Lumazine synthase and riboflavin synthase	Synthesis of the xylene ring of riboflavin (vitamin B_2)	[94]
2-oxo acid dehydrogenase complexes	Central metabolism checkpoints	[95–97]
2-enoyl-CoA hydratase (ECH), L-3-hydroxyacyl-CoA dehydrogenase (HACD) and 3-ketoacyl-CoA thiolase (KACT)	Fatty acid β-oxidation cycle	[28, 98]
Glycine decarboxylase multienzyme system	Oxidative decarboxylation and deamination of glycine in the C_2 cycle	[99]
Tryptophan synthase	Final two steps in L-tryptophan biosynthesis	[24]
Pyruvate carboxylase (PC)	Biotin-dependent production of oxaloacetate	[100]

The first insights into a multienzyme complex were provided by the structure of tryptophan synthase from *Salmonella typhimurium* determined in 1988 [23, 24]. Several other multienzyme assemblies followed (Table 2), including large complexes such as transcarboxylase, a 1.2 MDa multienzyme assembly [25], the catalytic core of an archaeal 2-oxoacid dehydrogenase multienzyme complex, a 42-mer protein assembly [26], and pyruvate decarboxylase [27]. Among the multienzyme assemblies, the structure of the fatty acid beta-oxidation multienzyme complex at 3.4 Å resolution [28] revealed domain rearrangements upon the binding of 3-hydroxyhexadecanoyl-CoA or nicotinamide adenine dinucleotide.

3 Mechanistic Details of Multienzyme Assemblies

Early X-ray crystallographic studies in multienzyme complexes suggested a direct transfer ("channeling") of a metabolic intermediate between the active sites of enzymes that catalyze sequential reactions in a biosynthetic pathway, a process known as substrate channeling [29]. The active sites can be located either on separate domains in a multifunctional enzyme or on separate subunits in a multienzyme complex. Although channeling can offer some potential benefits, such as enhanced reaction rates and protection of unstable intermediates [30], there is still considerable debate whether it is advantageous or not [31]. Recently, the idea of super channeling was introduced, suggesting that in some complexes the final modification could take place independently of the preceding steps if it is not mechanistically dependent on them [32]. Nevertheless, the precise control of substrate channeling outside the cellular environment has been of particular importance in areas such as synthetic biology, synthetic chemistry, and biomedicine [33]. Thus, the concept of substrate channeling has been explored in building artificial multienzyme assemblies using a variety of approaches, with or without the use of artificial scaffolds [32, 33]. Substrate channeling can proceed in three different ways as discussed below.

3.1 Electrostatic Guidance

Electrostatic channeling was first proposed based on studies of the bifunctional thymidylate synthase (TS)-dihydrofolate reductase (DHFR) enzyme from *Leishmania major* [34]. The negatively charged dihydrofolate produced by thymidylate synthase was proposed to be handed over to DHFR via electrostatic interactions formed by a series of lysine and arginine residues on the surface of the protein. A 40 Å "electrostatic highway" created across the surface of the protein was suggested as an explanation for how channeling may occur [35]. The mechanism has also been proposed for other protozoa bifunctional DHFR-TS enzymes, but it appears absent in *Cryptosporidium hominis* [36] (Fig. 1).

Electrostatic channeling has been suggested for a fusion protein of yeast mitochondrial citrate synthase (CS) and yeast

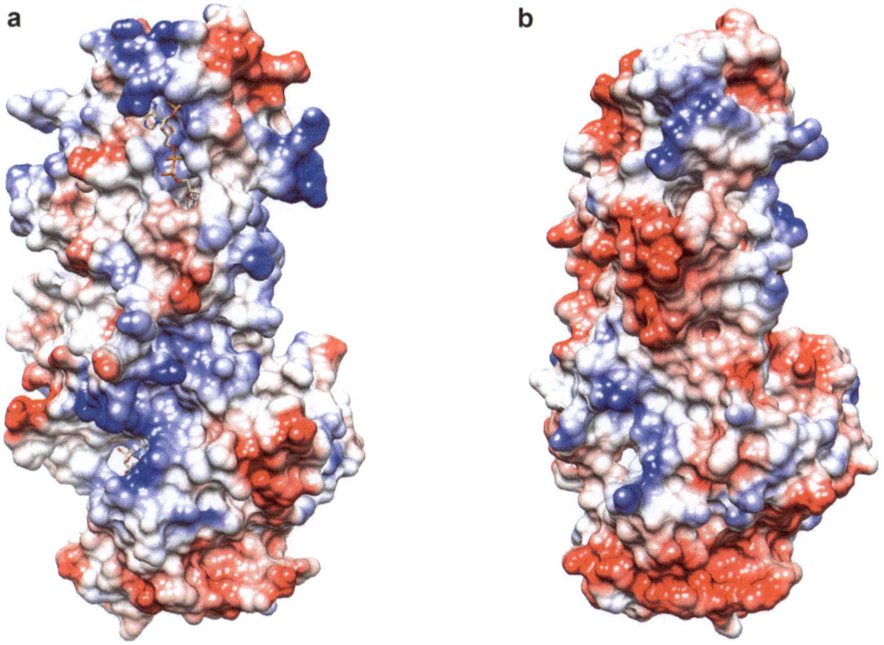

Fig. 1 Electrostatic guidance in DHFR-TS. (**a**) *Plasmodium falciparum* DHFR-TS (PDB id 1j3i) (**b**) *Cryptosporidium hominis* DHFR-TS (PDB id 1qzf). Coulombic surfaces are shown (red = negative charges; blue = positive charges). Ligands are shown as sticks. All figures were created with Chimera [101]

mitochondrial malate dehydrogenase (MD) created by connection of the C-terminal of CS through a 3-residue linker to the N-terminal of MD [37]. The kinetics of the fusion protein indicated that oxaloacetate was channeled between the active sites of MD and CS. A positive electrostatic surface between the two active sites was proposed based on molecular modeling and molecular dynamics simulations. Structural analysis using in vitro cross linking and mass spectrometry of the malate dehydrogenase–citrate synthase–aconitase complex has provided insights into the potential mechanism by which substrate channeling could occur [38]. The formation of a continuous electrostatic channel with positive charge was shown using surface analysis of the complex created by constraint protein docking.

A mechanism of electrostatic pivoting of the CoA moiety in PaaZ, a bifunctional enzyme that catalyzes the ring opening of oxepin-CoA and its conversion to 3-oxo-5,6-dehydrosuberyl-CoA, was recently suggested [39]. PaaZ is used by many bacteria to degrade various environmental pollutants and its structure was determined by cryo-electron microscopy, revealing a trilobed architecture with a dimer in each lobe. A set of conserved positively charged residues was proposed to be involved in the transfer of a key enzyme intermediate between the active sites of PaaZ.

3.2 Tunneling

A number of enzymes utilize molecular tunnels to move reactive intermediates between active sites [40]. Physical tunnels that transfer the product of one enzyme directly to the active site of a second enzyme are found in some multienzyme assemblies [30]. The presence of tunnels in multienzyme complexes offers several beneficial effects, such as acceleration of overall reaction rates, protection of unstable substrates–intermediates as well as circumvention of both unfavorable equilibria and substrate competition between different pathways. Because of their importance, such tunnels have been proposed as drug targets for the design of new inhibitors [41].

Typical examples of substrate tunneling are found in tryptophan synthase (TRS), carbamoyl phosphate synthase (CPS), and in several other enzymes [42]. In TRS, a metabolic intermediate (indole) from the active site of the alpha subunit is channeled to the active site of the beta subunit through a hydrophobic channel [43], resulting in an increase of the reaction rate by 1–2 orders of magnitude. The enzyme is a stable $\alpha_2\beta_2$ heterotetramer of 143 kDa and a tunnel connects the two active sites which are ~25 Å apart (Fig. 2). The tunnel is large enough to accommodate up to four molecules of indole in transit. Although the active sites are initially exposed to the solvent, upon substrate binding they undergo conformational changes that restrict solvent access to them and the tunnel. The hydrophobic channel prevents diffusion of indole to the solvent and provides a safe passage for indole to reach the

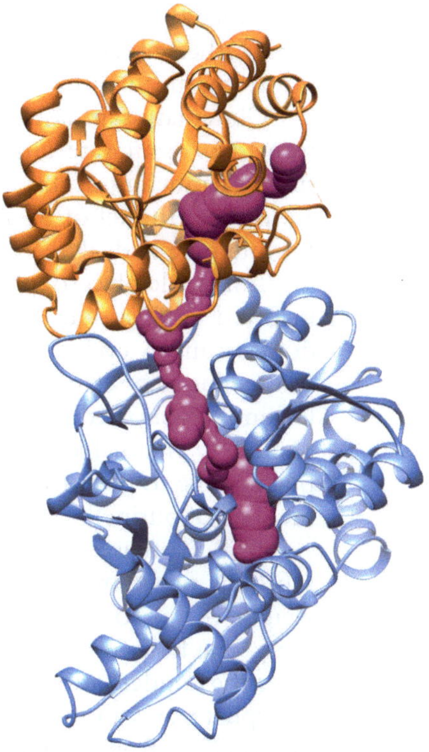

Fig. 2 Substrate channeling in TRS (PDB id 2trs). Chain A is in orange and chain B in cornflower blue. The tunnel that connects the active sites is shown (magenta). All channel calculations were carried out using MOLEonline [102]

catalytic site where it is condensed with serine to yield tryptophan with the help of the cofactor pyridoxal phosphate [44].

Structural information obtained for the fatty-acid-oxidation multienzyme complex suggested that the complex undergoes a quaternary-structural conversion for substrate channeling, resulting in the direct transfer of the intermediate product among the enzymes without diffusion into the bulk solvent [28].

Carbamoyl phosphate synthetase (CPS) utilizes a tunnel that connects three active sites over a distance of 100 Å. CPS catalyzes the formation of carbamoyl phosphate, a key intermediate in the biosynthesis of arginine and pyrimidines in prokaryotes and eucaryotes and in the biosynthesis of urea in most terrestrial vertebrates. CPS consists of a small and a large subunit. The small subunit of CPS hydrolyzes glutamine to release ammonia, which travels partway through the tunnel to an active site in the N-terminal domain of the large subunit [45]. At this site, ammonia attacks carboxyphosphate formed by phosphorylation of bicarbonate by ATP. The resulting carbamate travels through the tunnel to the third active site in the C-terminal domain of the large subunit, where it is phosphorylated by ATP to provide carbamoyl phosphate.

CPS is one of several enzymes that channel NH_3 between active sites. Other enzymes that use NH_3 as a gas reactant and utilize a tunneling mechanism include:

Glutamate synthase: It catalyzes the synthesis of L-glutamate from 2-oxoglutarate and L-glutamine and contains three active sites. Channeling of the ammonia intermediate is accomplished by an internal tunnel of 31 Å in length, which runs from the site of L-glutamine hydrolysis to the site of L-glutamate synthesis [46].

Imidazole glycerol phosphate synthase (IGPS): This enzyme is involved in histidine biosynthesis and combines a glutaminase domain and a cyclase domain close to each other for ammonia channeling. The structure of yeast IGPS determined at 2.1 Å resolution [47] showed that ammonia generated by hydrolysis of glutamine is channeled to a second active site where it acts as a nucleophile to produce imidazole glycerol phosphate and an intermediate of purine biosynthesis.

PLP synthase: It is a 24-protein complex required for the synthesis of vitamin B_6. These complexes have a total molecular weight of up to 750 kDa and contain nearly 6000 amino acids [48]. The holoenzyme consists of two subunits: YaaD which catalyzes the condensation of ribulose 5-phosphate, glyceraldehyde-3-phosphate, and ammonia and YaaE which catalyzes the production of ammonia from glutamine. An ammonia channel that connects the active site of YaaE with the ribulose 5-phosphate binding site was identified. This channel is similar to the one found in imidazole glycerol phosphate synthase; however, the structural superposition of the two complexes shows that the glutaminase domains are rotated by about 180° to each other.

Glucosamine 6-phosphate synthase: It channels ammonia to catalyze the synthesis of glucosamine-6-P from fructose-6P and glutamine. During the enzymatic reaction, ammonia is transferred through a channel from the glutaminase active site to the synthase active site of the enzyme [49].

Asparagine synthetase: The enzyme catalyzes the ATP-dependent biosynthesis of L-asparagine in cells from L-aspartic acid using L-glutamine as a nitrogen source. Its structure consists of two domains [50]: an N-terminal domain which contains the site for the hydrolysis of glutamine and a C-terminal domain responsible for asparagine synthesis. The two sites are connected by a 20 Å long hydrophobic channel made by hydrophobic and non-polar amino acid residues (Fig. 3).

Glutamine amidotransferase CAB (GatCAB): The X-ray structure of GatCAB revealed that the two identified active sites of GatA and GatB are remarkably distant but are connected in the complex by an ammonia channel of 30 Å in length [51]. This channel couples the glutaminase and transamidase reactions, as the ammonia generated by GatA is transferred through the channel to the transamidase reaction center of GatB (Fig. 4). Neutron diffraction

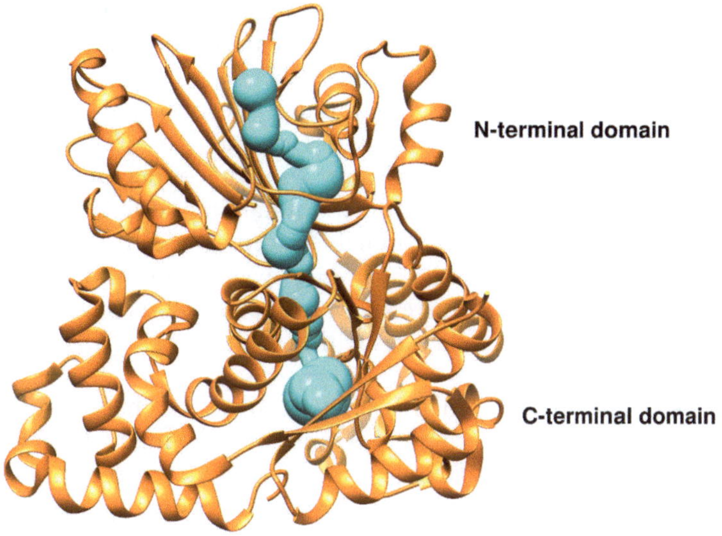

Fig. 3 Ribbon diagram of asparagine synthetase (PDB id 1ct9). The channel is depicted in cyan color. The two domains of the enzyme are labeled

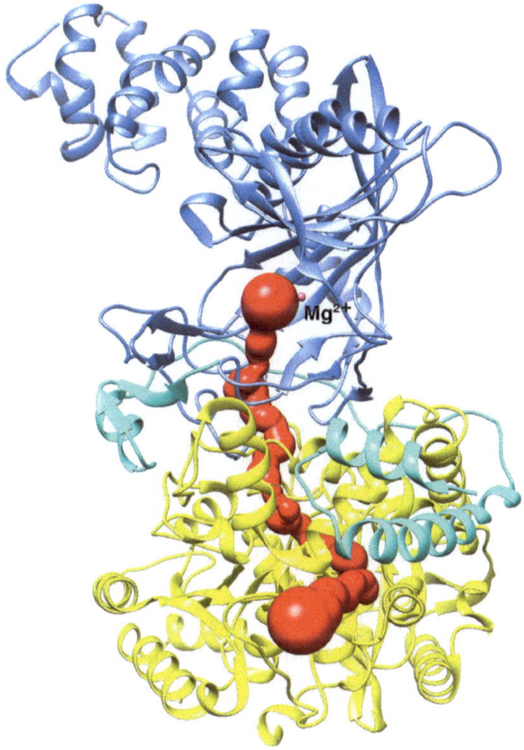

Fig. 4 Ribbon diagram of GatCAB (PDB id 2f2a). The channel that connects the two remote active sites is shown in red. The three domains are depicted in yellow (GatA), cornflower blue (GatB), and cyan (GatC)

experiments have been carried out to visualize ammonia by using large crystals GatCAB but they have been proved so far to be challenging [52].

Glutamine phosphoribosylpyrophosphate amidotransferase (GPATase): Contrary to previous examples, GPATase has no pre-existent tunnel connecting the active sites between the N-terminal glutamine domain, an Ntn hydrolase and the C-terminal acceptor domain, a phosphoribosyltransferase (PRTase). However, upon binding of appropriate ligands to each active site, a tunnel of ~20 Å in length is formed by conformational changes within each domain [53].

3.3 Swinging Arms

In this type of substrate channeling, a swinging arm, typically a chemical component connected to an enzyme complex through a flexible link, allows the direct transfer of substrate molecules between the active sites within the complex [54]. Three different swinging arms for shuttling around substrates or intermediates in multienzyme complexes have been identified, namely lipoyl-lysine, biotinyl-lysine, and phosphopantetheine.

Pyruvate dehydrogenase complex (PDC) is one of the largest multienzyme assemblies with a MW of 4.5–6 $\times 10^6$ and comprises multiple copies of three different enzymes (E1, pyruvate dehydrogenase; E2, dihydrolipoyl acetyltransferase; E3, dihydrolipoyl dehydrogenase). Lipoyl–lysine residues in PDC act as swinging arms, carrying substrate between the catalytic centers of the three enzymes. The Lys residue which becomes lipoylated belongs to a lipoid domain, a small (~80 aa) protein of a unique fold [55]. This domain is part of the multidomain-and-linker structure of the E2 component of the 2-oxo acid dehydrogenase complexes and its number per E2 subunit ranges from 1 to 3.

Pyruvate carboxylase (PC) presents a classical example of a swinging arm mechanism using biotinyl-lysine for the transfer of an activated carboxyl group between distant active sites. The biotinyl-lysine is located in a biotinyl domain which is part of the biotin carboxy carrier protein (BCCP). The biotinyl domain of BCCP of acetyl-CoA carboxylase was determined in 1995 and the biotinylated Lys residue was found at a hairpin β-strand of the structure (Fig. 5) [56]. The structure of PC from *Rhizobium etli* (RePC) was determined by single-wavelength anomalous diffraction at 2.0 Å resolution and revealed details of the mechanism [57, 58] (Fig. 6). The crystals contain a dimer in the asymmetric unit but only monomer A includes the complete polypeptide chain, composed of the three functional domains (carboxylase BC domain, carboxyltransferase CT domain, and BCCP), and a central allosteric domain (AD) that had not been previously described. The biotin cofactor on the BCCP is carboxylated via a MgATP-dependent reaction in the BC domain. The BCCP domain then swings to the CT domain where the carboxyl group is transferred

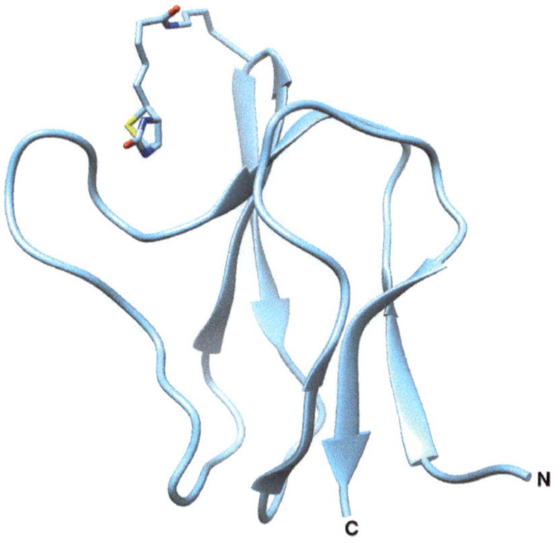

Fig. 5 The structure of the biotinyl domain of BCCP of acetyl-CoA carboxylase (PDB id 1bdo). The biotinylated lysine is shown as sticks

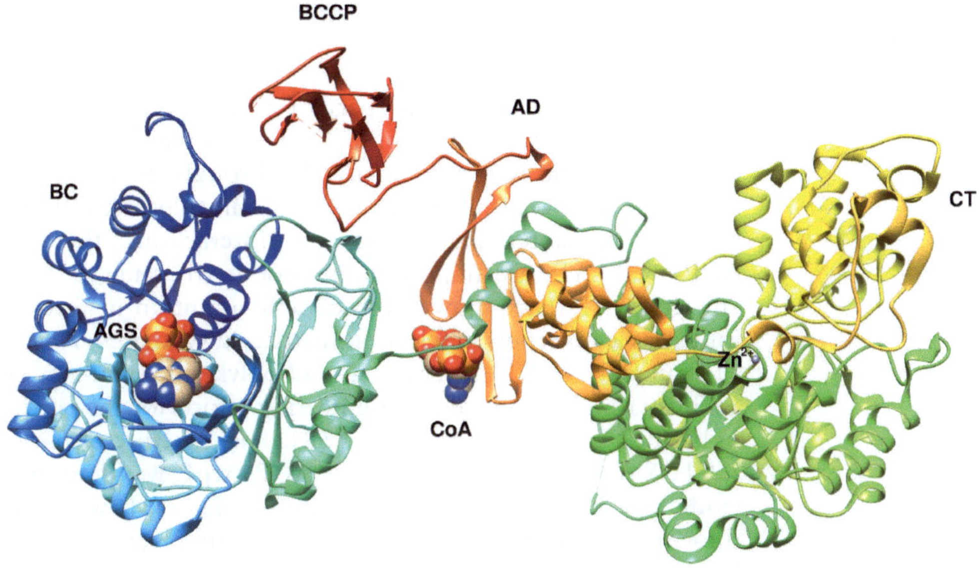

Fig. 6 Ribbon diagram of pyruvate carboxylase (PDB id 2qf7). The structure is rainbow-colored from blue at the N-terminal to red at the C-terminal. Bound ligands (AGS and CoA) are shown in space-filling representation

from the biotin cofactor to pyruvate, generating oxaloacetate. BCCP domain itself must move as it carries biotin between the active sites. The 34 Å linker connecting BCCP to the allosteric domain is therefore highly flexible in order to facilitate domain movement.

Phosphopantetheine provides a 20-Å long swinging arm attached to a Ser residue of a small ~10 kDa protein termed as acyl carrier protein (ACP). ACPs have a characteristic four-helix

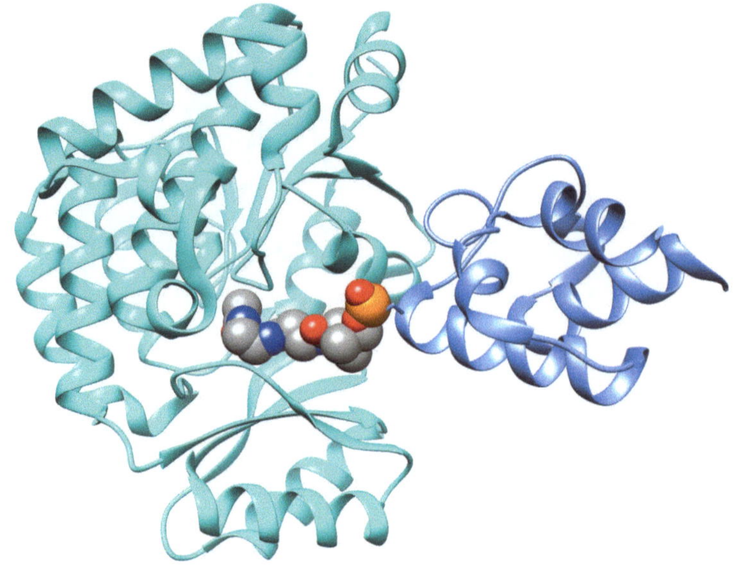

Fig. 7 Overall organization of the AcpP–FabD complex (PDB id 6u0j). Phosphopantetheine attached to the Ser residue of the AcpP protein is shown as space-filling model

bundle structure and can be part of a multidomain polypeptide (type I ACPs) or part of a multiprotein complex (Type II ACPs) [57]. Fatty acid synthesis proceeds through an acyltransferase (AT) that adds C2 units to an acyl chain in thioester linkage with the phosphopantetheine swing arm. Interactions, therefore, of ACPs with ATs are important in productive catalysis and high efficiency. ATs have drawn considerable attention in recent years for use as gatekeepers in selection. Fatty acid synthases (FASs) Type I (in animals) and II (in bacteria) and polyketide synthases (PKSs) are increasingly employed in engineering efforts for the production of biofuels and natural products. [58]. The structure of *Escherichia coli* FAS AT, FabD, and its cognate ACP, AcpP, was determined to study type II FAS ACP-AT interactions (Fig. 7). It was shown in these structural studies that AcpP adopts multiple, productive conformations at the AT binding interface, thus allowing the complex to sustain high transacylation rates [58].

In addition to the above swinging arm types, a swing arm mechanism using only lysine residues has also been reported in various enzymes. Pyridoxine synthase (PDX1) of the vitamin B_6 biosynthesis machinery has a single lysine swinging arm that helps substrate transferring between the two active sites, P2 and P1, of the enzyme [59]. The enzyme, as also the majority of its homologs, is a dodecamer of $(\beta/\alpha)_8$ barrels. Structural details from the *Arabidopsis thaliana* enzyme in the apo form and in complex with a substrate adduct showed structural snapshots of the reaction (Fig. 8). A Lys residue (Lys165) was identified to move from the

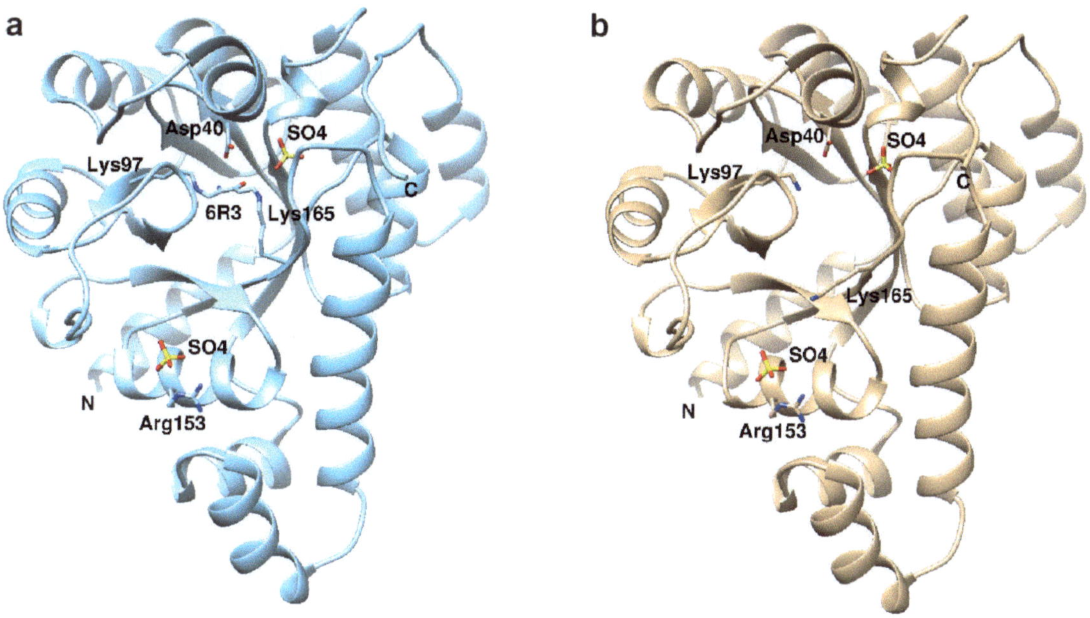

Fig. 8 PDX1 in complex (**a**) and in the apo form (**b**) (PDB id 5k2z for complex, 5k3v for apo). 6R3 corresponds to the chromophoric intermediate attached to Lys165

P2 active site to the P1 where it is held by a molecular catch and pin and is positioned for the transfer of the reaction intermediate back to the P2 active site where the reaction is completed.

4 Artificial Multienzyme Assemblies

Substrate channeling can be used as the basis to construct artificial multienzyme complexes by bringing enzymes together to enhance cascade enzymatic activity by the introduction of linkers [60, 61], enzyme co-localization [6, 62], or scaffolds [63] (Fig. 9).

Domain fusion using linkers can combine two or more active sites that participate in successive catalytic steps in a pathway. In a xylanase-laccase enzyme fusion, a laccase surface loop was used to insert xylanase with minimum disruption of the parent enzymes [64]. The interdomain interface was found to cause the dislocation of the loop comprising residues Leu-558 to Lys-573 in the laccase domain, resulting in a more accessible active site and exposing the type I Cu^{2+} ion to the solvent. Fusion of alcohol dehydrogenase with an aminotransferase through a spacer consisting of multiple Pro-Ala-Ser (PAS) sequences has been explored [65]. Three PAS spacers of 20, 40, and 60 residues were used and PAS20 was found to produce a fusion protein with a two-fold higher catalytic activity compared to a mix of the individual enzymes.

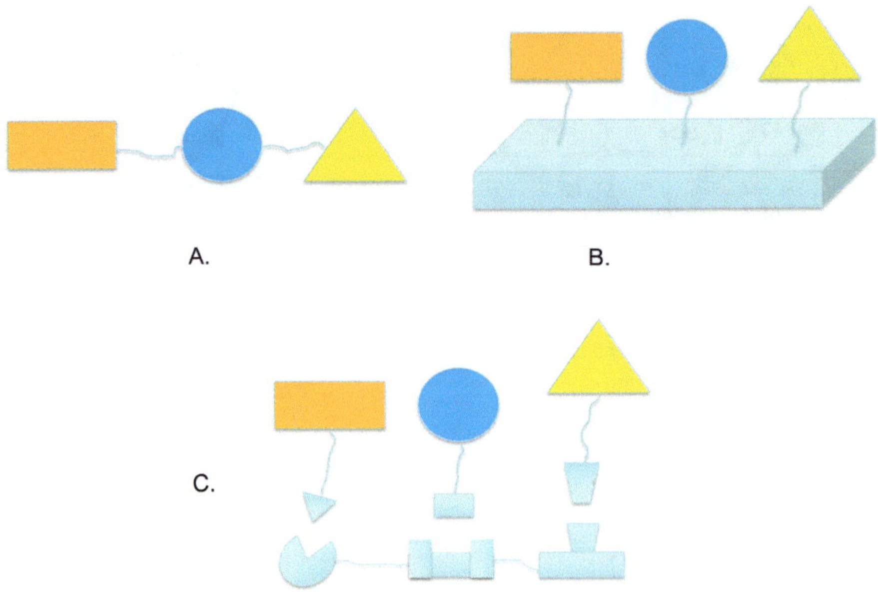

Fig. 9 Schematic illustration of multienzyme complexes created by different routes to bring enzymes in close proximity. (**a**) Fusion protein, (**b**) Multienzyme immobilization, (**c**) Scaffold unit. The enzymes are shown as rectangle, circle, and triangle in orange, blue, and yellow colours, respectively

A two-enzyme system for plastic depolymerization has been studied and the structures of the parent enzymes were recently determined [66]. *Ideonella sakaiensis* PETase depolymerizes polyethylene terephthalate (PET), liberating soluble products, including mono(2-hydroxyethyl) terephthalate (MHET), which is subsequently cleaved by an MHETase to terephthalic acid and ethylene glycol. MHETase follows an induced-fit mechanism as shown in its structure in the presence of a nonhydrolyzable MHET analog [67]. MHETase:PETase chimeric proteins of varying linker lengths were constructed and found to exhibit improved PET and MHET turnover relative to the free enzymes.

Based on the electrostatic guidance as a model, a charged oligopeptide bridge has been envisioned as a molecular construct to facilitate restricted diffusion of a charged intermediate between two sequential catalytic sites. Using a poly(lysine) peptide chain as a cationic bridge enzyme complex, a complex between glucose-6-phosphate dehydrogenase and hexokinase was constructed [68]. Molecular dynamics simulations were applied to test the concept of electrostatic guidance to facilitate substrate diffusion between the two sequential catalytic sites.

Substrate channeling has also inspired the development of artificial swinging arms. A formate:malate oxidoreductase system was constructed using a PEGylated nicotinamide adenine dinucleotide (NAD(H)) swinging arm [69]. A 2D enzyme network consisting of glucose-6- phosphate dehydrogenase (G6PDH) and lactate dehydrogenase (LDH) was positioned on a square lattice-like DNA origami. A double-stranded DNA swinging arm carrying an NAD^+ cofactor at the tip was positioned halfway between each pair of G6PDH and LDH [70]. The network achieved higher reaction efficiency compared to a single pair of enzymes.

DNA and RNA scaffolds have also been promising tools as the distance between the enzymes can be adjusted for optimal activity [63]. A two-enzyme cascade consisting of glucose-6-phosphate dehydrogenase (G6pDH) and malic dehydrogenase (MDH) using DNA nanostructures has been reported [71]. A photoresponsive molecule-based synthetic substrate channeling system on DNA origami to regulate enzyme cascade activity has been developed [33]. G6pDH and MDH were used as the enzyme cascade with nicotinamide adenine dinucleotide (NAD^+) acting as the cofactor. A light-driven Holliday junction was employed as a swing arm for substrate channeling.

An example of multienzyme assembly that may be useful in future bioengineering efforts is provided by nonribosomal peptide synthetases (NRPS) [72]. These are megaenzymes that make small molecules which can be used, for example, as antimicrobials, immunosuppressants, or antitumor agents. Nonribosomal peptide synthesis proceeds with assembly-line logic, where each station on the NRPS assembly line is a multi-domain unit called a module. Linear gramicidin synthetase subunit A (LrgA) is an NRPS with 16 modules that are physically attached to each other by flexible linkers or small docking domains. The use of crystallographic and SAXS data suggested that LgrA is highly flexible [73]. Besides, various structures revealed different relative positions of the modules. The lack of strong interactions between modules could have implications in future design efforts and facilitate module swapping efforts for better performances and altered products.

5 Emerging X-ray Crystallography Techniques

Recent developments in structural biology have opened up many new opportunities, especially in the studies of enzymatic reactions and protein dynamics [74, 75]. It is now widely recognized that proteins are not static entities but instead they are able to fluctuate between alternative conformations defined by complex energy landscapes [76]. Moreover, structural studies of enzymes over the years have clearly demonstrated a flexibility around active sites and multiple conformational stages during the catalytic cycle [86, 87]. Thus, investigation of the dynamical aspects of macromolecular function and the intermediates that form and decay during macromolecular reactions are of great interest [77].

Powerful X-rays in microfocus beamtimes at synchrotrons combined with fast photon-counting detectors offer new possibilities in understanding enzymatic processes, including studies of multienzyme assemblies. Rapid, shutterless X-ray detector technology on a synchrotron MX beamlines has been exploited on a single copper nitrite reductase crystal to perform low-dose serial crystallography. A single crystal survived long enough for 45 consecutive 100 K X-ray structures to be collected at 1.07–1.62 Å resolution, all sampled from the same crystal volume [78]. This serial crystallography approach revealed the gradual conversion of the substrate bound at the catalytic type 2 Cu center from nitrite to nitric oxide, following reduction of the type 1 Cu electron-transfer center by X-ray-generated solvated electrons. Besides, a dioxygen species was identified in data collected to 1.6 Å [79].

Moreover, the emergence of X-ray free-electron lasers (XFELs) has enabled time-resolved serial femtosecond (fs) crystallography (TR-SFX), where the X-ray damage is outpaced by ultra-short fs X-ray pulses. In SFX experiments, microcrystals suspended in a liquid jet are delivered to the beam of an X-ray free-electron laser with a gas dynamic virtual nozzle [80]. Data are collected from randomly oriented microcrystals which are intersected by 40-fs X-ray pulses. In this way, diffraction can be recorded before the crystals are destructed ("diffraction before destruction"). In one of the first demonstrations of this method, high-resolution intermediates of photoactive yellow protein (PYP) were revealed in time-resolved crystallography at 1.6 Å resolution using X-FEL [81]. Importantly, crystals measuring only a few microns or less and at near-physiological conditions can be used for time-resolved studies for structural and functional analysis [82]. The small size of the crystals facilitates also a quick diffusion of the substrates into the crystal lattice. In cytochrome c oxidase (C*c*O) crystals, for example, diffusion of O_2 occurred within 2 ms [83]. The microcrystals used were of average dimensions ~20 × 20 × 4 μm and soaked with O_2-saturated buffer in a mixer. Intermediates in the reaction of O_2 with C*c*O had never been determined before, limiting therefore the understanding of the coupling between O_2 reduction and proton translocation. With TR-SFX, the structure of a catalytic intermediate was obtained in the reaction of reduced microcrystals of bovine C*c*O with O_2.

The new techniques offer the possibility of using different temperatures to record the dynamics that otherwise may be suppressed or altered by cryocooling in standard protein crystallography experiments [84] as well as the dynamic transitions that structures may undergo at cryogenic temperatures above 100 K and the study of allosteric mechanisms [85].

Notably, the new techniques have allowed data collection at room temperatures and without the presence of cryoprotectants and the need for cryogenic temperatures [74]. Microcrystals of the enzyme β-lactamase from *Mycobacterium tuberculosis* (BlaC) were soaked with ceftriaxone (CEF), a third-generation antibiotic, using mix-and-inject serial crystallography (MISC). The reaction was followed from 30 ms to 2 s [86]. Electron density at 30 ms revealed that diffusion of the substrate was successful and the covalently bound CEF adduct was observed after 100 ms and it was the dominant species at 500 ms. Nucleophilic attack by Ser70 to open the lactam ring was observed at times longer than 100 ms. Also, serial femtosecond X-ray crystallography with simultaneous X-ray emission spectroscopy using multi-flash visible laser excitation at room temperature was employed to study all (meta)stable states of Kok's cycle in photosystem II. Various high-resolution structures (2.04–2.08 Å) were obtained [87].

Enzymatic reactions can also be triggered by using inactive (caged) substrates that could be activated by a laser pulse [88]. The idea of using caged compounds was born by the use of the so-called time-resolved Laue crystallography method [89–91]. The first proof of concept for time-resolved Laue crystallography triggered by a caged substrate was achieved using crystals of p21(H-ras) [92]. While this method has shown great potential, its application has so far been limited due to significant experimental challenges, one of them is the need for large crystals ($>6 \times 10^5$ μm^3) and therefore lack of uniform reaction initiation owing to the large size crystal. Time-resolved crystallography using synchrotron radiation was recently applied to fluoroacetate dehalogenase (FAcD) [93]. Data on FAcD crystals at resolutions between 1.7 and 1.9 Å and time points ranging from 30 ms to ~30 s were collected after initiation of the catalytic reaction by light. An unreactive, caged substrate soaked into the solvent channels of the FAcD crystals was uncaged by an ultra-short ultraviolet-laser pulse in a photochemical reaction that was orders of magnitude faster than the enzymatic reaction. The data revealed molecular-breathing motions but interestingly, no significant changes were observed during the catalytic cycle, which was found consistent with the idea of "allostery without changes." Nevertheless, only a few time-resolved experiments have been reported so far where highly reactive, caged substrates are readily available [75].

6 Conclusions

Efficient use of multienzyme assemblies requires a detailed understanding of their component enzymes. Biochemical analysis can provide insights into the enzyme mechanisms and kinetics. However, structural information is also needed to give mechanistic details of the catalytic reaction, conformational changes, and dynamics for a better characterization of their catalytic properties and the design of modifications to improve their potential in applied processes. The advent of modern diffraction techniques, powerful X-rays, and ultra-fast X-ray detectors could offer new avenues in studying the dynamics of enzymatic reactions in many multienzyme systems.

References

1. Sheldon RA, Woodley JM (2018) Role of biocatalysis in sustainable chemistry. Chem Rev 118:801–838. https://doi.org/10.1021/acs.chemrev.7b00203
2. Cipolatti EP, Cerqueira Pinto MC, Henriques RO et al (2019) Enzymes in green chemistry: the state of the art in chemical transformations. In: Advances in enzyme technology. Elsevier, Amsterdam, pp 137–151
3. Heckmann CM, Paradisi F (2020) Looking back: a short history of the discovery of enzymes and how they became powerful chemical tools. ChemCatChem 12: 6082–6102. https://doi.org/10.1002/cctc.202001107
4. Giannakopoulou A, Gkantzou E, Polydera A, Stamatis H (2020) Multienzymatic nanoassemblies: recent progress and applications. Trends Biotechnol 38:202–216. https://doi.org/10.1016/j.tibtech.2019.07.010
5. Ricca E, Brucher B, Schrittwieser JH (2011) Multi-enzymatic cascade reactions: overview and perspectives. Adv Synth Catal 353: 2239–2262. https://doi.org/10.1002/adsc.201100256
6. Hwang ET, Lee S (2019) Multienzymatic cascade reactions via enzyme complex by immobilization. ACS Catal 9:4402–4425. https://doi.org/10.1021/acscatal.8b04921
7. Jiang Y, Zhang X, Yuan H et al (2021) Research progress and the biotechnological applications of multienzyme complex. Appl Microbiol Biotechnol 105:1759–1777. https://doi.org/10.1007/s00253-021-11121-4
8. Blow D (2000) So do we understand how enzymes work? Structure 8:R77–R81. https://doi.org/10.1016/S0969-2126(00)00125-8
9. Pauling L (1948) Nature of forces between large molecules of biological interest. Nature 161:707–709. https://doi.org/10.1038/161707a0
10. Blake CC, Koenig DF, Mair GA et al (1965) Structure of hen egg-white lysozyme. A three-dimensional Fourier synthesis at 2 Å resolution. Nature 206:757–761. https://doi.org/10.1038/206757a0
11. Johnson LN, Phillips DC (1965) Structure of some crystalline lysozyme-inhibitor complexes determined by X-ray analysis at 6 Angstrom resolution. Nature 206:761–763. https://doi.org/10.1038/206761a0
12. Steitz TA, Ludwig ML, Quiocho FA, Lipscomb WN (1967) The structure of carboxypeptidase A: V. studies of enzyme-substrate and enzyme-inhibitor complexes at 6 Å resolution. J Biol Chem 242:4662–4668. https://doi.org/10.1016/S0021-9258(18)99508-6
13. Kartha G, Bello J, Harker D (1967) Tertiary structure of ribonuclease. Nature 213: 862–865. https://doi.org/10.1038/213862a0
14. Kraut J (1977) Serine proteases: structure and mechanism of catalysis. Annu Rev Biochem 46:331–358. https://doi.org/10.1146/annurev.bi.46.070177.001555
15. Schneider G, Lindqvist Y, Brändén CI, Lorimer G (1986) Three-dimensional structure of ribulose-1,5-bisphosphate carboxylase/oxygenase from Rhodospirillum rubrum at 2.9 A resolution. EMBO J 5:3409–3415
16. Andersson I, Knight S, Schneider G et al (1989) Crystal structure of the active site of ribulose-bisphosphate carboxylase. Nature 337:229–234. https://doi.org/10.1038/337229a0
17. McPhalen CA, Vincent MG, Picot D et al (1992) Domain closure in mitochondrial aspartate aminotransferase. J Mol Biol 227: 197–213
18. Fletterick RJ, Sprang SR (1982) Glycogen phosphorylase structures and function. Acc Chem Res 15:361–369. https://doi.org/10.1021/ar00083a004
19. Irvine HS, Shaw SM, Paton A, Carrey EA (1997) A reciprocal allosteric mechanism for efficient transfer of labile intermediates between active sites in CAD, the mammalian pyrimidine-biosynthetic multienzyme polypeptide. Eur J Biochem 247:1063–1073. https://doi.org/10.1111/j.1432-1033.1997.01063.x
20. Krause KL, Volz KW, Lipscomb WN (1987) 2.5 A structure of aspartate carbamoyltransferase complexed with the bisubstrate analog N-(phosphonacetyl)-L-aspartate. J Mol Biol 193:527–553. https://doi.org/10.1016/0022-2836(87)90265-8
21. Evans PR, Farrants GW, Hudson PJ (1981) Phosphofructokinase: structure and control. Philos Trans R Soc Lond Ser B Biol Sci 293: 53–62. https://doi.org/10.1098/rstb.1981.0059
22. Hajdu J, Acharya KR, Stuart DI et al (1987) Catalysis in the crystal: synchrotron radiation

studies with glycogen phosphorylase b. EMBO J 6:539–546

23. Hyde CC, Ahmed SA, Padlan EA et al (1988) Three-dimensional structure of the tryptophan synthase alpha 2 beta 2 multienzyme complex from Salmonella typhimurium. J Biol Chem 263:17857–17871
24. Hyde CC, Miles EW (1990) The tryptophan synthase multienzyme complex: exploring structure-function relationships with X-Ray Crystallography and mutagenesis. Nat Biotechnol 8:27–32
25. Hall PR, Zheng R, Antony L et al (2004) Transcarboxylase 5S structures: assembly and catalytic mechanism of a multienzyme complex subunit. EMBO J 23:3621–3631. https://doi.org/10.1038/sj.emboj.7600373
26. Marrott NL, Marshall JJT, Svergun DI et al (2012) The catalytic core of an archaeal 2-oxoacid dehydrogenase multienzyme complex is a 42-mer protein assembly. FEBS J 279:713–723. https://doi.org/10.1111/j.1742-4658.2011.08461.x
27. Frank RAW, Pratap JV, Pei XY et al (2005) The molecular origins of specificity in the assembly of a multienzyme complex. Structure 13:1119–1130. https://doi.org/10.1016/j.str.2005.04.021
28. Tsuchiya D, Shimizu N, Ishikawa M et al (2006) Ligand-induced domain rearrangement of fatty acid beta-oxidation multienzyme complex. Structure 14:237–246. https://doi.org/10.1016/j.str.2005.10.011
29. Miles EW, Rhee S, Davies DR (1999) The molecular basis of substrate channeling*. J Biol Chem 274:12193–12196. https://doi.org/10.1074/jbc.274.18.12193
30. Zhang Y-HP (2011) Substrate channeling and enzyme complexes for biotechnological applications. Biotechnol Adv 29:715–725. https://doi.org/10.1016/j.biotechadv.2011.05.020
31. Sweetlove LJ, Fernie AR (2018) The role of dynamic enzyme assemblies and substrate channelling in metabolic regulation. Nat Commun 9:2136. https://doi.org/10.1038/s41467-018-04543-8
32. Oh J-H, Hyun J-Y, Chen S-J, Varshavsky A (2020) Five enzymes of the Arg/N-degron pathway form a targeting complex: the concept of superchanneling. Proc Natl Acad Sci U S A 117:10778–10788. https://doi.org/10.1073/pnas.2003043117
33. Chen Y, Ke G, Ma Y et al (2018) A synthetic light-driven substrate channeling system for precise regulation of enzyme cascade activity based on DNA Origami. J Am Chem Soc 140:8990–8996. https://doi.org/10.1021/jacs.8b05429
34. Knighton DR, Kan CC, Howland E et al (1994) Structure of and kinetic channelling in bifunctional dihydrofolate reductase-thymidylate synthase. Nat Struct Biol 1:186–194. https://doi.org/10.1038/nsb0394-186
35. Atreya CE, Johnson EF, Williamson J et al (2003) Probing electrostatic channeling in protozoal bifunctional thymidylate synthase-dihydrofolate reductase using site-directed mutagenesis. J Biol Chem 278:28901–28911. https://doi.org/10.1074/jbc.M212689200
36. Anderson KS (2017) Understanding the molecular mechanism of substrate channeling and domain communication in protozoal bifunctional TS-DHFR. Protein Eng Des Sel 30:253–261. https://doi.org/10.1093/protein/gzx004
37. Shatalin K, Lebreton S, Rault-Leonardon M et al (1999) Electrostatic channeling of oxaloacetate in a fusion protein of porcine citrate synthase and porcine mitochondrial malate dehydrogenase. Biochemistry 38:881–889. https://doi.org/10.1021/bi982195h
38. Wu F, Minteer S (2015) Krebs cycle metabolon: structural evidence of substrate channeling revealed by cross-linking and mass spectrometry. Angew Chem Int Ed Engl 54:1851–1854. https://doi.org/10.1002/anie.201409336
39. Sathyanarayanan N, Cannone G, Gakhar L et al (2019) Molecular basis for metabolite channeling in a ring opening enzyme of the phenylacetate degradation pathway. Nat Commun 10:4127. https://doi.org/10.1038/s41467-019-11931-1
40. Weeks A, Lund L, Raushel FM (2006) Tunneling of intermediates in enzyme-catalyzed reactions. Curr Opin Chem Biol 10:465–472. https://doi.org/10.1016/j.cbpa.2006.08.008
41. Marques SM, Daniel L, Buryska T et al (2017) Enzyme tunnels and gates as relevant targets in drug design. Med Res Rev 37:1095–1139. https://doi.org/10.1002/med.21430
42. Raushel FM, Thoden JB, Holden HM (2003) Enzymes with molecular tunnels. Acc Chem Res 36:539–548. https://doi.org/10.1021/ar020047k
43. Anderson KS, Miles EW, Johnson KA (1991) Serine modulates substrate channeling in tryptophan synthase. A novel intersubunit triggering mechanism. J Biol Chem 266:

8020–8033. https://doi.org/10.1016/S0021-9258(18)92934-0

44. Dunn MF, Niks D, Ngo H et al (2008) Tryptophan synthase: the workings of a channeling nanomachine. Trends Biochem Sci 33: 254–264. https://doi.org/10.1016/j.tibs.2008.04.008
45. Thoden JB, Wesenberg G, Raushel FM, Holden HM (1999) Carbamoyl phosphate synthetase: closure of the B-domain as a result of nucleotide binding. Biochemistry 38: 2347–2357. https://doi.org/10.1021/bi982517h
46. Binda C, Bossi RT, Wakatsuki S et al (2000) Cross-talk and ammonia channeling between active centers in the unexpected domain arrangement of glutamate synthase. Structure 8:1299–1308. https://doi.org/10.1016/S0969-2126(00)00540-2
47. Chaudhuri BN, Lange SC, Myers RS et al (2001) Crystal structure of imidazole glycerol phosphate synthase: a tunnel through a (β/α) 8 barrel joins two active sites. Structure 9: 987–997
48. Zein F, Zhang Y, Kang YN et al (2006) Structural insights into the mechanism of the PLP synthase holoenzyme from Thermotoga maritima. Biochemistry 45:14609–14620. https://doi.org/10.1021/bi061464y
49. Mouilleron S, Badet-Denisot M-A, Golinelli-Pimpaneau B (2006) Glutamine binding opens the ammonia channel and activates glucosamine-6P synthase. J Biol Chem 281: 4404–4412. https://doi.org/10.1074/jbc.M511689200
50. Larsen TM, Boehlein SK, Schuster SM et al (1999) Three-dimensional structure of Escherichia coli asparagine synthetase B: a short journey from substrate to product. Biochemistry 38:16146–16157. https://doi.org/10.1021/bi9915768
51. Nakamura A, Yao M, Chimnaronk S et al (2006) Ammonia channel couples glutaminase with transamidase reactions in GatCAB. Science 312:1954–1958. https://doi.org/10.1126/science.1127156
52. Li L, Adachi M, Yu J et al (2019) Neutron crystallographic study of heterotrimeric glutamine amidotransferase CAB. Acta Crystallogr F Struct Biol Commun 75:193–196. https://doi.org/10.1107/S2053230X19000220
53. Krahn JM, Kim JH, Burns MR et al (1997) Coupled formation of an amidotransferase interdomain ammonia channel and a phosphoribosyltransferase active site. Biochemistry 36:11061–11068. https://doi.org/10.1021/bi9714114
54. Perham RN (2000) Swinging arms and swinging domains in multifunctional enzymes: catalytic machines for multistep reactions. Annu Rev Biochem 69:961–1004. https://doi.org/10.1146/annurev.biochem.69.1.961
55. Dardel F, Davis AL, Laue ED, Perham RN (1993) Three-dimensional structure of the lipoyl domain from Bacillus stearothermophilus pyruvate dehydrogenase multienzyme complex. J Mol Biol 229:1037–1048. https://doi.org/10.1006/jmbi.1993.1103
56. Athappilly FK, Hendrickson WA (1995) Structure of the biotinyl domain of acetyl-coenzyme A carboxylase determined by MAD phasing. Structure 3:1407–1419. https://doi.org/10.1016/s0969-2126(01)00277-5
57. Farmer R, Thomas CM, Winn PJ (2019) Structure, function and dynamics in acyl carrier proteins. PLoS One 14:e0219435. https://doi.org/10.1371/journal.pone.0219435
58. Misson LE, Mindrebo JT, Davis TD et al (2020) Interfacial plasticity facilitates high reaction rate of E. coli FAS malonyl-CoA: ACP transacylase, FabD. Proc Natl Acad Sci U S A 117:24224–24233. https://doi.org/10.1073/pnas.2009805117
59. Robinson GC, Kaufmann M, Roux C, Fitzpatrick TB (2016) Structural definition of the lysine swing in Arabidopsis thaliana PDX1: intermediate channeling facilitating vitamin B6 biosynthesis. Proc Natl Acad Sci U S A 113:E5821–E5829. https://doi.org/10.1073/pnas.1608125113
60. Bülow L, Mosbach K (1991) Multienzyme systems obtained by gene fusion. Trends Biotechnol 9:226–231. https://doi.org/10.1016/0167-7799(91)90075-s
61. Arai R, Ueda H, Kitayama A et al (2001) Design of the linkers which effectively separate domains of a bifunctional fusion protein. Protein Eng 14:529–532. https://doi.org/10.1093/protein/14.8.529
62. Schoffelen S, van Hest JCM (2013) Chemical approaches for the construction of multienzyme reaction systems. Curr Opin Struct Biol 23:613–621. https://doi.org/10.1016/j.sbi.2013.06.010
63. Wei X, Han P, You C (2020) Facilitation of cascade biocatalysis by artificial multi-enzyme complexes — A review. Chin J Chem Eng 28:

2799–2809. https://doi.org/10.1016/j.cjche.2020.05.022
64. Ribeiro LF, Furtado GP, Lourenzoni MR et al (2011) Engineering bifunctional laccase-Xylanase chimeras for improved catalytic performance. J Biol Chem 286:43026–43038. https://doi.org/10.1074/jbc.M111.253419
65. Lerchner A, Daake M, Jarasch A, Skerra A (2016) Fusion of an alcohol dehydrogenase with an aminotransferase using a PAS linker to improve coupled enzymatic alcohol-to-amine conversion. Protein Eng Des Sel 29:557–562. https://doi.org/10.1093/protein/gzw039
66. Knott BC, Erickson E, Allen MD et al (2020) Characterization and engineering of a two-enzyme system for plastics depolymerization. Proc Natl Acad Sci U S A 24: 202006753. https://doi.org/10.1073/pnas.2006753117
67. Palm GJ, Reisky L, Böttcher D et al (2019) Structure of the plastic-degrading Ideonella sakaiensis MHETase bound to a substrate. Nat Commun 10:1717–1710. https://doi.org/10.1038/s41467-019-09326-3
68. Liu Y, Hickey DP, Guo J-Y et al (2017) Substrate channeling in an artificial metabolon: a molecular dynamics blueprint for an experimental peptide bridge. ACS Catal 7: 2486–2493. https://doi.org/10.1021/acscatal.6b03440
69. Ozbakir HF, Garcia KE, Banta S (2018) Creation of a formate: malate oxidoreductase by fusion of dehydrogenase enzymes with PEGylated cofactor swing arms. Protein Eng Des Sel 31:103–108
70. Yang YR, Fu J, Wootten S et al (2018) 2D enzyme cascade network with efficient substrate channeling by swinging arms. Chembiochem 19:212–216. https://doi.org/10.1002/cbic.201700613
71. Fu J, Yang YR, Johnson-Buck A et al (2014) Multi-enzyme complexes on DNA scaffolds capable of substrate channelling with an artificial swinging arm. Nat Nanotechnol 9: 531–536. https://doi.org/10.1038/nnano.2014.100
72. Chiocchini C, Linne U, Stachelhaus T (2006) In vivo biocombinatorial synthesis of lipopeptides by COM domain-mediated reprogramming of the surfactin biosynthetic complex. Chem Biol 13:899–908. https://doi.org/10.1016/j.chembiol.2006.06.015
73. Reimer JM, Eivaskhani M, Harb I et al (2019) Structures of a dimodular nonribosomal peptide synthetase reveal conformational flexibility. Science 366:eaaw4388. https://doi.org/10.1126/science.aaw4388
74. Chapman HN, Fromme P, Barty A et al (2011) Femtosecond X-ray protein nanocrystallography. Nature 469:73–77. https://doi.org/10.1038/nature09750
75. Schmidt M (2020) Reaction initiation in enzyme crystals by diffusion of substrate. Crystals 10:116. https://doi.org/10.3390/cryst10020116
76. Keedy DA, Kenner LR, Warkentin M et al (2015) Mapping the conformational landscape of a dynamic enzyme by multitemperature and XFEL crystallography. elife 4: e07574. https://doi.org/10.7554/eLife.07574
77. Schmidt M (2019) Time-resolved macromolecular crystallography at pulsed X-ray sources. Int J Mol Sci. https://doi.org/10.3390/ijms20061401
78. Horrell S, Antonyuk SV, Eady RR et al (2016) Serial crystallography captures enzyme catalysis in copper nitrite reductase at atomic resolution from one crystal. IUCrJ 3:271–281
79. Halsted TP, Yamashita K, Hirata K et al (2018) An unprecedented dioxygen species revealed by serial femtosecond rotation crystallography in copper nitrite reductase. IUCrJ 5:22–31
80. Boutet S, Lomb L, Williams GJ et al (2012) High-resolution protein structure determination by serial femtosecond crystallography. Science 337:362–364
81. Tenboer J, Basu S, Zatsepin N et al (2014) Time-resolved serial crystallography captures high-resolution intermediates of photoactive yellow protein. Science 346:1242–1246. https://doi.org/10.1126/science.1259357
82. Orville AM (2020) Recent results in time resolved serial femtosecond crystallography at XFELs. Curr Opin Struct Biol 65: 193–208. https://doi.org/10.1016/j.sbi.2020.08.011
83. Ishigami I, Lewis-Ballester A, Echelmeier A et al (2019) Snapshot of an oxygen intermediate in the catalytic reaction of cytochrome C oxidase. Proc Natl Acad Sci U S A 116: 3572–3577. https://doi.org/10.1073/pnas.1814526116
84. Horrell S, Kekilli D, Sen K et al (2018) Enzyme catalysis captured using multiple structures from one crystal at varying temperatures. IUCrJ 5:283–292. https://doi.org/10.1107/S205225251800386X
85. Keedy DA (2019) Journey to the center of the protein: allostery from multitemperature

multiconformer X-ray crystallography. Acta Crystallogr D Struct Biol 75:123–137. https://doi.org/10.1107/S2059798318017941
86. Olmos JL, Pandey S, Martin-Garcia JM et al (2018) Enzyme intermediates captured "on the fly" by mix-and-inject serial crystallography. BMC Biol 16:1–15. https://doi.org/10.1186/s12915-018-0524-5
87. Kern J, Chatterjee R, Young ID et al (2018) Structures of the intermediates of Kok's photosynthetic water oxidation clock. Nature 563:421–425
88. Klán P, Šolomek T, Bochet CG et al (2013) Photoremovable protecting groups in chemistry and biology: reaction mechanisms and efficacy. Chem Rev 113:119–191. https://doi.org/10.1021/cr300177k
89. Moffat K, Szebenyi D, Bilderback D (1984) X-ray laue diffraction from protein crystals. Science 223:1423–1425. https://doi.org/10.1126/science.223.4643.1423
90. Hajdu J, Machin PA, Campbell JW et al (1987) Millisecond X-ray diffraction and the first electron density map from Laue photographs of a protein crystal. Nature 329: 178–181. https://doi.org/10.1038/329178a0
91. Hajdu J, Johnson LN (1990) Progress with Laue diffraction studies on protein and virus crystals. Biochemistry 29:1669–1678. https://doi.org/10.1021/bi00459a001
92. Scheidig AJ, Sanchez-Llorente A, Lautwein A et al (1994) Crystallographic studies on p21 (H-ras) using the synchrotron Laue method: improvement of crystal quality and monitoring of the GTPase reaction at different time points. Acta Crystallogr D Biol Crystallogr 50:512–520. https://doi.org/10.1107/S090744499301443X
93. Mehrabi P, Schulz EC, Dsouza R et al (2019) Time-resolved crystallography reveals allosteric communication aligned with molecular breathing. Science 365:1167–1170. https://doi.org/10.1126/science.aaw9904
94. Ladenstein R, Fischer M, Bacher A (2013) The lumazine synthase/riboflavin synthase complex: shapes and functions of a highly variable enzyme system. FEBS J 280: 2537–2563. https://doi.org/10.1111/febs.12255
95. Ævarsson A, Seger K, Turley S et al (1999) Crystal structure of 2-oxoisovalerate and dehydrogenase and the architecture of 2-oxo acid dehydrogenase multienzyme complexes. Nat Struct Mol Biol 6:785–792. https://doi.org/10.1038/11563
96. Artiukhov AV, Graf AV, Bunik VI (2016) Directed regulation of multienzyme complexes of 2-Oxo acid dehydrogenases using phosphonate and phosphinate analogs of 2-Oxo acids. Biochemistry (Mosc) 81: 1498–1521. https://doi.org/10.1134/S0006297916120129
97. Frank RAW, Titman CM, Pratap JV et al (2004) A molecular switch and proton wire synchronize the active sites in thiamine enzymes. Science 306:872–876
98. Ishikawa M, Tsuchiya D, Oyama T et al (2004) Structural basis for channelling mechanism of a fatty acid β-oxidation multienzyme complex. EMBO J 23:2745–2754. https://doi.org/10.1038/sj.emboj.7600298
99. Douce R, Bourguignon J, Neuburger M, Rébeillé F (2001) The glycine decarboxylase system: a fascinating complex. Trends Plant Sci 6:167–176
100. St Maurice M, Reinhardt L, Surinya KH et al (2007) Domain architecture of pyruvate carboxylase, a biotin-dependent multifunctional enzyme. Science 317:1076–1079
101. Pettersen EF, Goddard TD, Huang CC et al (2004) UCSF Chimera--a visualization system for exploratory research and analysis. J Comput Chem 25:1605–1612
102. Pravda L, Sehnal D, Svobodová Vařeková R et al (2018) ChannelsDB: database of biomacromolecular tunnels and pores. Nucleic Acids Res 46:D399–D405. https://doi.org/10.1093/nar/gkx868

Chapter 5

Assessing Protein Interactions for Clustering of Mitochondrial Urea Cycle Enzymes

Ljubica Caldovic, Shivaprasad Bhuvanendran, and Jyoti Jaiswal

Abstract

Enzyme clustering is a phenomenon that involves partitioning of proteins that function together in a common subcellular or sub-organellar compartment. Traditional genetic, biochemical, and biophysical approaches for studying protein–protein interactions in complexes with defined stoichiometry yield inconclusive results when applied to clustered proteins. This chapter describes a combination of approaches to study clustered proteins including co-immunoprecipitation, biochemical co-localization in purified mitochondria, and super resolution imaging of endogenous proteins in situ. These approaches can be used to study interactions among proteins that form clusters. We will illustrate this approach by using the urea cycle enzymes that localize in the mitochondrial matrix, and form clusters at the inner mitochondrial membrane.

Key words Protein–protein interactions, Protein cluster, Mitochondria, Urea cycle, Inner mitochondrial membrane, Co-immunoprecipitation, Mitochondrial fractionation, Immunofluorescence, Confocal microscopy, Super resolution microscopy, gSTED

1 Introduction

Stable and transient protein–protein interactions are critical for all forms of protein functions ranging from signaling, metabolism, cellular communication, intracellular trafficking, and even transcription, translation, or protein degradation. Approaches to detect and characterize protein–protein interactions can be classified into—(A) Genetic approaches, such as yeast and bacterial two-hybrid screens [1, 2]. (B) Biophysical approaches, such as surface plasmon resonance [3], light scattering [4], fluorescence resonance transfer (FRET) [5–8], nuclear magnetic resonance [9], and hydrogen exchange [10]. (C) Biochemical approaches, such as co-immunoprecipitation coupled with immunoblotting or tandem mass spectroscopy [11–13], chemical crosslinking [14, 15], and

Ljubica Caldovic and Shivaprasad Bhuvanendran contributed equally with all other contributors.

Haralambos Stamatis (ed.), *Multienzymatic Assemblies: Methods and Protocols*,
Methods in Molecular Biology, vol. 2487, https://doi.org/10.1007/978-1-0716-2269-8_5,
© The Author(s), under exclusive license to Springer Science+Business Media, LLC, part of Springer Nature 2022

enzymatic tagging [16]. Omics-based high throughput approaches are suitable for large scale discovery of protein–protein interactions [17–19] while biochemical and biophysical methods can be used for characterization of interacting surfaces between specific proteins [20–22] in complexes with defined stoichiometry and structure [14–16, 23, 24].

Assembly of enzymes in a metabolic pathway into multi-protein complexes enables channeling of pathway intermediates that presumably enhances flux through the pathway, protects unstable intermediates from degradation, and prevents diffusion of toxic intermediates into other cellular compartments [25, 26]. One such example is the urea cycle. Urea cycle operates in mammalian liver to convert ammonia, a product of protein catabolism, into urea. Ammonia is a potent neurotoxin, while urea is non-toxic and excreted through the kidneys [27]. Six urea cycle enzymes are partitioned between mitochondria and cytoplasm and two transporters shuttle intermediates between them. Mitochondrial N-acetylglutamate synthase (NAGS; EC 2.3.1.1) catalyzes formation of N-acetylglutamate (NAG) which activates carbamoyl phosphate synthetase 1 (CPS1; EC 6.3.4.16) [27–29]. CPS1 and ornithine transcarbamylase (OTC; EC 2.1.3.3) collectively convert ammonia, bicarbonate, ATP, and ornithine into citrulline in the mitochondria. Citrulline is converted into urea by the cytosolic enzymes argininosuccinate synthetase 1 (ASS1; EC 6.3.4.5), argininosuccinate lyase (ASL; EC 4.3.2.1), and arginase 1 (ARG1; EC 3.5.3.1) [27]. Ornithine transporter (ORNT encoded by the *SLC25A15/Slc25A15* gene) and mitochondrial aspartate-glutamate carrier 2 (also called citrin or ALARAL2 and encoded by the *SLC25A13/Slc25A13* gene) shuttle urea cycle intermediates between mitochondria and cytoplasm [27, 30, 31].

Substrate channeling of urea cycle intermediates among urea cycle enzymes [32, 33] and requirement for arginine supplementation in patients with urea cycle disorders who received liver transplant [34] suggest that urea cycle enzymes interact. Presumably, this prevents breakdown of chemically unstable carbamoyl phosphate and enables efficient ureagenesis to keep concentration of plasma ammonia below 30–50 μM. Abundance of urea cycle enzymes in liver cells differs by orders of magnitude [35–38]. This makes existence of a multienzyme complex with defined composition and stoichiometry unlikely and suggests that these enzymes may exist as clusters that can assemble into a functional complex.

Enzyme clustering is a concept that explains interactions between proteins of highly disparate abundance, that do not catalyze consecutive reactions of a metabolic pathway and assemble into a super-complex or cluster on demand [39–42]. Enzymes of the coenzyme Q biosynthesis form a super-complex only while catalyzing coenzyme Q formation inside mitochondria [42], while enzymes in the de novo purine biosynthesis cluster in a cell cycle dependent manner [39, 40]. Direct evidence of interactions among proteins in such clusters is often inconclusive because interactions are not as specific and strong as in multienzyme complexes with defined stoichiometry [39]. Similar issues exist with mitochondrial urea cycle enzymes. While NAGS, CPS1, and OTC can be co-immunoprecipitated from isolated liver mitochondria [43], stable interactions among the three enzymes are not evident in vitro [44]. Further, attempts to crosslink NAGS, CPS1, and OTC in isolated liver mitochondria remain inconclusive. This suggests that either the conditions used in vitro do not mimic in vivo environment or that interactions between these enzymes are not as strong and specific as is expected for a stable multienzyme complex [14–16, 23, 24]. Additional obstacles for studying protein–protein interactions among urea cycle enzymes include weak expression even in liver-derived human cell lines and poor long-term viability of primary hepatocyte cultures. These challenges are not unique to the urea cycle enzymes and the approaches we describe below are aimed at studying interactions among proteins that partition into a common subcellular compartment but do not form stable and defined complexes. Our efforts help overcome these obstacles by using a multipronged approach involving a combination of protein co-immunoprecipitation, biochemical co-localization in the purified mitochondria, and in situ imaging co-localized NAGS, CPS1, and OTC in the primary hepatocytes to provide complimentary evidence for clustering of these proteins in vivo. These methods can be applied for assessing interactions among other proteins that cluster, provided that reliable antibodies are available for immunoblotting and immunofluorescence.

2 Materials

2.1 Isolation of Mitochondria by Differential Fractionation (See Note 1)

1. Liver homogenization buffer (LHB): 10 mM HEPES, 1 mM EDTA (*see* **Note 2**), 10 mM KCl, 50 mM sucrose, 0.2 M mannitol; adjust pH to 7.4 with KOH (*see* **Note 3**).
2. Equipment: Dissecting tools, Potter-Elvehjem homogenizer (~0.09 mm clearance; 25 mL working volume) equipped with overhead high-torque electric motor (thyristor-controlled), low-speed centrifuge with swinging bucket rotor and appropriate tubes, high-speed centrifuge with fixed-angle rotor and 40- to 50-mL polycarbonate tubes, vacuum pump, Dounce homogenizer with loose-fitting pestle.

2.2 Protein Determination

1. Bio-Rad Protein Assay Dye Reagent Concentrate.

2.3 Co-immunoprecipitation

1. Mitochondrial lysis buffer (MLB): 2% (w/v) CHAPS in phosphate-buffered saline (PBS; 0.0100 M Na_2HPO_4, 0.0018 KH_2PO_4, 0.1370 NaCl, 0.0027 KCl, pH 7.4; *see* **Note 4**).
2. Co-IP wash buffer: 0.05% (v/v) Triton X-100 in PBS.
3. Pierce™ Protein A/G Magnetic Agarose Beads.
4. Antibodies against NAGS, CPS1 and OTC (*see* **Note 5**).
5. ImmunoPure IgG Elution Buffer.
6. DynaMag™-2 Magnet.

2.4 Fractionation of Rat Liver Mitochondria

1. Mitochondrial fractionation buffer (MFB): 5 mM Tris–HCl, 250 mM sucrose, 1 mM EDTA, pH 7.2.
2. Mitochondria resuspension buffer (MRB): 20 mM HEPES, pH 8.0.
3. Triton stock solution: 6% (v/v) Triton X-100 in deionized H_2O.
4. MFB with digitonin: 4 mg/mL digitonin in MFB.
5. Equipment: ultrasonic cell disruptor/homogenizer, ultracentrifuge with tubes and rotor suitable for centrifugation at 144,000 × *g*.

2.5 Immunoblotting

1. Precast 4–15% polyacrylamide gels or precast 4–12% bis-Tris protein gels.
2. 5× SDS-PAGE sample loading buffer: 250 mM Tris–HCl, pH 6.8, 10% (w/v) SDS, 30% (v/v) glycerol, 10 mM DTT, 0.05% (w/v) Bromophenol Blue.
3. 10× SDS-PAGE gel running buffer: 250 mM Tris–HCl, 1.92 M Glycine, 1% (w/v) SDS, pH 8.3.
4. Pierce™ Blocking buffer containing 0.5% (v/v) Surfact-Amps 20.
5. Tris buffered saline with Tween-20 (TBST): 20 mM Tris–HCl, 150 mM NaCl, 0.005% (v/v) Tween-20, pH 7.6.
6. ECL Western Blotting Substrate.
7. Fast Western Kit, SuperSignal West Pico Substrate.
8. Primary antibodies against NAGS, CPS1, OTC (*see* **Note 5**), CoxIV, Grp75, and VDAC.
9. Secondary antibodies: horseradish peroxidase (HRP) conjugated goat anti-mouse IgG (H+L) and donkey anti-rabbit IgG (H+L) cross-adsorbed with HRP.
10. Equipment: Mini-PROTEAN®, Criterion™ or equivalent vertical electrophoresis gel apparatus, Trans-Blot Turbo Transfer System or equivalent system for protein transfer from polyacrylamide gel to either nitrocellulose or nylon filters,

ChemiDoc or equivalent imaging system for visualization and quantification of protein bands.

2.6 Culturing Mouse Primary Hepatocytes on Coverslips

1. Coverslip cleaning solutions: 1 M HCl and 100% ethanol.
2. Collagen solution for coating coverslips: Prepare 3 mg/mL collagen stock solution in 50 mM HCl; Incubate at 37 °C until collagen is completely dissolved. Dilute 3 mg/mL collagen stock solution with 50 mM HCl for final concentration 1 mg/mL. Dilute 3 parts of the 1 mg/mL collagen solution with 50 parts sterile H_2O for the final collagen concentration 60 μg/mL. This solution is stable at 4 °C for 6 months.
3. Phosphate-buffered saline (PBS): 0.0100 M Na_2HPO_4, 0.0018 KH_2PO_4, 0.1370 NaCl, 0.0027 KCl, pH 7.4.
4. Trypan Blue stain: 0.4% (w/v) trypan blue in PBS.
5. MP100 hepatocyte plating medium.
6. MM250 hepatocyte maintenance medium.

2.7 Immuno-fluorescence

1. Fixing solution: 4% (w/v) PFA in PBS.
2. Permeabilization solution: 0.1% (v/v) Triton X-100 in PBS.
3. Blocking solution: 5% (v/v) goat serum in PBS.
4. DAPI staining solution: Dissolve DAPI (4′,6-diamidino-2-phenylindole; Thermo Fisher Scientific Cat. No. D1306) in deionized H_2O according to manufacturer's instructions to obtain 300 μM DAPI intermediate dilution. Dilute DAPI intermediate solution with H_2O to a 1 μM final concentration. Mix 28.5 μL of 1 μM DAPI with 200 μL PBS.
5. Primary antibodies against NAGS, CPS1, and OTC (*see* **Note 5**) and TOMM20.
6. Secondary antibodies: goat anti-rabbit IgG (H+L) antibody highly cross-adsorbed with Alexa Fluor Plus 647 and chicken anti-rabbit IgG (H+L) antibody cross-adsorbed with Alexa Fluor 594.
7. ProLong™ Diamond Antifade Mounting solution.
8. Equipment and other supplies: Whatman filter paper, hematocytometer, tissue culture incubator.

2.8 Confocal and Super Resolution (gSTED) Microscopy

1. Equipment: Leica TCS SP8 microscope equipped with.
   (a) 405 nm laser,
   (b) a white light laser (WLL) with a range of 470–670 nm,
   (c) two depletion lasers (660 nm and 775 nm),
   (d) acousto-optical beam splitter (AOBS),
   (e) two PMTs (one would be sufficient for the method described here),

(f) two high-sensitive hybrid detectors,

(g) HC PL APO CS2 100×/1.40 oil STED objective,

(h) a low magnification objective preferably oil immersion (HC PL APO 20×/0.75 multi-immersion objective was used),

(i) fluorescence illumination (a metal-halide illumination source with intensity control to minimize photobleaching was used),

(j) DAPI and TRITC filters (FITC long-pass was available but not used),

(k) stage with motorized z-focus (preferably also with X and Y positioning ability),

(l) a workstation with simultaneous image acquisition and processing capabilities (HP Z8G4 was used. Specifications are available at https://www8.hp.com/h20195/v2/GetPDF.aspx/c05527763.pdf).

2. Software: Leica Application Suite (LAS X) and Huygens Professional version 17.04 (Scientific Volume Imaging; https://svi.nl).

3 Methods

When describing experimental procedures that were used to show that mitochondrial urea cycle genes cluster at the IMM we assumed that investigators have been trained to harvest liver tissue from euthanized animals, are familiar with performing SDS-PAGE, balancing ultracentrifuge tubes, and cell counting with hemocytometer.

3.1 Co-immunoprecipitation of Mitochondrial Urea Cycle Enzymes from Liver Mitochondria

NAGS, CPS1, and OTC are co-immunoprecipitated from isolated mouse liver mitochondria that were lysed with mitochondrial lysis buffer (MLB). Co-immunoprecipitated proteins were resolved using SDS-PAGE and visualized using immunoblotting.

1. Sacrifice a mouse using method approved by the Institutional Animal Care and Use Committee (*see* **Note 6**). Open the abdominal cavity and transfer the liver to a chilled beaker containing 1–2 mL ice cold LHB. All solution volumes are for purification of mitochondria from a mouse liver weighing 1–2 g. Transfer liver onto ice cold Petri dish and mince the liver with razor blade.
2. Transfer minced liver into Dounce homogenizer, agitate minced tissue in approx. 3 mL of ice cold LHB, allow the pieces to settle, decant the LHB, and replace it with 4–5 mL of ice cold LHB.

3. Homogenize the minced liver in Potter-Elvehjem homogenizer using five to six up-and-down strokes of the pestle, rotating at ~500 rpm (*see* **Note 7**). Decant the homogenate into 15 screw cap polypropylene tube and keep on ice.
4. Centrifuge liver homogenate for 10 min at 1000 × *g* and 4 °C in a refrigerated low-speed centrifuge. Transfer supernatant into fresh tube (*see* **Note 8**) and centrifuge for 10 min at 3000 × *g* and 4 °C, in high-speed centrifuge. Remove the supernatant and as much of the floating lipid layer as possible with a Pasteur pipette; remove as much as possible of the loose-packed pinkish layer that overlies the brown mitochondria. Remove remaining lipids adhering to the wall of the tube with a cotton swab (*see* **Note 9**).
5. Gently resuspend the pellet in 4–5 mL of ice cold LHB and centrifuge for 10 min at 3000 × *g* at 4 °C. Remove the supernatant and repeat resuspension of the pellet and centrifugation two more times.
6. Resuspend purified mitochondria in MLB. Determine protein concentration using Protein Assay Dye Reagent Concentrate according to manufacturer's instructions. Dilute mitochondrial lysate to 1 mg/mL protein for co-immunoprecipitation with antibodies against CPS1 and OTC, or to 5 mg/mL protein for co-immunoprecipitation with anti-NAGS antibodies.
7. Attach antibodies against NAGS, CPS1, or OTC to protein A/G magnetic agarose beads according to manufacturer's instructions. Attach non-specific IgG to protein A/G magnetic agarose beads and use as negative control in co-immunoprecipitations.
8. Incubate mitochondrial lysates and beads at 20 °C for 10 min with gentle agitation, separate magnetic beads from the lysate using DynaMag™-2 Magnet according to manufacturer's instructions. Wash the beads five times with co-IP wash buffer. Elute co-immunoprecipitated proteins with ImmunoPure IgG Elution Buffer according to manufacturer's instructions.
9. Determine protein concentration using Protein Assay Dye Reagent Concentrate according to manufacturer's instructions.
10. To determine whether NAGS co-immunoprecipitates with CPS1 and/or OTC resolve 0.5–1 μg of proteins immunoprecipitated with anti-CPS1 and anti-OTC antibodies using SDS-PAGE and transfer proteins onto either nylon or nitrocellulose membrane (*see* **Note 10**). Block the membrane with Pierce™ Blocking buffer according to manufacturer's instructions. Probe the filter with anti-NAGS primary antibody diluted 1: 5000 in TBST. Incubate for 1 h at room temperature and wash three times for 5 min with TBST. Incubate the filter for 1 h at

room temperature with HRP-conjugated donkey anti-rabbit secondary antibody at 1:50,000 dilution. NAGS bands are visualized using SuperSignal™ West Pico Substrate according to the manufacturer's instructions.

11. To determine whether CPS1 co-immunoprecipitates with NAGS and/or OTC resolve 0.5–1 μg of proteins immunoprecipitated with anti-NAGS and anti-OTC antibodies using SDS-PAGE and transfer proteins onto either nylon or nitrocellulose membrane. Block the membrane with Pierce™ Blocking buffer. Probe the filter with anti-CPS1 primary antibody diluted 1:5000 in TBST. Incubate for 1 h at room temperature and wash three times for 5 min with TBST. Incubate the filter for 1 h at room temperature with HRP-conjugated goat anti-mouse secondary antibody at 1:10,000 dilution. CPS1 bands are visualized using ECL Western Blotting Substrate according to the manufacturer's instructions.
12. To determine whether OTC co-immunoprecipitates with NAGS and/or CPS1 resolve 0.5–1 μg of proteins immunoprecipitated with anti-NAGS and anti-CPS1 antibodies using SDS-PAGE and transfer proteins onto either nylon or nitrocellulose membrane. Block the membrane with Pierce™ Blocking buffer. Probe the filter with anti-OTC primary antibody diluted 1:5000 in TBST. Incubate for 1 h at room temperature and wash three times for 5 min with TBST. Incubate the filter for 1 h at room temperature with HRP-conjugated donkey anti-rabbit secondary antibody at 1:10,000 dilution. OTC bands are visualized using ECL Western Blotting Substrate according to the manufacturer's instructions.

3.2 Biochemical Co-localization of Mitochondrial Urea Cycle Enzymes at the IMM

CPS1 and OTC associate with IMM of rat mitochondria [45] and we assessed whether NAGS co-localizes with the two urea cycle enzymes. Accordingly, we purified rat liver mitochondria and prepared IMM vesicles using either solubilization of the outer mitochondrial membrane with digitonin or three rounds of freezing and thawing. IMM vesicles were probed for the presence of NAGS, CPS1, and OTC and for the presence of glucose related protein 75 (GRP75), subunit IV of the cytochrome c oxidase (COX IV), and voltage-gated soluble ion channel (VDAC) as markers of the mitochondrial matrix, inner and outer mitochondrial membrane, respectively. Unless indicated otherwise in the protocol, all solutions and containers should be pre-chilled for isolation and fractionation of rat liver mitochondria.

1. Rat liver mitochondria were purified using the same protocol as for mouse liver mitochondria. All solution volumes were scaled up for rat liver weighing approx. 10 g.

2. Resuspend purified mitochondria in 0.5 mL of 20 mM MFB that contains 4 mg/mL digitonin and incubate 20 min at room temperature. Add 1 mL of MFB and centrifuge for 10 min at 16,000 × *g* and 4 °C. Remove the supernatant, resuspend the pellet in 250 μL of mitochondrial resuspension buffer (MRB), and centrifuge again for 10 min at 16,000 × *g* and 4 °C. Resuspend pelleted material in 1 mL of MRB and sonicate for 20 s twice. Keep lysate on ice between the two rounds of sonication.
3. Alternatively, resuspend purified mitochondria in 1.5 mL of 20 mM MFB and subject to three rounds of freezing in the isopropanol-dry ice bath and thawing. Centrifuge for 10 min at 16,000 × *g* and 4 °C. Remove the supernatant, resuspend the pellet in 250 μL of MRB, and centrifuge again for 10 min at 16,000 × *g* and 4 °C.
4. Divide the solution containing IMM vesicles that resulted either from the sonication treatment or freezing and thawing in four equal portions and add Triton stock solution to 0, 0.1, 0.5, and 1.0% final concentration; incubate for 30 min at room temperature. Pellet IMM vesicles by ultracentrifugation at 144,000 × *g* for 60 min. Keep the supernatants for immunoblotting and determine protein concentration in them. Wash the pellets three times with MRB, resuspend the pellets in the same buffer, and determine protein concentration.
5. Use SDS-PAGE to resolve 5–10 μg of proteins in the supernatants and pellets after Triton X-100 treatment and transfer onto nylon or nitrocellulose filters.
6. Carry out NAGS, CPS1, and OTC immunoblotting as described in Subheading 3.1.
7. Verify the purity of IMM vesicles by immunoblotting proteins in supernatants and pellets using protocols described in Subheading 3.1 and primary antibodies against CoxIV at 1:5000 dilution, Grp75 at 1:1000 dilution, and VDAC at 1:1000 dilution and HPR-conjugated secondary antibody at 1: 10,000 dilution. Visualize CoxIV, Grp75, and VDAC bands using ECL Western Blotting Substrate according to manufacturer's instructions.
8. Determine intensity of each band using a GS-800 Calibrated Densitometer and the Quantity One software package. Calculate percent of NAGS, CPS1, and OTC in the mitochondrial matrix and associated with IMM vesicles from the known volumes of IMM vesicle suspensions that were treated with Triton X-100 and their protein concentrations.

3.3 Super Resolution Imaging of Mitochondrial Urea Cycle Enzymes at the IMM

Location of NAGS, CPS1, and OTC was imaged in primary mouse hepatocytes. Both commercially available primary mouse hepatocytes or cells prepared from freshly harvested mouse livers can be used for immunofluorescence.

1. Prepare coverslips for collagen coating. Place coverslips in 1 M HCl in a glass container and keep at 50–60 °C for 4–16 h with occasional agitation. Rinse the coverslips two to three times with deionized water followed by two rinses with double deionized water (*see* **Note 11**). Rinse coverslips in 100% ethanol, airdry them and keep in a clean and sterile container.
2. Collagen Coating of Coverslips. Place coverslips in 6-well plates in the laminar flow hood. Add 500 μL collagen solution to each well, cover with lids and incubate at 37 °C for 4 h. Rinse coverslips three times with 2 mL of PBS to neutralize the acid (*see* **Note 12**).
3. Harvest hepatocytes by gentle trypsinization and pellet by centrifuging at 76 × *g* for 6 min, followed by gente resuspending by inverting tube few times.
4. Use Trypan Blue exclusion assay to determine primary hepatocyte viability. Mix 400 μL of PBS with 50 μL of Trypan Blue stain. Pipet 50 μL of hepatocyte suspension into tube with diluted Trypan Blue stain using wide bore tip. Wait 1 min and gently flick and tap the tube.
5. Gently pipet 10 μL of cell suspension into each side of hemocytometer and count live cells (clear of yellowish) and dead cells (blue color) in the four outer quadrants of the hematocytometer.
6. Calculate percent cell viability as 100 × (number of live cells)/(total number of cells).
7. Determine total live cell density: number of cells/mL = (average number of live cells) × 10,000 × 10 (*see* **Note 13**).
8. Adjust cell density to 5 × 10^5 viable cells/mL in warm hepatocyte plating medium. Plate approx. 10^5 cells in 1 mL of MP100 hepatocyte plating medium per well of a 6-well collagen coated plate. Place the plate in a 37 °C/5% CO_2 incubator.
9. Disperse the cells by moving the plate, with your hand on top of it, parallel to the incubator shelf in a north-south, east-west motion. Without removing plate from the incubator, repeat the shaking motion at 15, 30, and 45 min post-seeding. After 15 more minutes, remove plate(s) from the incubator, carefully aspirate the medium, and replace with 1.5 mL of fresh warm MP100 hepatocyte plating medium per well. Incubate the cells for a minimum total of 3–5 h post-seeding. Replace plating medium with 1.5 mL of warm MM250 hepatocyte

maintenance medium and allow cells to recover overnight at 37 °C in 5% CO_2.

10. Fixing primary hepatocytes for immunofluorescence. 24 h after plating, aspirate the cell maintenance medium and wash cells twice with PBS. To fix the cells, add 1100 μL of fixing solution for 15 min at room temperature. Wash cells three times 5 min with PBS. Fixed cells can be kept temporarily at 4 °C until use.
11. Permeabilize fixed hepatocytes with permeabilization solution for 10 min at room temperature.
12. Incubate permeabilized cells in blocking solution for 30 min at room temperature.
13. Apply anti-NAGS, CPS1, OTC, or TOMM20 primary antibody diluted with PBS for either 2 h at room temperature or overnight at 4 °C. Anti-NAGS, CPS1, and OTC primary antibodies are diluted 1:200 while anti-TOMM20 primary antibody is diluted 1:250. Wash cells three times 5 min with PBS.
14. Apply secondary antibody diluted 1:500 with PBS for 1 h at room temperature in the dark.
15. Stain with DAPI for 10 min at room temperature followed by three 5 min washes with PBS and a wash with double deionized water.
16. Mount the slides with Prolong Diamond without DAPI.
17. Hepatocytes can be stained for single proteins to verify immunofluorescence conditions or can be sequentially stained for two different proteins. In latter case stain the cells for less abundant protein first followed by staining for the more abundant protein (e.g., stain for NAGS first and then for CPS1 or OTC or TOMM20, or stain for OTC first and then for TOMM20 or CPS1). CPS1 is the most abundant liver protein and it should always be the second protein to stain.

3.4 Confocal Microscopy

This can be performed using any commercially available confocal microscope, but here we describe imaging hepatocytes labeled with DAPI, Alexa Fluor 532, and Alexa Fluor 647 using a Leica TCS SP8 confocal microscope.

1. Switch on the confocal microscope and start the LAS-X software (*see* **Note 14**).
2. In the Configuration tab, switch on the required laser lines—405 nm and White Light Laser (WLL). It is recommended to switch on the depletion lasers (660 nm and 775 nm) at this time, if planning to do STED microscopy during the same imaging session. Also, under Hardware Settings set the bit-depth to 12-bit (the software defaults to 8-bit).

Table 1
Three-fluorochrome sequential acquisition settings

Sequence	Fluorochrome	Excitation (nm)	AOBS setting (nm)	Detector used
Seq 1	DAPI	405	(420–480)	PMT
Seq 2	AF 532	515	(520–590)	HyD1
Seq 3	AF 647	645	(650–720)	HyD2

3. For a x/y/z scan, set up the image acquisition parameters: scan speed of 400 Hz, 1024 × 1024 image format, and pinhole at 1 Airy Unit.
4. Under the Acquire tab, set up a three fluorochrome sequential scan as shown in Table 1.
5. Set up the slide on the stage and focus on the imaging area using any low magnification objective, e.g., 20× objective, if available.
6. Using the DAPI filter, focus on the nuclei of the hepatocytes through the oculars.
7. Switch to the 100×/1.40 oil objective and adjust the focus.
8. Using a TRITC filter, focus on the urea cycle enzyme labeled with AF 532 (Excitation maxima - 532 nm, Emission maxima - 554 nm).
9. Switch to the confocal scanning (CS) mode.
10. In the LAS X software using Seq 1, do a real-time scan (live) to center and focus the sample area to be imaged. To start with, have the 405 nm laser between 1% and 5% power, Smart Gain at 800, and Smart Offset at 0. Check the focus and centricity for the AF 532 and AF 647 using Seq 2 and Seq 3, respectively, keeping the laser power as low as possible and the Smart Gain at 100%.
11. Using the over/underexposure (Lookup table) view, adjust each sequence for optimal dynamic range, ensuring no pixels are saturated. For Seq 1 (uses a PMT), the laser power, Smart Gain, and Smart Offset can be adjusted so that there is no over-exposure and minimal underexposure (*see* **Note 15**). For Seq 2 and Seq 3, adjust laser power and Smart Gain to eliminate over-exposure; the Smart Offset adjustment is disabled for the hybrid detectors used in these sequences.
12. To avoid photobleaching, avoid averaging particularly in areas where STED imaging will be done. But, if averaging is required, line averaging is preferred.
13. Acquire the confocal image.

Table 2
Two-fluorochrome confocal-gSTED combination

Sequence	Fluorochrome	Excitation laser (nm)	Depletion laser	AOBS setting (nm)	Detector used
Seq 1 (Simultaneous confocal)	AF 532	515	–	(520–590)	HyD1
	AF 647	645	–	(650–720)	HyD2
Seq 2 (gSTED)	AF 647	645	775 nm	(650–720)	HyD2
Seq 3 (gSTED)	AF 532	515	660 nm	(520–590)	HyD1

3.5 Confocal–gSTED Microscopy

1. Ensure that the WLL is switched on first and then the depletion lasers (660 nm and 775 nm) (*see* **Note 16**). Set the laser power to desired levels (*see* **Note 17**). Bit-depth should also be set to 12-bit.
2. Under the acquire tab, set up sequential scan with three sequences as shown in Table 2.
3. Sequential acquisition should be done as tabulated in Table 2. Simultaneous confocal images should be in the first sequence, gSTED image of AF647 should be in the second sequence (irrespective of the enzyme labeled by it) and the final sequence should be for gSTED of AF532 (again irrespective of the enzyme labeled by it).
4. For the second sequence, start with lower (~20%) laser power (setting under the Acquire tab) for the 775 nm laser (*see* **Note 17**). Since 775 nm is a pulsed laser, time-gating may not be necessary in this sequence. Time-gating can be set to 0.5–6.0 ns as the 775 nm laser is usually calibrated with a delay time of 0.5 ns. Note that the 645 nm laser may require twice the power than used in confocal.
5. It is recommended that the lasers are left on for a while (usually 30 - 60 minutes) before imaging so that they reach the required operating temperature. If the 775 nm laser has reached the required operating temperature, then it's STED delay time can be left at default (0 ps). If the 775 nm depletion laser is being used in the warm-up stage, then a STED delay time should be set up as per Leica instructions.
6. For the third sequence, set the laser power in the mid-range for the 660 nm laser (*see* **Note 18**). Since 660 nm is CW-laser, higher time-gating (say 2.0–6.0 ns) can be helpful (*see* **Note 19**). Here again the 532 nm laser may require twice the power than used in confocal imaging.
7. Use the first sequence to acquire a confocal image.
8. Based on this confocal image, one of the enzyme clusters can be selected and centered for gSTED imaging.

9. Using the confocal sequence, set up a z-stack for the region of interest by marking the beginning and end of the stack. For the z-step size, select System Optimized. It is preferred to over-sample with a step size of ~ 0.15 nm. Move to the center of the stack.
10. Staying in the confocal sequence, zoom on to the region of interest.
11. Move to the second sequence and select "Optimal XY format." Usually, this would give a pixel size of 20–30 nm. You may also try different combinations of image format and suitable zoom factors to get optimal pixel size.
12. For 3D STED acquisition, STED laser needs to be configured for lateral as well as axial directions. The STED-3D slider needs to be adjusted for distributing the laser in both directions. The schematic display of effective PSF is available under the "Sketch of estimated effective PSF" dialogue box. Please note that when using 3D STED, laser power would need to be adjusted to compensate for the redistribution in the lateral and axial directions.
13. In the Configuration tab under STED, do a Áuto Beam Alignment' to optimally superimpose the excitation and depletion lasers (*see* **Note 20**).
14. Acquire the image of all three sequences (*see* **Notes 21** and **22**).
15. Following the initial acquisition, experiment with various settings—depletion laser power, time-gating, excitation laser power, Smart Gain, etc. to optimizing the image.
16. Export the image to Huygens Professional for deconvolution and analysis. Use this initial image processing to optimize the acquisition settings.
17. It is recommended to do calibrations at regular intervals to verify the working of STED modality (*see* **Note 23**).
18. While here just two fluorochromes were chosen based on need and hardware limitations, one should also explore other new fluorochromes that are being added to the STED-capable list (*see* **Note 24**).

3.6 Image Processing

1. The acquired images can be deconvolved using either the deconvolution module available in LAS software or exported to Huygens Professional for deconvolution. (Fig. 1).
2. Here gSTED and confocal microscopy images were deconvolved using the Huygens Professional version 17.04 from Scientific Volume Imaging.

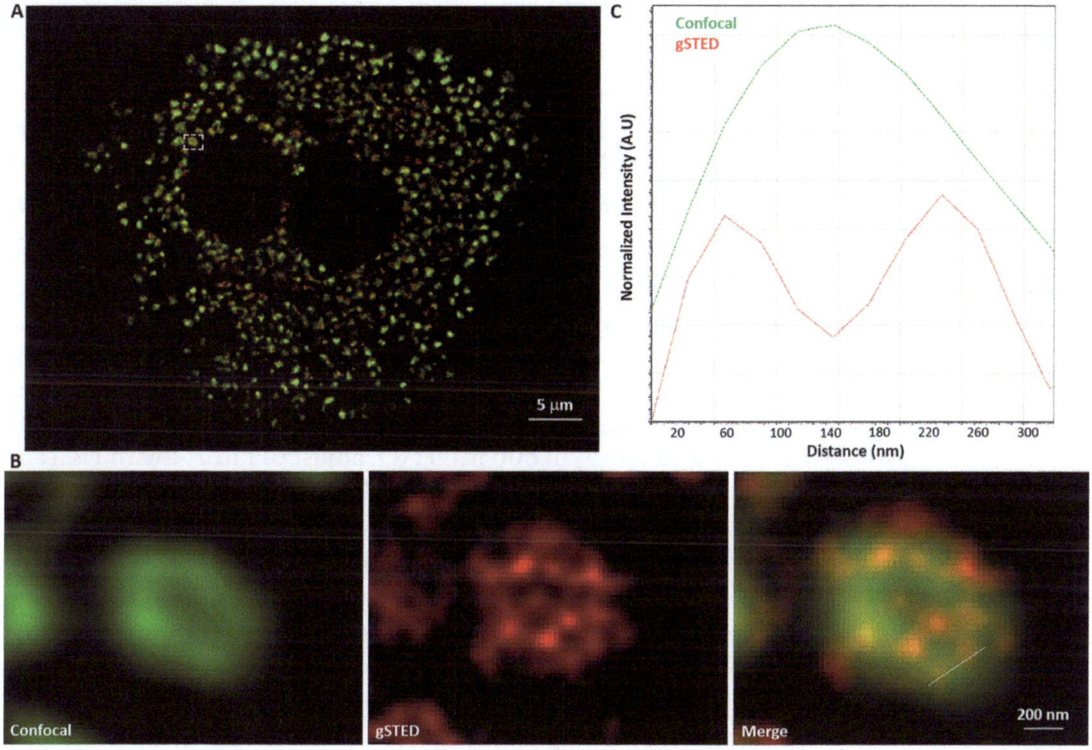

Fig. 1 Confocal and gSTED imaging of CPS1 clusters in mitochondria. Primary mouse hepatocytes were immune-stained using anti-CPS1 primary and Alexa Fluor 647 conjugated secondary antibody. The cells were imaged using confocal and gSTED modalities and the images were deconvolved. Images show (**a**) a whole cell and (**b**) zoomed image of an individual mitochondrion marked in panel A by dotted lines. (**c**) Normalized intensity profile plot showing the pixel intensity along the line indicated on the merged image in panel (**b**). Images are pseudo-colored to show Confocal (green) and gSTED (red) images for the same region. Scale bars are as indicated on the image

3. The intensity profiles plots can be created within the LAS-X software or by using other software such as Image J, MetaMorph, etc.
4. In the LAS-X software, open an image and draw a Region of Interest (ROI) where the plot of intensity profile is required.
5. Under Quantify-Tools tabs choose "Line Profile" and then "Intensity." For comparing, confocal and STED select the respective channels (Fig. 1c).
6. Better publication quality intensity profiles plots can be obtained by exporting the data from the intensity profile for graphing using Excel, R, Python, or other software.

4 Notes

1. Protease inhibitors can be included in all solutions used for purification of liver mitochondria. Any protease inhibitor cocktail is suitable for this purpose.
2. EDTA is added to LHB as 100 mM Na_2EDTA stock solution.
3. EDTA can be replaced with the same concentration of EGTA and HEPES can be replaced with MOPS.
4. PBS is prepared as 10x stock solution and diluted to 1× as needed. 10× PBS is 1.37 M NaCl, 27 mM KCl, 100 mM Na_2HPO_4, 18 mM KH_2PO_4, pH 7.4 solution. To prepare 10× PBS dissolve 80 g NaCl, 2 g KCl, 14.4 g Na_2HPO_4, and 2.4 g KH_2PO_4 in 800 mL H_2O, adjust pH to 7.4 with HCl and add H_2O to 1 L. Divide 10× PBS into aliquotes, sterilize them either by filter sterilization or in the autoclave (liquid cycle for 20 min at 15 psi) and store at room temperature.
5. Primary antibodies against recombinant mouse NAGS and human OTC proteins were raised in rabbits. Antibodies against NAGS and OTC that were validated and used in the Human Protein Atlas Project [46, 47] can be used. Both commercially available antibodies should be validated by immunoblotting purified recombinant NAGS and OTC proteins.
6. Use of live animals for liver harvesting must be reviewed and approved by an Institutional Animal Care and Use Committee (IACUC) or equivalent and must conform to governmental regulations regarding the care and use of laboratory animals.
7. The Potter-Elvehjem homogenizer motor should be securely mounted either to a wall or to a bench via a G clamp. Attachment to a free-standing stand is not adequate.
8. A 20- to 30-mL syringe attached to a metal filling cannula (i.d. 0.8–1.0 mm) can be used to transfer supernatants that are to be recentrifuged.
9. Lipid removal is essential because free fatty acids are potent electron transport uncouplers.
10. Transfer of proteins can be carried out using wet or semi-dry methods, depending on the equipment available to investigators.
11. Distilled and double distilled H_2O can be used instead of deionized and double deionized H_2O. Make sure to rinse the acid between coverslips that are stuck together.
12. Coverslips can be used immediately or stored at 4 °C for 2 weeks.
13. 10,000 is the surface area of hematocytometer; 10 is dilution factor.

14. The STED option should be selected at this stage of starting the software if planning to do STED microscopy following the confocal microscopy.
15. Opinion on underexposure adjustment varies among experts. Many equate adjustment of offset, particularly to get rid of background, to loss of signal and would prefer post-acquisition scaling of images.
16. For the pulsed STED, Leica supplies more than one combination of lasers that act as trigger master and trigger slaves. The trigger master emits the pulses and the trigger slave is synchronized with the trigger master using a time delay. Please follow the Leica instruction in terms of switching on the lasers based on available combination.
17. The general rule is that the higher the laser power, the higher the resolution of the STED image. The 775 nm laser is a pulsed laser which usually gives better resolution even in comparatively lower laser power.
18. Testing for anti-Stokes's fluorescence should be done when setting up a STED sequence. Presence of anti-Stokes's fluorescence can be tested by imaging with the excitation wavelength to 0% and depletion laser at 100%.
19. Higher resolution can be obtained by increasing the time-gating without having to increase the STED depletion power. Signal-to-noise ratio (SNR) is compromised as gating is increased.
20. The beam alignment should be done after the lasers have warmed up (30–60 min after switching on the lasers). It may also be needed every 2–3 h or at shorter intervals if the room temperature fluctuates.
21. Sequential acquisition should be done as tabulated in Table 2. Simultaneous confocal images should be in the first sequence, gSTED image of AF647 should be in the second sequence (irrespective of which enzyme is labeled by it) and final sequence should be for gSTED of AF532.
22. To maximize STED resolution, one has to experiment with the depletion laser power and time-gating. Before acquiring images in regions of interest, different depletion power settings can be tried on other areas of similar intensity.
23. STED calibration—GATTA-beads can be used for STED calibration. 3D-STED PSF measured using these beads can improve the deconvolution output. These are also helpful in post-acquisition correction of sample drift https://www.gattaquant.com/products/gatta-beads.html. Maintain room temperature within ± 1 °C to minimize sample drift. Also refer to https://www.leica-microsystems.com/science-lab/

measuring-the-3d-sted-psf-with-a-new-type-of-fluorescent-beads/ for additional details.

24. Leica's Guide to STED Sample Preparation (https://www.leica-microsystems.com/science-lab/the-guide-to-sted-sample-preparation/) provides a good list of various fluorochromes/combinations that work with specific depletion lasers. The STAR dyes from Abberior (https://abberior-labels.com/products/dyes-lables/) are new addition to the family of brilliant STED dyes.

References

1. Battesti A, Bouveret E (2012) The bacterial two-hybrid system based on adenylate cyclase reconstitution in Escherichia coli. Methods 58: 325–334
2. Young KH (1998) Yeast two-hybrid: so many interactions, (in) so little time. Biol Reprod 58: 302–331
3. Rich RL, Myszka DG (2007) Higher-throughput, label-free, real-time molecular interaction analysis. Anal Biochem 361:1–6
4. Attri AK, Minton AP (2005) Composition gradient static light scattering: a new technique for rapid detection and quantitative characterization of reversible macromolecular hetero-associations in solution. Anal Biochem 346: 132–138
5. Martin SF, Tatham MH, Hay RT, Samuel ID (2008) Quantitative analysis of multi-protein interactions using FRET: application to the SUMO pathway. Protein Sci 17:777–784
6. Matsumoto S, Hammes GG (1975) Fluorescence energy transfer between ligand binding sites on aspartate transcarbamylase. Biochemistry 14(2):214–224
7. Pollok BA, Heim R (1999) Using GFP in FRET-based applications. Trends Cell Biol 9: 57–60
8. Sekar RB, Periasamy A (2003) Fluorescence resonance energy transfer (FRET) microscopy imaging of live cell protein localizations. J Cell Biol 160(5):629–633
9. Vinogradova O, Qin J (2012) NMR as a unique tool in assessment and complex determination of weak protein-protein interactions. Top Curr Chem 326:35–45
10. Englander SW (2006) Hydrogen exchange and mass spectrometry: a historical perspective. J Am Soc Mass Spectrom 17:1481–1489
11. Keilhauer EC, Hein MY, Mann M (2015) Accurate protein complex retrieval by affinity enrichment mass spectrometry (AE-MS) rather than affinity purification mass spectrometry (AP-MS). Mol Cell Proteomics 14:120–135
12. Morris JH, Knudsen GM, Verschueren E, Johnson JR, Cimermancic P, Greninger AL, Pico AR (2014) Affinity purification-mass spectrometry and network analysis to understand protein-protein interactions. Nat Protoc 9: 2539–2554
13. Rigaut G, Shevchenko A, Rutz B, Wilm M, Mann M, Seraphin B (1999) A generic protein purification method for protein complex characterization and proteome exploration. Nat Biotechnol 17:1030–1032
14. Herzog F, Kahraman A, Boehringer D, Mak R, Bracher A, Walzthoeni T, Leitner A, Beck M, Hartl FU, Ban N, Malmstrom L, Aebersold R (2012) Structural probing of a protein phosphatase 2A network by chemical cross-linking and mass spectrometry. Science 337:1348–1352
15. Wu F, Minteer S (2015) Krebs cycle metabolon: structural evidence of substrate channeling revealed by cross-linking and mass spectrometry. Angew Chem Int Ed Engl 54:1851–1854
16. Rhee HW, Zou P, Udeshi ND, Martell JD, Mootha VK, Carr SA, Ting AY (2013) Proteomic mapping of mitochondria in living cells via spatially restricted enzymatic tagging. Science 339:1328–1331
17. Floyd BJ, Wilkerson EM, Veling MT, Minogue CE, Xia C, Beebe ET, Wrobel RL, Cho H, Kremer LS, Alston CL, Gromek KA, Dolan BK, Ulbrich A, Stefely JA, Bohl SL, Werner KM, Jochem A, Westphall MS, Rensvold JW, Taylor RW, Prokisch H, Kim JP, Coon JJ, Pagliarini DJ (2016) Mitochondrial protein interaction mapping identifies regulators of respiratory chain function. Mol Cell 63:621–632
18. Huttlin EL, Bruckner RJ, Paulo JA, Cannon JR, Ting L, Baltier K, Colby G, Gebreab F, Gygi MP, Parzen H, Szpyt J, Tam S, Zarraga G, Pontano-Vaites L, Swarup S,

White AE, Schweppe DK, Rad R, Erickson BK, Obar RA, Guruharsha KG, Li K, Artavanis-Tsakonas S, Gygi SP, Harper JW (2017) Architecture of the human interactome defines protein communities and disease networks. Nature 545:505–509

19. Schweppe DK, Huttlin EL, Harper JW, Gygi SP (2018) BioPlex display: an interactive suite for large-scale AP-MS protein-protein interaction data. J Proteome Res 17:722–726
20. Jie J, Lohr F, Barbar E (2015) Interactions of yeast dynein with dynein light chain and dynactin: general implications for intrinsically disordered duplex scaffolds in multiprotein assemblies. J Biol Chem 290:23863–23874
21. Nyarko A, Song Y, Barbar E (2012) Intrinsic disorder in dynein intermediate chain modulates its interactions with NudE and dynactin. J Biol Chem 287:24884–24893
22. Nyarko A, Song Y, Novacek J, Zidek L, Barbar E (2013) Multiple recognition motifs in nucleoporin Nup159 provide a stable and rigid Nup159-Dyn2 assembly. J Biol Chem 288:2614–2622
23. Velot C, Mixon MB, Teige M, Srere PA (1997) Model of a quinary structure between Krebs TCA cycle enzymes: a model for the metabolon. Biochemistry 36:14271–14276
24. Bulutoglu B, Garcia KE, Wu F, Minteer SD, Banta S (2016) Direct evidence for metabolon formation and substrate channeling in recombinant TCA cycle enzymes. ACS Chem Biol 11:2847–2853
25. Castellana M, Wilson MZ, Xu Y, Joshi P, Cristea IM, Rabinowitz JD, Gitai Z, Wingreen NS (2014) Enzyme clustering accelerates processing of intermediates through metabolic channeling. Nat Biotechnol 32:1011–1018
26. De La Fuente IM, Martinez L, Perez-Samartin AL, Ormaetxea L, Amezaga C, Vera-Lopez A (2008) Global self-organization of the cellular metabolic structure. PLoS One 3:e3100
27. Brusilow SW, Horwich AL (2001) Urea cycle enzymes. In: Scriver CR, Beaudet AL, Sly WS, Valle D (eds) The metabolic & molecular bases of inherited disease, vol 2. McGraw-Hill, New York, NY, pp 1909–1963
28. Grisolia S, Cohen PP (1952) The catalytic role of carbamyl glutamate in citrulline biosynthesis. J Biol Chem 198:561–571
29. Grisolia S, Cohen PP (1953) Catalytic role of glutamate derivatives in citrulline biosynthesis. J Biol Chem 204:753–757
30. Bradford NM, McGivan JD (1980) Evidence for the existence of an ornithine/citrulline antiporter in rat liver mitochondria. FEBS Lett 113:294–298
31. Kobayashi K, Sinasac DS, Iijima M, Boright AP, Begum L, Lee JR, Yasuda T, Ikeda S, Hirano R, Terazono H, Crackower MA, Kondo I, Tsui LC, Scherer SW, Saheki T (1999) The gene mutated in adult-onset type II citrullinaemia encodes a putative mitochondrial carrier protein. Nat Genet 22:159–163
32. Cheung CW, Cohen NS, Raijman L (1989) Channeling of urea cycle intermediates in situ in permeabilized hepatocytes. J Biol Chem 264:4038–4044
33. Cohen NS, Cheung CW, Sijuwade E, Raijman L (1992) Kinetic properties of carbamoyl-phosphate synthase (ammonia) and ornithine carbamoyltransferase in permeabilized mitochondria. Biochem J 282:173–180
34. Tuchman M (1989) Persistent acitrullinemia after liver transplantation for carbamylphosphate synthetase deficiency. N Engl J Med 320:1498–1499
35. Cohen NS, Cheung CW, Kyan FS, Jones EE, Raijman L (1982) Mitochondrial carbamyl phosphate and citrulline synthesis at high matrix acetylglutamate. J Biol Chem 257: 6898–6907
36. Raijman L (1976) Enzyme and reactant concentrations and the regulation of urea synthesis. In: Grisolia S, Baguena R, Mayor F (eds) The urea cycle. John Wiley & Sons, New York, pp 243–259
37. Sonoda T, Tatibana M (1983) Purification of N-acetyl-L-glutamate synthetase from rat liver mitochondria and substrate and activator specificity of the enzyme. J Biol Chem 258:9839–9844
38. Wang Y, Palmfeldt J, Gregersen N, Makhov AM, Conway JF, Wang M, McCalley SP, Basu S, Alharbi H, St Croix C, Calderon MJ, Watkins S, Vockley J (2019) Mitochondrial fatty acid oxidation and the electron transport chain comprise a multifunctional mitochondrial protein complex. J Biol Chem 294: 12380–12391
39. An S, Kumar R, Sheets ED, Benkovic SJ (2008) Reversible compartmentalization of de novo purine biosynthetic complexes in living cells. Science 320(5872):103–106
40. Chan CY, Zhao H, Pugh RJ, Pedley AM, French J, Jones SA, Zhuang X, Jinnah H, Huang TJ, Benkovic SJ (2015) Purinosome formation as a function of the cell cycle. Proc Natl Acad Sci U S A 112:1368–1373
41. Lapuente-Brun E, Moreno-Loshuertos R, Acin-Perez R, Latorre-Pellicer A, Colas C, Balsa E, Perales-Clemente E, Quiros PM, Calvo E, Rodriguez-Hernandez MA, Navas P, Cruz R, Carracedo A, Lopez-Otin C, Perez-

Martos A, Fernandez-Silva P, Fernandez-Vizarra E, Enriquez JA (2013) Supercomplex assembly determines electron flux in the mitochondrial electron transport chain. Science 340:1567–1570

42. Subramanian K, Jochem A, Le Vasseur M, Lewis S, Paulson BR, Reddy TR, Russell JD, Coon JJ, Pagliarini DJ, Nunnari J (2019) Coenzyme Q biosynthetic proteins assemble in a substrate-dependent manner into domains at ER-mitochondria contacts. J Cell Biol 218: 1353–1369

43. Haskins N, Bhuvanendran S, Anselmi C, Gams A, Kanholm T, Kocher KM, LoTempio J, Krohmaly KI, Sohai D, Stearrett N, Bonner E, Tuchman M, Morizono H, Jaiswal JK, Caldovic L (2020) Mitochondrial enzymes of the urea cycle cluster at the inner mitochondrial membrane. Front Physiol 11:542950

44. Makris G, Lauber M, Rufenacht V, Gemperle C, Diez-Fernandez C, Caldovic L, Froese DS, Haberle J (2021) Clinical and structural insights into potential dominant negative triggers of proximal urea cycle disorders. Biochimie 183:89–99

45. Powers-Lee SG, Mastico RA, Bendayan M (1987) The interaction of rat liver carbamoyl phosphate synthetase and ornithine transcarbamoylase with inner mitochondrial membranes. J Biol Chem 262:15683–15688

46. Thul PJ, Akesson L, Wiking M, Mahdessian D, Geladaki A, Ait Blal H, Alm T, Asplund A, Bjork L, Breckels LM, Backstrom A, Danielsson F, Fagerberg L, Fall J, Gatto L, Gnann C, Hober S, Hjelmare M, Johansson F, Lee S, Lindskog C, Mulder J, Mulvey CM, Nilsson P, Oksvold P, Rockberg J, Schutten R, Schwenk JM, Sivertsson A, Sjostedt E, Skogs M, Stadler C, Sullivan DP, Tegel H, Winsnes C, Zhang C, Zwahlen M, Mardinoglu A, Ponten F, von Feilitzen K, Lilley KS, Uhlen M, Lundberg E (2017) A subcellular map of the human proteome. Science 356:aal3321

47. Uhlen M, Fagerberg L, Hallstrom BM, Lindskog C, Oksvold P, Mardinoglu A, Sivertsson A, Kampf C, Sjostedt E, Asplund A, Olsson I, Edlund K, Lundberg E, Navani S, Szigyarto CA, Odeberg J, Djureinovic D, Takanen JO, Hober S, Alm T, Edqvist PH, Berling H, Tegel H, Mulder J, Rockberg J, Nilsson P, Schwenk JM, Hamsten M, von Feilitzen K, Forsberg M, Persson L, Johansson F, Zwahlen M, von Heijne G, Nielsen J, Ponten F (2015) Proteomics. Tissue-based map of the human proteome. Science 347:1260419

Chapter 6

DNA Nanoscaffolds for Multienzyme Systems Assembly

Sung Won Oh, Zhicheng Wang, and Jinglin Fu

Abstract

Multienzyme reactions play an important role in cellular metabolic functions. The assembly of a metabolon is often observed, in which the position and the orientation of composite enzymes are optimized to facilitate the substrate transport. The recent progress of DNA nanotechnology is promising to organize the assembly of bimolecular complexes with precise controlled geometric patterns at nanoscale, such as enzyme cascades assembly, biomimetic substrate channeling, and compartmentalization. Here, we present detailed protocols of using DNA nanoscaffolds to assemble a multienzyme system with control over spatial interactions and arrangements of individual components. The protocols include the preparation and purification of DNA nanostructures, the bioconjugation of DNA with proteins and cofactors, the chromatography purification of DNA-conjugated biomolecules, the characterization of assemblies by routine gel electrophoresis and advanced AFM imaging, as well as the activity evaluation of multienzyme assemblies.

Key words DNA nanostructure, Multienzyme assembly, Nanoreactors, Self-assembly, Bioconjugation and purification, Gel electrophoresis, Protein chromatography

1 Introduction

DNA is a biopolymer that self-assembles into the double-helical structure following the Watson–Crick base pairing rule [1]. Starting with Ned Seeman's work of artificial "Holliday" junctions, structural DNA nanotechnology has progressed rapidly to develop approaches to fold DNA molecules into various rationally designed, multi-dimensional nanostructures [2, 3], such as DNA origami [4–6], DNA bricks [7], DNA crystals [8, 9], nanoscale polyhedrons [10], and polyhedral meshes [11, 12]. The self-assembled DNA nanostructures are useful in organizing and patterning biomolecular complex into controlled geometric arrangements for mimicking cellular spatial interactions and functions [1–3]. For example, multiple enzymes were assembled on DNA nanostructures with precise control over their spatial distance to govern the mass transport of substrates [13, 14]. A DNA-based biomimetic swinging arm was able to transfer electrons within a two-enzyme cascade system

Haralambos Stamatis (ed.), *Multienzymatic Assemblies: Methods and Protocols*,
Methods in Molecular Biology, vol. 2487, https://doi.org/10.1007/978-1-0716-2269-8_6,
© The Author(s), under exclusive license to Springer Science+Business Media, LLC, part of Springer Nature 2022

[15, 16]. DNA nanocages were used to encapsulate individual enzymes or enzyme cascades for improved stability and boosted activities [17, 18]. DNA nanomachines can regulate enzyme activities with trigger-actuated function [19, 20]. In this chapter, we present a detailed protocol of how to design, purify, conjugate, and assemble multienzyme systems on DNA nanostructures, as well as the characterization of structures and functions.

2 Materials

Prepare all oligonucleotides in DNA Grade water (DNase, Protease free) and store in −20 °C freezer. Prepare enzyme in 50 mM N-(2-Hydroxyethyl)piperazine-N′-(2-ethanesulfonic acid) sodium salt (HEPES·Na), pH 7.5 and store in 4 °C refrigerator. All salt buffers are filtered by 0.22 μm filter and are stored at 4 °C refrigerator.

2.1 Common Stock Buffer

Prepare a 1 L solution with distilled water, then filter the solution by 0.22 μm filter.

1. 50× Tris-acetate-EDTA (TAE): 242.2 g Tris base, 57.1 mL 1 M acetic acid, 37.2 g $EDTA{\cdot}Na_2{\cdot}2H_2O$.
2. 10× TAE-Mg (125 mM Mg^{2+}): 26.8 g $(CH_3COO)_2Mg{\cdot}4H_2O$, 200 mL 50× TAE.
3. 10× Tris-borate-EDTA (TBE): 108 g Tris base, 55 g boric acid, 40 mL 0.5 M EDTA, pH 8.0.
4. 0.5 M EDTA, pH 8.0: 186.1 g $EDTA{\cdot}Na_2$, 24 g NaOH.
5. 100 mM HEPES, pH 7.5: Add 26.03 g HEPES·Na into 1 L of distilled water. Adjust the pH to 7.5 with 5 M HCl.
6. 5 M HCl: 10 mL concentrated HCl (36.5% v/v) is added with 14 mL distilled water. Store in a chemical hood at room temperature.
7. 50 mM HEPES, pH 7.5: Add 500 mL of distilled water to 500 mL of 100 mM HEPES.
8. 50 mM HEPES-Na (1 M), pH 7.5: Add 58.5 g NaCl into 1000 mL of 100 mM HEPES buffer.

2.2 Gel Electrophoresis Supplies

1. Water circulating bath (ISOTEMP 500LCU).
2. Gel box (SE 600 Ruby Standard Electrophoresis Set).
3. Power supply (Power Pac 1000).
4. 50 mL and 10 mL pipettes.
5. 100 mL volumetric cylinder.
6. 1000 mL cylinder.

7. 250 mL Erlenmeyer flask.
8. Gel loading tips.

2.3 Denature Polyacrylamide Gel Electrophoresis (PAGE)

1. 20% Denature PAGE mix: 500 mL 40% acrylamide stock solution (19:1 Ac:Bis) is added with 500 g urea, 100 mL 10× TBE. Adjust the volume to a 1 L with distilled water (*see* **Note 1**). Store at room temperature.
2. 0% Denature PAGE mix: 500 g urea is added into 100 mL 10× TBE. Adjust the volume to 1 L with distilled water. Store at room temperature.
3. Ammonium persulfate (APS): 10% w/v solution in distilled water, 1 g APS and 9 mL distilled water (*see* **Note 2**).
4. *N*,*N*,*N*′,*N*′-tetramethylethane-1,2-diamine (TEMED): Store at room temperature.
5. Bromophenol Blue tracking dye: 90 mL Formamide, 40 mg NaOH, 37 mg EDTA·Na_2·$2H_2O$, 100 mg Bromophenol Blue. Store at room temperature.
6. DNA Elution buffer: 19.27 g NH_4AC, 1.07 g $Mg(AC)_2$, 2 mL 0.5 M EDTA, pH 8.0. Prepare a 500 mL solution with distilled water.
7. UV lamp.
8. Spin-X® Centrifuge Tube Filters.
9. Butyl alcohol.
10. Ethyl alcohol (>190 proof).
11. 70% ethyl alcohol.
12. Vacufuge.

2.4 DNA-Cofactor Bioconjugation

1. β-Nicotinamide-N^6-(2-aminoethyl) adenine dinucleotide (AE-NAD^+) stock: Prepare 50 mM in DNase-free water. Store at −20 °C freezer in the dark.
2. 200 nmol, 200 μL NAD^+: 4 μL 50 mM NAD^+, 20 μL 1 M HEPES, pH 8.0, and 176 μL DNase-free water.
3. Oligonucleotide with 5′-, 3′- or internal amine modification.
4. Disuccinimidyl suberate (DSS) cross linker.
5. Diethylaminoethyl (DEAE)-Sepharose resin: Store at 4 °C (*see* **Note 3**).
6. DNA-loading buffer: 10 mM acetate acid and 0.005% v/v Triton X-100. Prepare a 1 L solution with distilled water.
7. Sigma Prep column.
8. Amicon Ultra-0.5 mL Centrifugal Filters, 3KD and 10KD.
9. Agilent 1260 Bio-inert high-performance liquid chromatography (HPLC).

10. First-step organic buffer: Dry dimethylformamide (DMF), 2% (v/v) DIPEA.
11. Second-step aqueous buffer: 100 mM HEPES, pH 8.0.
12. Elution buffer: 50 mM HEPES, 1.5 M NaCl, pH 8.0. Add 87.75 g NaCl into 500 mL of 100 mM HEPES buffer and adjust the pH to 8.0. Then adjust volume to 1 L with distilled water.
13. Solvent A for HPLC: 100 mM Triethylammonium Acetate (TEAA) is made of by adding 100 mL of 1 M TEAA into 900 mL of distilled water.
14. Solvent B for HPLC: methanol, HPLC grade.
15. Column: Clarity 3u Oligo-RP.
16. Lyophilizer.

2.5 DNA-Enzyme Bioconjugation

1. Succinimidyl 3-(2-pyridyldithio) propionate (SPDP): Store SPDP solid in a-20 °C freezer (*see* **Note 4**). Prepare 20 mM solution stock in DMSO.
2. 1 M $NaHCO_3$, pH 8.5: Dissolve 8.4 g of $NaHCO_3$ into 100 mL distilled water and adjust the pH to 8.5 (*see* **Note 5**). Aliquot 1 M $NaHCO_3$ solution into 1 mL centrifuge tubes, and store in a −20 °C freezer.
3. Tris(2-carboxyethyl)phosphine hydrochloride (T-CEP): Store T-CEP solid in a −20 °C freezer. Prepare 20 mM solution in a 10× TBS buffer and use 5 N NaOH to adjust the pH between 7 and 8 (*see* **Note 6**).
4. 5 M NaOH: Dissolve 20 g of NaOH into 80 mL of distilled water. Make up to 100 mL with distilled water. Store at room temperature.
5. Amicon Ultra-4 Centrifugal Filters, 10KD and 30KD.
6. Agilent 1260 Bio-inert HPLC.
7. GE Ionic-exchange Column: Mono QTM, 4.6/100 PE.

2.6 Non-denature PAGE

1. 40% acrylamide stock solution (19:1 Ac:Bis).
2. Low-range DNA ladder.
3. Ethidium Bromide (EB) staining solution: Add 30 μL EB (1% m/v solution) into 100 mL of distilled water.
4. SYBR™ Green I (1:10,000 dilution): Add 10 μL of SYBR Green into 100 mL of 1× TAE-Mg buffer (12.5 mM Mg^{2+}).
5. UVP ChemiDoc-It^2 Imager.

2.7 AFM Characterization

1. Mica.
2. Metal plate.
3. Super glue (liquid).

4. Double-sided tape (Ted Pella).
5. Bruker Multi-mode 8 AFM.
6. 0.22 μm syringe filter.
7. Nickel sulfate.
8. ScanAsyst-Air probe/ScanAsyst-Fluid$^+$ probe.

2.8 Enzyme Activity Characterization

1. 96-well microplate reader: Multi-mode from Molecular Devices or BioTek Cytation 3.
2. 96-well plate.
3. 1× TBS-Mg (1 mM), pH 7.5: Add 1 mL of 1 M $MgCl_2$ into 1 L of 1× TBS, pH 7.5 buffer.

3 Methods

Carry out all protocols at room temperature unless otherwise specified.

3.1 Design of DNA Nanoscaffolds

The design of DNA nanostructures is mostly carried out by open-access programs, such as Cadnano [21], CanDo [22], and Tiamat [23]. Detailed instructions of these programs are available on YouTube and software downloading websites. Nanostructures assembled from B-form double-stranded DNA (dsDNA) have a helical structure with a helical diameter of ~2.4 nm, a helical repeat of ~3.4 nm for every 10.5 base pairs (bp), and a ~34.3° rotation angle between two adjacent base pairs [24]. DNA nanoscaffolds can be used to direct the assembly of enzymes with controlled parameters of spacing distance [13], relative anchor angles [15], geometric arrangement [14], and stoichiometry number [16].

3.2 Denature PAGE for Oligonucleotide purification

3.2.1 Gel Casting Assembly of Denature PAGE Gel

1. Turn on the water bath at 35 °C.
2. Prepare 1× TBE 5000 mL for the first time use or to replace the old buffer in the Lower Buffer Chamber (LBC) (*see* **Note 7**). Discard the buffer in the Upper Buffer Chamber (UBC) after each use.
3. Assemble the sandwich with two glass plates with 2 spacers and secure it with clamps.
4. Seal the gel sandwich on the casting stand using cam.

3.2.2 Preparation of Denature PAGE

1. Add the gel mixer solution using 50 mL pipette into a clean 50 mL Tube (*see* **Note 8**). Use Table 1 to make the solution (each gel needs ~35 mL volume of the gel mixer solution). Add 263 μL of 10% APS and 14.7 μL of TEMED.
2. Shake 2–3 times and quickly load the prepared gel mixture solution to the assembled gel casting cassettes in between the

Table 1
Denature PAGE Gel Recipe (in mL)

	5%	6%	8%	10%	12%	14%	20%
20% poly mix (mL)	8.8	10.5	14	17.5	21	24.5	35
0% mix (mL)	26.3	24.5	21	17.5	14	10.5	0
Separate range (bp)	70–300	45–70	35–45	25–35			8–25
Bromophenol blue	35	26	19	12	8		

glass plates using the 50 mL pipet carefully. Insert the combs (*see* **Note 9**).

3. Wait for an hour for the polymerization of the gel (*see* **Note 10**).
4. After the gel polymerizes, gently take out combs by pulling straight up. Sit the UBC on top of the gel assembly (*see* **Note 11**). Insert black cams into the holes on the sides of the UBC and turn 180° to seal the UBC with the gel sandwiches.
5. Rinse the wells twice with distilled water using 200 μL pipette. Fill in the wells with distilled water.

3.2.3 Preparation of Samples for Oligonucleotide Purification

1. Set the thermocycler to 90 °C.
2. Centrifuge the sample tubes containing the dry samples for 1 min (*see* **Note 12**).
3. Add distilled water to each dry DNA sample to make 0.5 OD/μL (*see* **Note 13**). Vortex each sample tubes for 20 s and spin at 2,000 rpm (376 rcf) for 30 s. Store the DNA solution in a –20 °C freezer.
4. Take 4 OD of each sample (8 μL) into a new labeled tube and add equal volume of 2× denaturing dye to each sample, vortex 10 s to mix and then spin at 2,000 rpm (376 rcf) for 30 s.
5. Heat the DNA/dye mixture at 90 °C in a thermal block for 5 min (*see* **Note 14**).
6. Once the gel is polymerized and ready, load 20 μL/well into the gel wells (*see* **Note 15**) using gel loading tips (*see* **Note 16**).
7. Run the gel at constant current ~ 30–40 mA per gel (or 60–80 mA for two gels) for around 2–3 h depending on the length of the interested oligonucleotide fragments (*see* **Note 17).**
8. Once the gel is done running, turn off the power supply, the stirrer, and the water circulating bath.

3.2.4 Collect Oligonucleotide from Denature PAGE

1. In a dark room, lift the gel and gently put it on the transilluminator. Set the UV lamp to 254 nm to view the DNA band by its native absorbance and cut out the major band. Turn off the UV lamp, chop the band into small pieces, and collect the small gel blocks into Spin-X® Centrifuge Tube Filters. Caution! UV face shield must be worn.
2. Add 500 μL DNA elution buffer into a filter and store it at −20 °C for 10 min, then store the filter in a 4 °C refrigerator overnight to let the oligonucleotide migrate out from the gel into the elution solution.
3. Centrifuge filters at 8,000 rpm (6,010 rcf) for 6 min to separate the gel block from the eluted oligonucleotide.
4. Add 200 μL of DNA elution buffer onto the gel blocks.
5. Shake on the shaker at 800 rpm for 30 min.
6. Centrifuge at 8,000 rpm (6,010 rcf) for 2 min. Discard the gel blocks.
7. Transfer eluted oligonucleotide into 2 mL tubes.
8. Add in 1000 μL of butyl alcohol, vortex each tube for 1 min and centrifuge at 2,000 rpm (376 rcf) for 1 min. After the spin, discard the upper layer of butyl alcohol with a pipette into a waste under a venting hood. This step is to extract any organic soluble from the oligonucleotide sample like tracking dyes.
9. Add in 1000 μL of ethyl alcohol (>190 proof) and vortex. Leave the mixture in the −20 °C freezer for 30 min (*see* **Note 18**). This will precipitate oligonucleotide as it has a lower solubility in 70% ethanol.
10. Spin at 13,000 rpm (15,871 rcf), 4 °C for 30 min. This step is to precipitate the oligonucleotide, since it is not soluble in 70% ethanol.
11. Pour out the ethyl alcohol, then add 1000 μL of ice cold 70% ethyl alcohol (*see* **Note 19**) and mix well. This is a washing step to remove water and ethanol solubles like buffer salts.
12. Centrifuge at 13,000 rpm (15,871 rcf), 4 °C for 10 min. Pour out the ethyl alcohol.
13. Use the vacufuge to dry the sample for 3 h at 30 °C (*see* **Note 20**).
14. Add 50 μL of DNase-free water and vortex for 1 min to dissolve the purified oligonucleotide strands. Store in a −20 °C freezer.

3.3 DNA-Cofactor Bioconjugation

AE-NAD^+ is conjugated to an amino-modified single stranded DNA (ssDNA) using a DSS linker (Fig. 1).

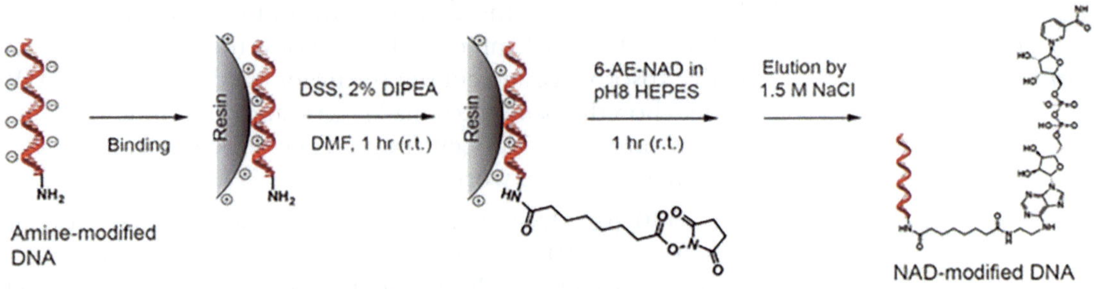

Fig. 1 The conjugation of aminoethyl NAD$^+$ to 5′ amine-modified DNA strands using DSS crosslinking chemistry (modified from [25] with permission from Springer Nature)

3.3.1 DNA-Cofactor Bioconjugation

1. Prepare DNA and AE-NAD$^+$ in DNase-free water.
2. Shake the resin bottle vigorously to emulsify the resin solution. Resin amount: 10 μL/1 nmol DNA. e.g., 200 μL resin is needed for 20 nmol DNA.
3. Add 200 μL of resin to Sigma Prep column then spin down at 2,000 rpm (376 rcf) for 20 s. Wash the resin three times with 500 μL of DNA-loading buffer.
4. Prepare the DNA in a loading buffer: 20 nmol DNA in 200 μL of loading buffer (0.1 mM); need 2 μL for 10 mM DNA stock.
5. Cap the bottom of the spin column with the red rubber. Add DNA to resin and incubate DNA-resin mixture on the shaker at 1,000 rpm, 25 °C for 15 min for the DNA to bind to the resin.
6. After the incubation, spin down at 2,000 rpm (376 rcf) for 20 s to remove extra solution. Calculate the DNA binding efficiency using the following equation:

$$\text{Binding efficiency} = \left(A_{260(\text{before})} - A_{260(\text{after})}\right)/A_{260(\text{before})}$$

7. Wash the resin with 500 μL of dry DMF twice to remove extra water.
8. Prepare and add DSS linker to DNA-resin. Cap the bottom of the spin column with the red rubber. Incubate on the shaker at 1,000 rpm, 25 °C for 1 h.
9. Wash the reactant 3 times with 500 μL of dry DMF. Wash once with 500 μL of DNA-loading buffer to re-stabilize the surface-bound DNA.
10. Use 10× excess 6-AE-NAD$^+$ for NAD$^+$ coupling. For 20 nmol DNA, you need 200 nmol NAD$^+$. Cap the bottom of the spin column with the red rubber. Incubate on the shaker at 1,000 rpm, 25 °C for 1 h.

11. Add 300 μL of DNA-loading buffer to the reaction mixture and incubate on the shaker at 1,000 rpm, 25 °C for 10 min to re-stabilize the DNA bound resin.
12. Spin down the mixture at 3,000 rpm (845 rcf) for 30 s to remove extra NAD^+, then wash 3 times with 200 μL of elution buffer. Collect the filtrate into new 0.5 mL or 1.5 mL tube.
13. Wash the eluted solution 3 times using the Amicon Ultra-0.5 mL Centrifugal Filter with DNase-free water at 10,000 rpm (9,391 rcf), 4 °C for 10 min to wash eluted solution. The washed solution is ready for HPLC purification.

3.3.2 HPLC Purification of Cofactor-Conjugated DNA

1. HPLC is used to carry out the purification of cofactor-conjugated DNA molecules with a Clarity 3u Oligo-RP column. The elution gradient is described in Table 2. Figure 2 shows the HPLC chromatogram of NAD^+-modified ssDNA molecule.
2. Collect fractions into 15 mL tubes for each identified peaks. Store at −80 °C overnight. Use the lyophilizer to remove solvent for dry compounds.
3. Dissolve the dry DNA with DNase-free water. Store in a −20 °C freezer.

3.4 DNA-Enzyme Conjugation

First three sub-protocols must be finished on the same day (~6–8 h work).

Thiol-modified DNA can be commercially synthesized by IDT or it can be converted from amine-modified DNA as published [25].

3.4.1 Conjugation of PROTEIN-SPDP with thiol-DNA (Fig. 3)

1. Wash the crude protein 3 times using the Amicon Ultra-4 30 KD Centrifugal Filter with 50 mM HEPES, pH 7.5 buffer at 4,000 rpm (2,934 rcf), 4 °C for 10 min to remove any amine impurities.
2. Prepare 20 mM SPDP in DMSO (*see* **Note 4**).
3. Mix the protein solution with SPDP in 1:5 molar ratio to react and use 1 M $NaHCO_3$ to adjust the pH to 8.5, as shown in Table 3 (*see* **Note 21**).
4. Incubate on the rocker for 1 h at room temperature (*see* **Note 22**).
5. Wash the protein-SPDP 3 times using the Amicon Ultra-4 30 KD Centrifugal Filter with 50 mM HEPES, pH 7.5 buffer 4,000 rpm (2,934 rcf), 4 °C for 10 min.
6. Add 20 μL of 50 mM HEPES, pH 7.5 buffer to rinse the filter after the last wash and then collect the rinsed solution.
7. Collect the protein from the filter into a centrifuge tube and wrap it with an aluminum foil to protect from the light.

Table 2
HPLC gradient for separating NAD⁺-modified DNA molecules

Step	Time (min)	% B (methanol)	Flow rate (mL/min)	Max. press. (bar)
1	0	10	1	400
2	10	25	1	400
3	60	35	1	400
4	63	100	1	400
5	68	100	1	400
6	70	10	1	400
7	75	10	1	400

Solvent A: 100 mM TEAA; Solvent B: methanol

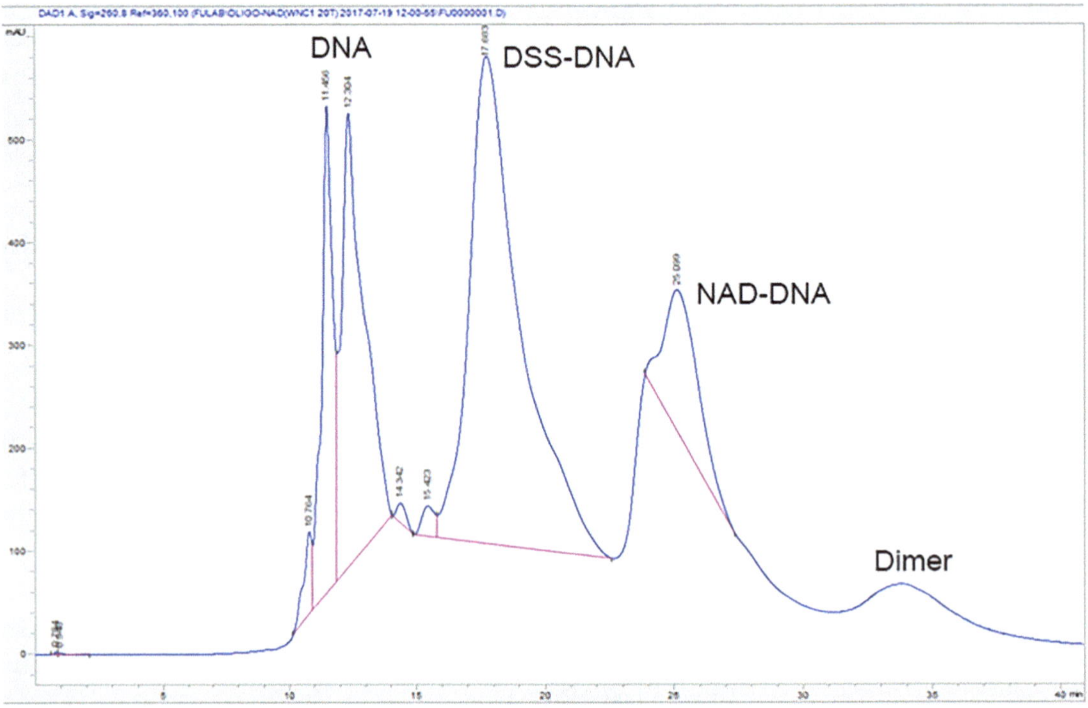

Fig. 2 HPLC purification of NAD^+-modified 6T loop hairpin (modified from [20] with permission John Wiley and Sons)

8. Measure the absorbance at 280 nm and 343 nm by using the NanoDrop to quantify the protein concentration.
9. Prepare 20 mM T-CEP (*see* **Note 6**).
10. Quantify the protein-SPDP labeling ratio by mixing 10 μL of protein-SPDP with T-CEP in 1:20 molar ratio.

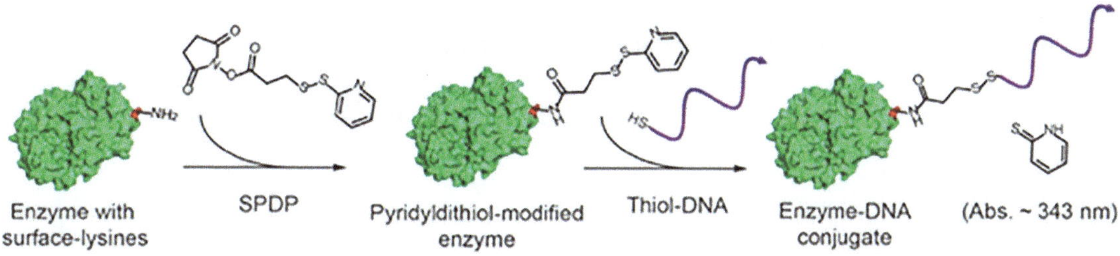

Fig. 3 Protein-SPDP with thiol-DNA conjugation (modified from [25] with permission from Springer Nature)

Table 3
Calculation example of protein-SPDP ratio

Protein concentration (μM)	Protein volume (μL)	SPDP excess ratio	SPDP stock (mM)	SPDP volume (μL)
50	100	5	20	1.25

11. Incubate for 30 min at a room temperature then measure the absorbance at 343 nm.

$$\text{Concentration (SPDP)} = \left(A_{343\ (\text{after T-CEP})} - A_{343\ (\text{before T-CEP})}\right)/8080$$

$$\text{SPDP labeling ratio} = \text{Concentration (SPDP)}/\text{Concentration (protein)}$$

3.4.2 DNA-protein Conjugation

1. Dilute the protein-SPDP to ~ 50 μM in 50 mM HEPES, pH 7.5 buffer. Mix the protein-SPDP with thiolated-DNA in a molar ratio of 1:8. Add 1 M $NaHCO_3$ to adjust the pH to 8.
2. Incubate on the rocker for 1 h at room temperature (*see* **Note 23**).
3. Estimate the protein-DNA conjugation yield by measuring the absorbance at 343 nm. Evaluate the DNA conjugation efficiency:

$$\text{Concentration (thiol} - \text{DNA on protein)} = A_{343}/8080$$

$$\text{DNA} - \text{Protein ratio} = \text{Concentration (thiol} - \text{DNA on protein)}/\text{Protein concentration}$$

4. Wash the DNA-protein 3 times using the Amicon Ultra-4 30 KD Centrifugal Filter with elution buffer at 4,000 rpm (2,934 rcf), 4 °C for 10 min.
5. Use 20 μL of 50 mM HEPES, pH 7.5 buffer to rinse the filter after the last wash and collect the rinse solution. Store the protein-DNA in a 4 °C refrigerator.

3.4.3 Anion-exchange Purification of DNA-conjugated Protein

1. HPLC or fast protein liquid chromatography (FPLC) is used to purify protein-DNA solution with an anion-exchange column. The elution gradient is described in Table 4. Figure 4 shows the FPLC purification of G6PDH-DNA conjugation.

Table 4
FPLC gradient for separating G6pDH-modified DNA molecules

Step	Time (min)	% B (50 mM HEPES-Na (1 M), pH 7.5)	Flow rate (mL/min)	Max. press. (bar)
1	0	0	1	40
2	5	0	1	40
3	10	20	1	40
4	60	60	1	40
5	62	100	1	40
6	70	100	1	40
7	72	0	1	40
8	85	0	1	40

Solvent A: 50 mM HEPES, pH 7.5; Solvent B: 50 mM HEPES-Na (1 M), pH 7.5

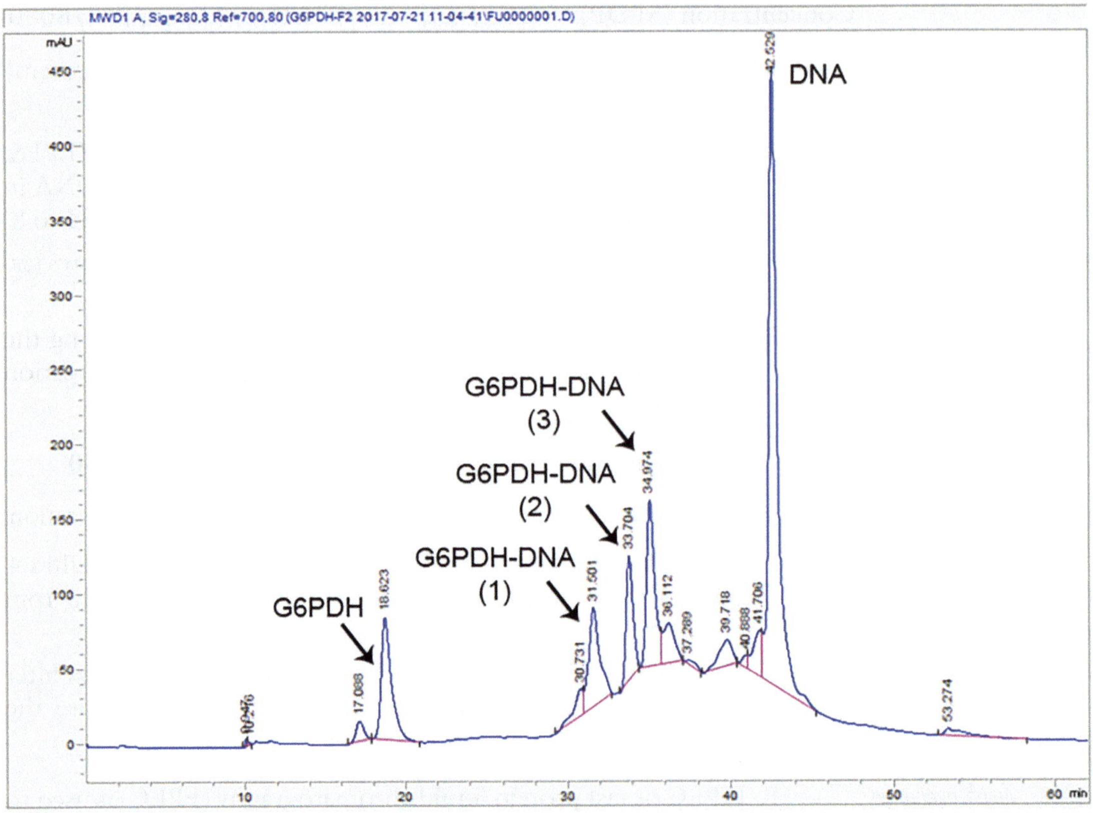

Fig. 4 Anion-exchange purification of enzyme-DNA with different numbers of labeled DNA molecules (modified from [20] with permission John Wiley and Sons)

Table 5
Calculation of making a DNA origami

DNA origami	Final concentration (μM)	Stock concentration (μM)	Volume (μL)
Core-1	0.1	1	10
Core-2	0.1	1	10
Core-3	0.1	1	10
M-13 Scaffold	0.02	0.4	5
TAE-Mg	1×	10×	10
distilled water			55
Total volume			100

Table 6
Detailed annealing program for assembling the DNA origami

Temperature (°C)	Gradient
90	30 s
86–71	1 min/step
70–60	10 min/step
59–30	15 min/step
29–26	10 min/step
25	25 min
4	Hold

2. Characterize the collected fractions of sample peak, concentrate the peak by using Amicon Ultra-4 30 KD Centrifugal Filter.

3.5 Assembly of Enzymes on DNA Nanoscaffolds

3.5.1 DNA Nanoscaffold Assembly

1. Prepare a stock solution of 1 μM staple strands and 0.4 μM M13 strands.
2. Use Table 5 to prepare 100 μL, 20 nM DNA origami with 3 cores of staple strands.
3. Anneal 100 μL of DNA origami solution using the temperature gradient shown in Table 6.
4. Wash the solution 3 times using the Amicon Ultra-0.5 mL 100 KD Centrifugal Filter with 1× TAE-Mg, pH 7.5 buffer 10,000 rpm (9,391 rcf), 4 °C for 10 min.
5. Use 20 μL of 1× TAE-Mg buffer to rinse the filter after the last wash and collect the rinse solution.

Table 7
Detailed annealing program for assembling the enzyme onto the DNA nanoscaffold

Temperature (°C)	Gradient
37	5 min
36–16	2 min/°C
15	Hold

6. Measure the DNA origami concentration at 260 nm absorbance using the NanoDrop, assuming an extinction coefficient of 109,119,009/M*cm.

3.5.2 Enzyme Assembly on DNA Nanoscaffold

1. Prepare 100 μL, 20 nM DNA origami solution as described in Table 5.
2. Mix the DNA origami with DNA-conjugated enzyme at a molar ratio of 1:3.
3. Anneal 100 μL of mixture using the temperature gradient shown in Table 7.

3.6 Non-denature PAGE to Characterize Small DNA Assemblies

1. Prepare a 5% non-denature PAGE gel at a room temperature and leave it to polymerize for 3 h, similarly as previously described. The gel components are shown in Table 8.
2. After loading the samples and a low-range DNA ladder, run the gel for 2.5–3 h at a constant voltage of 200 V.
3. Stain the gel with ethidium bromide for 5 min or SYBR™ Green I for 20 min and rinse in distilled water for another 5 min.
4. The gel is imaged using a UVP ChemiDoc-It^2 Imager instrument.

3.7 AFM Characterization

3.7.1 Mica Preparation

1. Mica is attached to a metal plate using super glue (liquid), or double-sided tape. Do not leave air bubbles between mica and metal plate.
2. Prepare a clean and flat mica surface by peeling off old mica layers with tape (*see* **Note 25**).

3.7.2 Image Nanostructures by SCAN AYST in Fluid

It is recommended to image DNA nanostructures in liquid mode to maintain the native structures. AFM imaging in air mode generally produces less quality than the liquid mode. The protocol below is developed for Bruker multi-mode AFM system.

1. Turn on the AFM system as described in the instrument setup. Avoid looking directly at the laser beam.

Table 8
Non-denature PAGE gel recipe

	3.5%	5%	8%	12%	15%
40% A-Stock (mL)	3.1	4.4	7	10.5	13.1
10× TAE/Mg (mL)	3.5	3.5	3.5	3.5	3.5
Distilled water (mL)	28.4	27.1	24.5	21	18.4
To separate bp	100–1000	75–500	50–400	35–250	20–150
Bromophenol blue (*see* **Note 24**)	100	65	45	20	15
Xylene cyanol FF (*see* **Note 24**)	460	260	160	70	60

2. All buffers must be filtered by 0.22 μm syringe filter right before the use.
3. Pipette ~2 μL, 5–10 nM DNA origami solution, or 100–200 nM of small tiles, onto the center of a freshly cleaved mica surface; evenly distribute solution around the mica surface. Let it incubate for 1 min (do not dry) for absorption.
4. Add 80 μL of 1× TAE-Mg buffer to the mica surface and 2 μL of 100 mM Ni^{2+} to enhance the DNA adsorption. Wait for 2 min, then transfer the mica to the scan head of the AFM.
5. Lower the stage by "Tip UP" until the surface of the mica is at the same level of the stage, or even lower than the stage.
6. Assemble a ScanAsyst-Fluid$^+$ AFM probe onto the liquid cell. A stand magnifier may be needed to view the probe clearly.
7. Add 40 μL of buffer to the top of tip on the liquid cell. The liquid droplets should cover the cantilever without air bubbles. Flip the fluid cell with the tip side down (the droplet should hang on the tip) and insert the fluid cell into the scan head with tilted angle to avoid the tip from touching the mica surface.
8. Assemble the liquid cell loaded with a tip for ScanAsyst in Fluid. Before the scan, adjust the horizontal and vertical biased intensity to less than 0.1.
9. Proceed the imaging in ScanAsyst mode. The peak force setpoint should be <0.02 V. The higher value may result in a strong force to destroy sample structures during the imaging.
10. After imaging, use a Kimwipe to wipe the liquid from the mica as much as possible. Remove the mica from the AFM.

3.7.3 AFM Image Analysis

1. Open the raw image file in the NanoScope Analysis software.
2. Execute a Flatten function to correct the image for tilt and bow.

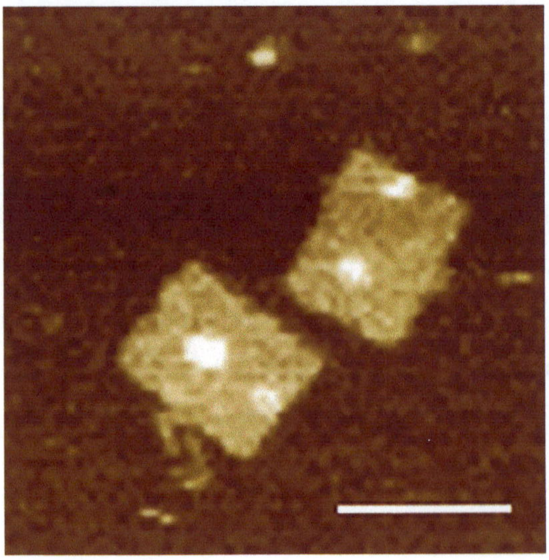

Fig. 5 AFM imaging of PK-HEK pairs assembled on a rectangular DNA origami tile, where the assembly of two enzymes results in the brighter spots due to the increased height compared to the surface of a DNA origami. Scale bar, 100 nm

3. The assembled enzymes typically result in brighter spots due to their increased height on the surface of the DNA structure (Fig. 5).
4. The researcher can zoom in to areas of interest for detailed analysis. The Section function can be used to measure parameters, such as height and distance.
5. Analyzed images should be converted to JPEG or PNG formats for future presentations.

3.8 Enzyme Activity Characterization

1. Perform the enzyme assay using a 96-well microplate reader (e.g., Multi-mode from Molecular Devices or BioTek Cytation 3). Set the instrument ready for kinetic measurements of absorbance or fluorescence.
2. For a typical assay, prepare all substrates in 1× TBS-Mg (1 mM), pH 7.5. First, add 50 μL of enzyme–DNA assemblies to each well. Then transfer the substrate to a 96-well plate with 50 μL of solution per well to initiate the reaction and start the reading immediately. Enzyme assays should be performed with at least three replicates.
3. An example of an enzyme/cofactor assay uses glucose-6-phosphate dehydrogenase (G6PDH) and a NAD cofactor assembled on DNA nanostructures, with the addition of a substrate mixture of 1 mM of glucose-6-phosphate (G6P), 500 μM of phenazine methosulfate (PMS), and 500 μM of resazurin in distilled water.

4. An example of an enzyme cascade reaction uses glucose oxidase (GOX) and horseradish peroxidase (HRP)-assembled DNA origami with the addition of a substrate mixture of 1 mM of glucose and 1 mM of 2,2′-Azinobis[3-ethylbenzothiazoline-6-sulfonic acid]-diammonium salt (ABTS) in 1× TBS-Mg (1 mM), pH 7.5. Another example is hexokinase (HEK) and pyruvate kinase (PK)-assembled DNA origami with the addition of a substrate mixture of 1 mM of pyruvate enol phosphate (PEP), 1 mM of glucose, and 100 μM of adenosine diphosphate (ADP).
5. Use the kinetical software carried by the instrument, or other data-analyzing software (e.g., GraphPad Prism or Origin) to fit the initial velocity of the enzyme-catalyzed reactions. Detailed kinetical analysis includes the measurement of Michaelis–Menten curves and standard curves of product molecules (*see* **Note 26**).

4 Notes

1. Filter with 0.22 μM filter and wrap the solution with aluminum foil.
2. APS, gel polymerization initiator. Prepare 10 mL of 10% APS solution, aliquot to 300 μL/tube, and store in −20 °C freezer. Take out when needed.
3. Shake well before use. The solution should be thoroughly mixed before use.
4. Take out the SPDP solid from its −20 °C freezer to room temperature 15 min prior to use. Check that there is no moisture present. The SPDP powder should not be clumped together. SPDP solution must be used within a few hours after it is dissolved in DMSO.
5. Prepare a large volume and aliquot to 1 mL/tube and store in −20 °C freezer. Take out when needed.
6. Prepare T-CEP solution right before the use. Add X μL of 5 N NaOH using the following equation: X mg T-CEP / (2.8 mg × 9 μL).
7. The buffer in the LBC can be reused many times within a month. Replace every month.
8. Wear gloves. Do not smell any of the solution. The acrylamide and bisacrylamide are severe neurotoxin.
9. Be careful not to introduce air bubbles.
10. Leave the gel casting assembly on the bench at room temperature undisturbed for the gel to polymerize (solidify).

11. Make sure the rubber gaskets at the bottom of the UBC are clean and dry.
12. Centrifuge just in case some oligonucleotide powders are on the walls of the tube.
13. Read the label on the sample tube or on the specification sheet online and calculate the amount of water to put in.
14. Do not overheat, both in time and temperature.
15. Generally, load 2 OD per lane.
16. You can leave one lane empty in between different samples, especially when their lengths (bp) are very close. If you are purifying strands with similar lengths, try to load the gel asymmetrically, i.e., leave one empty lane on one side or. This can help to identity the strands in the gel cut and purification process.
17. The starting voltage typically is ~100 V and the final voltage is ~250–400 V.
18. Shorter strands are harder to precipitate and hence need a lower temperature and a longer time.
19. Keep stored in the −20 °C freezer.
20. Load samples to the rotor symmetrically with all tube lids open to avoid the lids from tangling during the spin. If any tube lid is broken from the tube body, transfer the sample to a new tube. Vacufuge cover can only open when the power is on and standstill rotor. Do not choose a higher temperature. Only water and ethanol can be used in the vacufuge as other organic solvent may cause damage to the cover.
21. The pH must be controlled as the reaction is not efficient if the pH is less than 8.
22. Do not use the vortex as it is too strong and induces aggregation of the protein.
23. Make sure to protect the sample from the light.
24. Approximate the size of DNA fragment (bp) with which the dye would migrate.
25. To ensure a flat and new surface on the mica, the peeled surface on the tape must be smooth and a whole surface. Make small cracks on the edge of the mica to help peel off the surface.
26. It is strongly recommended to use electronic multiple-channel pipettes to increase the pipetting accuracy between multiple samples.

Acknowledgements

This publication was supported by a PECASE award to J.F. (W911NF1910240), an I-Corps award (W911NF1910414) and a DoD DURIP (W911NF2010107). J.F. also appreciate the support from the Rutgers TechAdvance Grant and NJ ACTS Pilot Award. The authors are also grateful to the Equipment Leasing Funds from the State of New Jersey.

References

1. Fu J, Liu M, Liu Y, Yan H (2012) Spatially-interactive biomolecular networks organized by nucleic acid nanostructures. Acc Chem Res 45:1215–1226
2. Fu J, Oh SW, Monckton K, Arbuckle-Keil G, Ke Y, Zhang T (2019) Biomimetic compartments scaffolded by nucleic acid nanostructures. Small 15:1900256
3. Fu J, Wang Z, Liang XH, Oh SW, St. Iago-McRae E, Zhang T (2020) DNA-scaffolded proximity assembly and confinement of multi-enzyme reactions. Top Curr Chem 378:38
4. Dietz H, Douglas SM, Shih WM (2009) Folding DNA into twisted and curved nanoscale shapes. Science 325:725–730
5. Han D, Pal S, Nangreave J, Deng Z, Liu Y, Yan H (2011) DNA origami with complex curvatures in three-dimensional space. Science 332: 342–346
6. Rothemund PW (2006) Folding DNA to create nanoscale shapes and patterns. Nature 440: 297–302
7. Ke Y, Ong LL, Shih WM, Yin P (2012) Three-dimensional structures self-assembled from DNA bricks. Science. 338:1177–1183
8. Ke Y, Ong LL, Sun W, Song J, Dong M, Shih WM, Yin P (2014) DNA brick crystals with prescribed depths. Nat Chem 6:994–1002
9. Zheng J, Birktoft JJ, Chen Y, Wang T, Sha R, Constantinou PE, Ginell SL, Mao C, Seeman NC (2009) From molecular to macroscopic via the rational design of a self-assembled 3D DNA crystal. Nature 461:74–77
10. Veneziano R, Ratanalert S, Zhang K, Zhang F, Yan H, Chiu W, Bathe M (2016) Designer nanoscale DNA assemblies programmed from the top down. Science 352:1534–1534
11. Benson E, Mohammed A, Gardell J, Masich S, Czeizler E, Orponen P, Högberg B (2015) DNA rendering of polyhedral meshes at the nanoscale. Nature 523:441–444
12. Zhang F, Jiang S, Wu S, Li Y, Mao C, Liu Y, Yan H (2015) Complex wireframe DNA origami nanostructures with multi-arm junction vertices. Nat Nanotechnol 10:779
13. Fu J, Liu M, Liu Y, Woodbury NW, Yan H (2012) Interenzyme substrate diffusion for an enzyme cascade organized on spatially addressable DNA nanostructures. J Am Chem Soc 134:5516–5519
14. Liu M, Fu J, Qi X, Wootten S, Woodbury NW, Liu Y, Yan H (2016) A three-enzyme pathway with an optimised geometric arrangement to facilitate substrate transfer. Chembiochem 17: 1097–1101
15. Fu J, Yang YR, Johnson-Buck A, Liu M, Liu Y, Walter NG, Woodbury NW, Yan H (2014) Multi-enzyme complexes on DNA scaffolds capable of substrate channelling with an artificial swinging arm. Nat Nanotechnol 9:531
16. Yang YR, Fu J, Wootten S, Qi X, Liu M, Yan H, Liu Y (2018) 2D enzyme cascade network with efficient substrate channeling by swinging arms. Chembiochem 19:212–216
17. Collins J, Zhang T, Oh SW, Maloney R, Fu J (2017) DNA-crowded enzyme complexes with enhanced activities and stabilities. Chem Commun 53:13059–13062
18. Zhao Z, Fu J, Dhakal S, Johnson-Buck A, Liu M, Zhang T, Woodbury NW, Liu Y, Walter NG, Yan H (2016) Nanocaged enzymes with enhanced catalytic activity and increased stability against protease digestion. Nat Commun 7: 1–9
19. Liu X, Wang D, Liu G, Srinivasan V, Liu Z, Hussain Z, Yang W (2013) Distinct charge dynamics in battery electrodes revealed by in situ and operando soft X-ray spectroscopy. Nat Commun 4:1–8
20. Oh SW, Pereira A, Zhang T, Li T, Lane A, Fu J (2018) DNA-mediated proximity-based assembly circuit for actuation of biochemical reactions. Angew Chem Int Ed 57: 13086–13090
21. Douglas SM, Marblestone AH, Teerapittayanon S, Vazquez A, Church GM, Shih WM (2009) Rapid prototyping of 3D

DNA-origami shapes with caDNAno. Nucleic Acids Res 37:5001–5006

22. Castro CE, Kilchherr F, Kim D-N, Shiao EL, Wauer T, Wortmann P, Bathe M, Dietz H (2011) A primer to scaffolded DNA origami. Nat Methods 8:221

23. Williams S, Lund K, Lin C, Wonka P, Lindsay S, Yan H (2008) Tiamat: a three-dimensional editing tool for complex DNA structures. In: International workshop on DNA-based computers. Springer, New York, pp 90–101

24. Douglas SM, Dietz H, Liedl T, Högberg B, Graf F, Shih WM (2009) Self-assembly of DNA into nanoscale three-dimensional shapes. Nature 459:414–418

25. Fu J, Yang YR, Dhakal S, Zhao Z, Liu M, Zhang T, Walter NG, Yan H (2016) Assembly of multienzyme complexes on DNA nanostructures. Nat Protoc 11:2243–2273

Chapter 7

Strategies for Multienzyme Assemblies

Qing Sun, Maryam Raeeszadeh-Sarmazdeh, Shen-Long Tsai, and Wilfred Chen

Abstract

Proteins are not designed to be standalone entities and must coordinate their collective action for optimum performance. Nature has developed through evolution the ability to co-localize the functional partners of a cascade enzymatic reaction in order to ensure efficient exchange of intermediates. Inspired by these natural designs, synthetic scaffolds have been created to enhance the overall biological pathway performance. In this chapter, we describe several DNA- and protein-based scaffold approaches to assemble artificial enzyme cascades for a wide range of applications. We highlight the key benefits and drawbacks of these approaches to provide insights on how to choose the appropriate scaffold for different cascade systems.

Key words Metabolon, Enzyme Cascade, Substrate Channeling, Protein Assembly, Bioconjugation

1 Introduction

In the past decade, synthetic multienzyme assemblies have been created to enable complex biological cascade reactions to proceed more efficiently [1–4]. The main advantage of these synthetic assemblies is that several enzymes can collaborate effectively through spatial confinement during the cascade process [5–8]. In these cascade reactions, enzymes are assembled into a larger macromolecular structure, and the reactivity can be adjusted through spatial organization, so that each intermediate formed from the reaction can be channeled as a substrate for the next transformation [9]. Since enzymes are close to each other, it overcomes diffusion limitations by transferring intermediates from one active site to another and maintaining a high local concentration [10–13]. Substrate channeling is especially critical to avoid accumulation of toxic metabolites [14] or to push a highly reversible reaction step toward

Qing Sun, Maryam Raeeszadeh-Sarmazdeh and Shen-Long Tsai contributed equally with all other contributors.

Haralambos Stamatis (ed.), *Multienzymatic Assemblies: Methods and Protocols*,
Methods in Molecular Biology, vol. 2487, https://doi.org/10.1007/978-1-0716-2269-8_7,
© The Author(s), under exclusive license to Springer Science+Business Media, LLC, part of Springer Nature 2022

the desired direction by coupling with irreversible sequestration steps [11].

While enzyme assemblies based on random clustering have been reported [11, 15, 16], a scaffold is often used to organize these enzymes to facilitate their spatial interactions [17]. We have previously demonstrated the use of either DNA-based or protein-based scaffold to create multienzyme assemblies [3, 4, 8, 12, 18–21]. We will highlight the benefits and drawbacks of these different strategies for a range of applications.

1.1 Cellulosome-Based Enzyme Assemblies

Some anaerobic microorganisms provide a unique scaffold-assisted enzyme supramolecular that is formed on the cell surface to degrade plant cell walls, called "cellulosome" [22]. Cellulosomes are megaton-sized extracellular enzyme complexes composed of multiple catalytic enzymes anchored on non-catalytic scaffold proteins, termed "scaffoldins." The enzymes are mainly composed of glycosyl hydrolases (mainly cellulases). Such hydrolytic enzyme complex is assembled by the high-affinity interaction between the cohesin domains on the scaffoldins and the dockerin domains of the hydrolytic enzymes. Such a supramolecular complex can hydrolyze cellulose faster than its non-complex cellulase mixture, especially on crystalline cellulose with low accessibility [19–21, 23, 24]. Inspired by nature cellulosome structures, many self-assembled supramolecular structures are mimetically constructed in order to accelerate other enzymatic reactions [3, 25, 26]. Nevertheless, the main concept of these studies is to fuse the catalytic enzymes with a specific protein or peptide, and then assemble them onto a scaffold protein through high-affinity protein–protein interactions (*see* **Note 1**). Herein, the protocol is based on the study published in 2009 using *E. coli* cells for cellulases production and *Saccharomyces cerevisiae* cells for the display of scaffoldin [19]. By incubating *E. coli* cell lysates with yeast cells, the assembly of dockerin and cohesin initiates automatically to form the resulting mini-cellulosome structure (Fig. 1).

1.2 Protein Nanoparticle Immobilization Strategies

Protein nanoparticle can be assembled using both naturally derived strategies (e.g., protein cages and virus-like particles (VLPs)) [27–29] and synthetic routes [30, 31] (e.g., amphiphilic nanoparticles and desolvation-driven nanoparticles). Protein nanoparticles, with advantages of self-assembly, uniform nano size, and environmentally friendly production process, have been extensively explored for sensor assembly, the delivery of therapeutic proteins, small molecule drugs, RNA, and DNA [32–34]. Modification of these nanoparticles with functional groups at the right density and right orientation is desired to achieve these functions. For example, unnatural amino acids incorporation has been used for the synthesis of site-specific antibody-drug conjugates to improve product homogeneity and for functionalizing the surface of protein

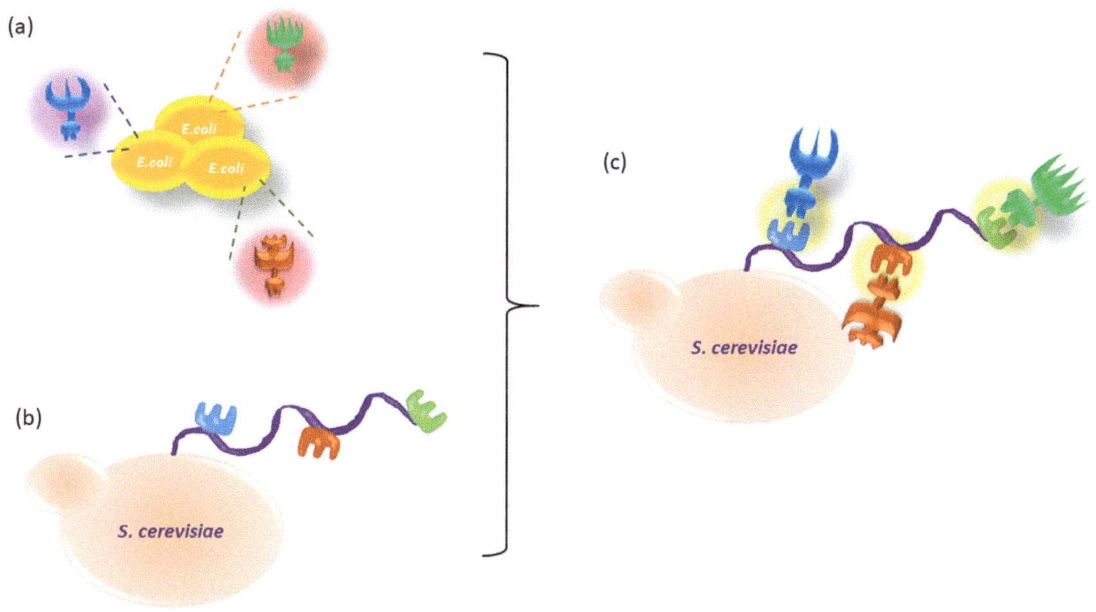

Fig. 1 Surface-displayed cellulosomes: (**a**) *E. coli* cells expressing dockerin-tagged cellulases. (**b**) *S. cerevisiae* surface displaying a scaffoldin. (**c**) Formation of mini-cellulosome

nanoparticles to control the orientation and density of targeting moieties [32–37]. Here, we will focus on more benign bioconjugation approaches such as sortase A-mediated ligation [38–40] and Spytag/SpyCatcher ligation [41–43] (Fig. 2a) to decorate functional moieties onto protein nanoparticles. Purification of functionalized protein nanoparticles is made possible by modifying with elastin-like-polypeptide (ELP), which enables the easy purification of the nanoparticles by inverse transition cycling (ITC) [44–46].

1.2.1 Sortase-Mediated Ligation

Sortase-mediated ligation (SML) and oligomerization of proteins were extensively used as a site-specific protein ligation technique [39, 40]. The benefit of SML for multienzyme assembly is the use of two short tags, LPXTG and GGG for common *Staphylococcus aureus* sortase A, which will not leave a significant scar after protein ligation and oligomerization [47]. This mechanism has been used as a molecular "stapler" to site-specifically ligate proteins with the C-terminal LPXTG motif to other proteins with glycine or aminomethylene motif [48, 49]. Because of their ability to site-specifically add complex polypeptide fragments, sortases have become an ideal tool for protein engineering to modify proteins at the N- and C-termini post-translationally. Because of the high site-specificity and the minor size (less than five amino acids) of the sorting motif, sortase-mediated transpeptidation (sortagging) has become one of the most frequently used methods for protein conjugation, cyclization, labeling/functionalization, purification, and/or

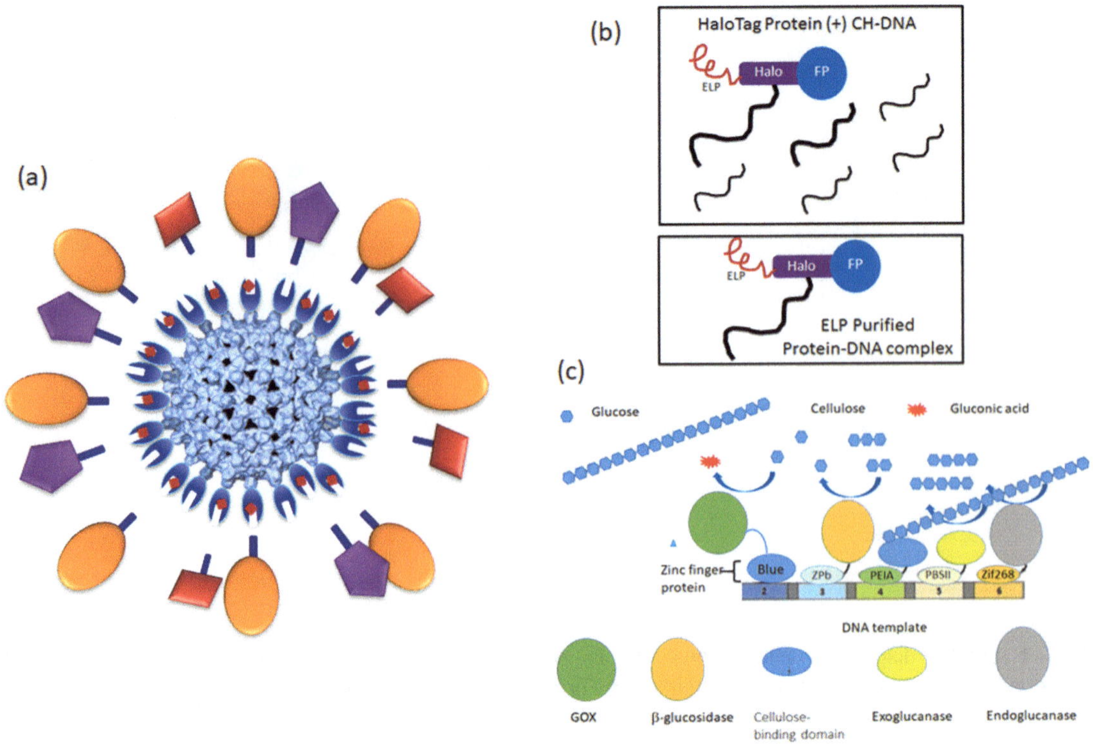

Fig. 2 Nanoparticle and DNA scaffolds. (**a**) Decoration of protein nanoparticles with different enzymes using Spytag/SpyCatcher chemistry. (**b**) Conjugation of a DNA oligo onto a HaloTag fusion protein. Conjugated proteins are separated using ITC. (**c**) Assembly of a functional enzyme cascade using different ZFPs

immobilization. This section describes SML ligation of enzymes, endoglucanase (CelA), and β-galactosidase (β-gal), on the surface of GGG functionalized E2 protein nanoparticle surface [2] that can be used for multienzyme assembly.

1.2.2 Spytag/SpyCatcher Bioconjugation

Spytag/SpyCatcher pair was first developed by splitting the *Streptococcus pyogenes* fibronectin-binding protein FbaB domain with a spontaneous amide bond between Lys and Asp reaction [41]. Spytag peptide could be fused at either terminus or internally [29, 50]. This section describes Spytag/SpyCatcher of enzymes, on the surface of viral nanoparticles.

1.3 Protein–Nucleic Acid Interaction Strategies

The ability to exploit the base-pairing property of nucleic acids has resulted in the creation of various DNA or RNA-based 3-D nanostructures as scaffolds for multienzyme assemblies structures [51]. We have utilized HaloTag mediated protein-DNA conjugation [12, 52] and zinc finger protein [18, 53] to enable protein organization onto DNA scaffolds (Fig. 2b). Toehold-mediated strand displacement was used to achieve dynamic enzyme function programming [54, 55].

1.3.1 HaloTag Mediated Protein-DNA Labeling

To assemble enzymes in a position-specific manner on DNA template, we proposed to conjugate single-stranded DNA onto each protein component to enable enzyme complex assembly through DNA hybridization. Since direct chemical modification of a DNA linker to proteins can sometimes lead to functional loss, the self-labeling HaloTag was used for conjugation. HaloTag is a 34 kDa monomeric mutant haloalkane dehalogenase designed to covalently bind to synthetic chlorohexane (CH) modified molecules, including DNA, fluorescent dyes, and solid supports [56]. The covalent attachment is highly specific and occurs rapidly under mild physiological conditions. By genetically tethering HaloTag to the different cellulosomal components, CH-modified DNA oligonucleotides can be site-specifically conjugated for cellulosome assembly [12, 13, 54].

1.3.2 Zinc Finger Protein Enabled Protein Immobilization on DNA Template

Zinc finger proteins (ZFPs) have been widely used for both in vitro and in vivo DNA binding applications [18, 57]. ZFPs are attractive as they can be engineered to specifically recognize any three base-pair (bp) sequence of interest with very high affinity (Kd in the nM range) [58] with chemical modifications (Fig. 2c). By taking advantage of the modularity of the individual zinc finger domain, multiple zinc finger motifs with nanomolar affinity have been designed to sketch the target sequences from 9 to 18 bp while retaining the exquisite specificity [59]. A potential drawback is the relative modest binding affinity that might lead to low organization efficiency [60].

1.3.3 Strand Displacement for Enzyme Function Programming

The use of dynamic DNA devices whose operation is based on toehold-mediated strand displacement can produce reconfigurable and autonomous DNA nanostructures to allow for dynamic protein assembly. We have recently demonstrate the use of strand displacement as a generalizable principle for constructing dynamic protein–DNA nanoassemblies suitable for modulating protein proximity and activities [54]. Using the HaloTag mediated DNA–protein complex, we assembled two proteins onto a DNA scaffold and used strand displacement to disassemble the complex. The activity and complex formation/disassemble were evaluated by enzyme activity assay and native acrylamide gel. The modularity of the design enables the creation of a universal set of input-guided logic devices for regulated dynamic protein assembly.

2 Materials

2.1 Cellulosome-Based Enzyme Assemblies

2.1.1 Culture Medium

1. Luria–Bertani (LB) medium: Dissolve 10.0 g of tryptone, 5.0 g yeast extract, and 10.0 g of NaCl in about 600 mL of ultrapure water (a specific resistance 18.2 MΩ·cm at 25 °C). Mix well by a stirrer and make up to 1 L with ultrapure water. Autoclave for 15 min at 15 psi at 121 °C. Depending on the selection marker

employed, either 100 μg/mL ampicillin or 50 μg/mL kanamycin may be supplemented.

2. YPD medium: Dissolve 10 g of yeast extract and 20 g of peptone in about 600 mL of ultrapure water and autoclave for 15 min at 15 psi at 121 °C. Dissolve 20 g of dextrose in 200 mL of ultrapure water and sterile-filtered with a 0.22 μm filter (*see* **Note 2**). Mix the two solutions and make up to 1 L with sterilized water. This medium is non-selective for *S. cerevisiae*.
3. SDC medium: Dissolve 6.7 g yeast nitrogen base without amino acids and 5.0 g of Casamino acids in about 600 mL of ultrapure water. Dissolve 20 g of dextrose in 200 mL of ultrapure water and sterile-filtered with a 0.22 μm filter. Well mix the solutions by a stirrer and make up to 1 L with ultrapure water. Autoclave for 15 min at 15 psi at 121 °C. This medium is tryptophan deficient as tryptophan will be hydrolyzed after autoclave.
4. SGC medium: The procedure and ingredients of SGC medium are similar to SDC medium, except for dextrose is replaced by galactose. This medium is used for the surface display of the mini-scaffoldin (*see* **Note 3**).

2.1.2 Protein Expression and Cell Lysis

1. 0.5 M Isopropyl β-d-1-thiogalactopyranoside (IPTG): Add 120 mg of IPTG powder in 1 mL of ultrapure water. After filtering with a 0.22 μm filter, make aliquots and stored at −80 °C.
2. LB-glycerol broth: Mix 10 g of tryptone, 5 g of yeast extract, and 10 g of NaCl in 900 mL of ultrapure water. Add 15 mL of glycerol to the solution and bring the volume to 1 L with ultrapure water. Use autoclave to sterilize.
3. 1 M Tris–HCl, pH 8.0: Dissolve 121.1 g of Tris-base in 500 mL water. Add concentrated HCl to adjust the pH to 8.0, then add ultrapure water to a final volume of 1 L.
4. Binding buffer: Dilute the 1 M Tris–HCl buffer to 50 mM, supplement with 100 mM NaCl and 10 mM $CaCl_2$ (*see* **Note 4**).

2.1.3 Immunofluorescence Microscope

1. Primary antibody solution: Add 0.5 μg of anti-c-Myc or anti-c-His immuno- globulin G (IgG) to 1 mL of binding buffer containing 1 mg of protease-free bovine serum albumin.
2. Secondary antibody solution: Add 0.5 μg of anti-mouse IgG conjugated with Alexa 488 to 1 mL of binding buffer containing 1 mg of protease-free bovine serum albumin (*see* **Note 5**).
3. Wash buffer: Supplement binding buffer with 1 g/L protease-free bovine serum albumin, store at 4 °C.

2.2 Sortase-Mediated and Spytag/SpyCatcher Bioconjugation

2.2.1 Protein Expression and Purification

1. Chemically competent cells BL21(DE3).
2. 5 M NaCl: Dissolve 292 g of NaCl in 800 mL of water. Adjust the volume to 1 L with water. Sterilize by autoclaving. Store the NaCl solution at room temperature.
3. 1 M KCL: Dissolve 74.55 g of KCl in 900 mL of water. Adjust the volume to 1 L with water. Sterilize by autoclaving. Store the KCl solution at room temperature.
4. 1 M $MgCl_2$: Dissolve 203.3 g of $MgCl_2 \cdot 6H_2O$ in 800 mL of water. Adjust the volume to 1 L with water. Sterilize by autoclaving. Store the $MgCl_2$ solution at room temperature.
5. 1 M $MgSO_4$: Dissolve 120 g of $MgSO_4$ (anhydrous) or 246 g of Mg $SO_4 \cdot 6H_2O$ in 800 mL of water. Adjust the volume to 1 L with water. Sterilize by autoclaving. Store the solution at room temperature.
6. 1 M glucose: Dissolve 180.16 g glucose in 800 mL of water. Adjust the volume to 1 L with water. Filter-sterilize and store at room temperature.
7. SOC media: 2g Bacto Tryptone, 0.5 g Bacto Yeast Extract, 0.2 mL of 5 M NaCl, 0.25 mL of 1 M KCl, 1 mL of 1 M $MgCl_2$, 1 mL of 1 M $MgSO_4$, 2 mL of 1 M glucose in 100 mL milliQ ultrapure water. Sterilize by autoclaving.
8. LB (Luria Broth) media: 10.0 g of tryptone, 5.0 g yeast extract, and 10.0 g of NaCl in 1 L of milliQ ultrapure water; and agar plates (same as LB media with 12 g of Agar) with antibiotic (Ampicillin or Kanamycin at final concentration of 50–100 μg/mL). Sterilize by autoclaving.
9. TB media: Add 24 g yeast extract, 20 g tryptone, 4 mL glycerol, 100 mL phosphate buffer (0.17 M KH_2PO_4, 0.72 M K_2HPO_4) to 1 L milliQ ultrapure water. Sterilize by autoclaving.
10. Phosphate buffer saline (PBS) buffer (pH 7.4): Add 8 g NaCl, 2 g KCl, 1.44 g Na_2HPO_4, and 0.24 g KH_2PO_4 to 1 L milliQ ultrapure water.
11. IPTG (Isopropyl β-d-1-thiogalactopyranoside) 10 mM stock solution:
12. DNaseI.
13. 1.5 μL microfuge tube (sterile).
14. 50 mL conical tubes (sterile).
15. 250 mL containers (sterile) for pelleting large quantity of cell culture (optional).
16. Water bath (42 °C).

17. Incubator shaker 37 °C.
18. Spectrophotometer for measuring cell growth (OD600 absorbance).

2.2.2 Sortase Reaction

1. Reaction buffer: 50 mM Tris, 150 mM NaCl, 50 mM KCl, 10 mM $CaCl_2$, pH 7.5

2.2.3 Spytag/SpyCatcher Reaction

1. Reaction buffer: 50 mM Tris, 150 mM NaCl at pH 8.0

2.2.4 Inverse Transition Cycling (ITC)

1. ELP precipitation buffer: 3 M ammonium sulfate: 3.972 g of ammonium sulfate in 10 mL $ddIH_2O$ stock (Millipore ultrapure water).
2. ELP solubilization buffer: 1.5 mL cold 50 mM Tris NaCl pH 7.6: dissolve 6.05 g Tris and 8.76 g NaCl in 800 mL of H_2O. Adjust pH to 7.6 with 1 M HCl and make volume up to 1 L with H_2O.

2.2.5 SDS-PAGE

1. 4× separating gel buffer: Dissolve 91 g Tris-base in 300 mL water. Adjust pH to 8.8. Adjust volume to 500 mL. Autoclave.
2. 4× stacking buffer: Dissolve 6.06 g Tris-base in 40 mL of water. Adjust pH to 6.8. Adjust final volume to 100 mL. Autoclave.
3. Separating gel (20 mL) by mixing 4.6 mL milliQ ultrapure water, 10 mL 30% acrylamide/Bis, 5 mL 4× separating gel buffer, 200 μL 10% SDS, 200 μL 10% APS (add right before pouring gel), 8 μL TEMED (add right before pouring gel). Pour the separating gels first (about ¾ of the total volume), add butanol on top and let dry for about 15–20 min.
4. Stacking gel (10 mL) by mixing milliQ ultrapure water, 1.7 mL 30% Acrylamide/Bis, 2.5 mL 4× stacking buffer, 100 μL 10% SDS, 100 μL 10% APS (add right before pouring gel), 10 μL TEMED (add right before pouring gel).
5. Sample loading buffer: Mix 50 mM Tris–HCl pH 6.8, 2% SDS, 10% glycerol, 12.5 mM EDTA, 0.02% bromophenol blue, and 1% β-mercaptoethanol (β-ME) (*see* **Note 6**).
6. Staining buffer: Add 100 mL of glacial acetic acid to 500 mL of ddH_2O. Add 400 mL of methanol and mix. Add 1 g of Coomassie R250 dye and mix.
7. Destaining buffer: Add 100 mL of glacial acetic acid to 500 mL of ddH_2O. Add 400 mL of methanol and mix.

2.2.6 Equipment

1. Table top centrifuge (Eppendorf).
2. Standard native acrylamide gel equipment, 10% native acrylamide gel.

2.3 HaloTag Mediated Protein-DNA Labeling

2.3.1 Equipment

1. Table top centrifuge (Eppendorf).
2. Ultrafiltration column (Vivaspin 500, Sartorius Stedim Biotech with 3000 Da cutoff).
3. Standard SDS-PAGE equipment, 10% SDS-PAGE gels, 4.5% native acrylamide gel.

2.3.2 Buffers and Reagents

1. HaloTag succinimidyl (O_4) chlorohexane ligand (Promega P6751)

2.4 Zinc Finger Protein Enabled Protein Immobilization on DNA Template

2.4.1 Equipment

1. Standard native acrylamide gel equipment, 10% native acrylamide gel

2.4.2 Buffers and Reagents

1. Lightshift Chemiluminescent EMSA kit (Pierce Chemical Co., Rockford, IL)

2.5 Strand Displacement for Enzyme Function Programming

2.5.1 Equipment

Standard native acrylamide gel equipment, 10% native acrylamide gel.

2.5.2 Buffers and Reagents

1. DNA hybridization buffer (20 mM Tris, 150 mM NaCl, 5 mM $MgCl_2$, pH 7.4).
2. 5× detergent buffer (25 mM DTT, 50% glycerol, and 0.5% IGEPAL).

3 Methods

3.1 Cellulosome-Based Enzyme Assemblies

3.1.1 Cellulase Expression and Cell Lysis

1. Add 3 mL of LB supplemented with the appropriate antibiotics into a 15 mL test tube. Pick single colonies expressing the cellulases as inoculates to each tube.
2. Shake the tubes at 250 rpm in a standard incubator at 37 °C overnight or until OD600 in the culture is 2.5.
3. Freshly transfer 2 mL of pre-culture to 200 mL of LB-glycerol medium containing appropriate antibiotics.
4. Grow the culture at 37 °C until OD600 reaches 1.0–1.2. Reduce the temperature in the shaker to 16 °C and induce the cellulase expression with 0.2 mM IPTG. Continue incubation overnight or until the OD600 is 4.5 at 16 °C and 250 rpm.

5. Harvest the cells at 6000 × *g* for 20 min, discard the supernatant, and store cell pellets at −20 °C or −80 °C until purification.
6. Resuspend the pellet in around 20 mL of bind buffer in a 50 mL conical tube by pipetting up and down slowly with a 1 mL disposable pipette until the cell suspension is homogeneous.
7. Lyse the cells by sonication (medium intensity, 50% duty cycle, 5× 30 s, 60 s, pausing after each cycle).
8. Centrifuge supernatant at 100,000 × *g* for 15 min at 4 °C and filter the supernatant through a 0.45 μm syringe filter into a new tube placed in an ice-cold water bath.

3.1.2 Display of Scaffoldin on Yeast Surface

1. Grow the yeast at 30 °C for 16–24 h until colonies are visible.
2. Pick up single colony and transfer to 3 mL of SDC media to start a liquid culture. Grow liquid cultures for 12 h shaking at 250 rpm at 30 °C.
3. Expand the culture by transferring cells into 20 mL SDC medium in a 100 mL Erlenmeyer flask. Grow cultures overnight (12–16 h) at 30 °C with shaking.
4. The following morning, dilute the cells in 200 mL of fresh SDC medium to the initial OD600 of 0.3. Continue to grow the cells in SDC at 30 °C until OD600 reaches 1.5.
5. Centrifuge the cells and remove the medium. Resuspend the cells in 200 mL SGC and grow at 25 °C for 36–48 h with constant shaking to induce the display of mini-scaffoldin.
6. Pellet the required volume of yeast at 2000 × *g* for 5 min in a conical centrifuge tube. Aspirate the supernatant and wash cells once with binding buffer.

3.1.3 Mini-cellulosome Assembly and Immunofluorescence Microscopy

1. Incubate yeast cells displaying the mini-scaffoldin with the as prepared cell lysates containing dockerin-tagged cellulases for 1 h at 4 °C with gentle mixing.
2. After incubation, wash the yeast cells twice with 10 mL of ice-cold binding buffer to remove the unbound cellulases.
3. Harvest the cells by centrifugation (2000 × *g*, 10 min) at 4 °C and resuspended in the binding buffer for further use.
4. To check the successful assembly of the mini-cellulosomes on the yeast surface, resuspend the yeast cells in 1 mL of primary antibody solution for 4 h with occasional mixing.
5. Pellet and wash the yeast cells twice with wash buffer.
6. Resuspend the cells in secondary antibody solution for 2 h with occasional mixing.

7. Pellet and wash the yeast cells twice with wash buffer, then resuspend them in binding buffer to an OD600 of 1.
8. Spot the cell suspensions onto a glass slide and add a coverslip.
9. Observe and capture images of Alexa 488 signal with appropriate microscope filter.

3.2 Sortase-Mediated and Spytag/SpyCatcher Conjugation

3.2.1 Transformation of pET Plasmid into BL21 E. coli Cells

1. Pre-warm SOC media and LB agar plates at room temperature.
2. Thaw the chemically competent cells BL21(DE3) on ice for ~20 min.
3. Transfer 50 μL of competent cells to 1.5 mL microcentrifuge tubes.
4. Add 50–100 ng of DNA plasmid to the microcentrifuge containing 50 μL cells. Mix the cells and DNA gently by pipetting up and down or flicking the tube 4–5 times.
5. Incubate on ice for 30 min.
6. Heat shock at 42 °C water bath for 30 s.
7. Transfer the tubes on the ice and incubate for 2 min.
8. Add 950 μL of warm SOC media to the tube.
9. Incubate the tube at 37 °C incubator shaker (250 rpm) for 60 min.
10. Spread 100 μL of the cells onto the LB agar plate with appropriate antibiotics (kanamycin).
11. Incubate the plate overnight at 37 °C.

3.2.2 Sortase and Target Protein Expression (IPTG Induction)

1. Grow a culture from a single isolated colony on the agar plate in LB-Kan media at 37 °C incubator shaker at 225 rpm.
2. Measure the OD600 of the overnight starter culture.
3. Start a 500 mL LB culture from the starter culture with starting OD600 of 0.05–0.1.
4. Incubate at 37 °C while shaking at 225 rpm.
5. Inoculate the culture with IPTG to final concentration of 0.1 mM, when the OD600 of the culture reaches 0.4–0.5, usually after about three to 4 h.
6. Grow the induced culture for 3–4 more hours at 37 °C while shaking at 225 rpm.
7. Harvest the induced cells via centrifugation (5000 × *g* for 15 min at 4 °C).
8. Discard the supernatant, and store the pellets at −80 °C overnight.
9. Resuspend the pellets in 1× PBS buffer by ultrasonic disruption.

10. Centrifuge the sample to separate the soluble fraction from the insoluble cell debris by centrifugation (15,000 × *g*, 4 °C, 15 min).

3.2.3 ELP-LPETG Protein Expression (Leaky Expression)

1. Start the culture by adding a single colony to 50 mL TB Ampicillin media and incubate at 37 °C overnight while shaking at 225 rpm.
2. Measure the OD600 of the overnight starter culture.
3. Once the OD600 of the starter culture is 0.05–0.1, add 10 mL starter culture to 500 mL culture (TB Ampicillin media).
4. Incubate the 500 mL inoculated media at 37 °C while shaking at 225 rpm.
5. Harvest the expression cultures via centrifugation (4000 × *g*, 15 min, 4 °C).
6. Discard the supernatant, and store the pellets at −80 °C until further cell lysis.
7. Cell lysis: Thaw the cell pellets and resuspend in 17.5 mL 50 mM Tris NaCl pH 7.6. Add 10 μL with 10 μL DNaseI to per 1 L of culture.
8. Lyse the cells by sonication (10 s on and 20 s off, total on time is 3 min, power output of 85 W.) Repeat the process twice.
9. Centrifuge the cell lysate at 15,000 × *g* for 15 min at 4 °C to separate cell debris from soluble protein fraction.
10. Retain the supernatant for the first round of ITC.

3.2.4 Sortase Reaction

1. Incubate sortase 5–20 μM, sortase reaction buffer, both substrates (ELP-LPETG and GGG-E2 fusion proteins) 5–100 μM at 37 °C incubator or water bath for 2–4 h in a 1.5 mL centrifuge tube (*see* **Notes** 7 and **8**).
2. Setup a negative control sample which does not include sortase, only the two substrates.
3. Proceed with ELP purification (Subheading 3.2.6).
4. Repeat the sortase ligation for the CelA-LPETG, beta-gal-LPETG or any desired enzyme modified with a C-terminal LPETG tag on the ELP-E2 nanoparticle decorated with GGG peptide following optimization of the reaction in **step 2**.

3.2.5 Spytag/SpyCatcher Bioconjugation

1. Express and purify Spytag- and SpyCatcher-fusion protein (enzymes) in *E. coli* using pET vector plasmids as described in the previous section (sortase and substrate #X).
2. Mix purified Spytag- and SpyCatcher-fusion proteins (5–25 μM) in PBS buffer and incubate at 25 °C for 2 h (*see* **Notes 9** and **10**).

3. Heat the samples in SDS loading buffer to stop the reactions on a thermal cycler or heating block at 95 °C for 7 min.
4. Run the products including negative controls on SDS-PAGE gels.
5. Follow the SDS-PAGE as described above.

3.2.6 ELP Purification

1. Precipitation: Add concentrated ammonium sulfate solution (stock solution concentration: 3 M) to the supernatant to the final concentration of 0.5 M. The solution should get cloudy.
2. Incubate the cloudy ELP solution in a water bath at 37 °C for 15 min.
3. "Hot" spin: Collect the protein by centrifuging at 16,000 × *g* at room temperature (25 °C) for 10 min.
4. Transfer the supernatant to another tube.
5. Precipitation: Add ammonium sulfate to precipitate the protein.
6. Resolubilization: After removing the supernatant, add solubilization buffer (1.5 mL cold 50 mM Tris NaCl pH 7.6) to the pellet and let sit in an ice bath for 10 min - 1 h before disturbing the pellet with the pipette.
7. Wait for the opaque pellet to turn translucent, indicating that the ELP has phase transitioned.
8. "Cold" spin: Centrifuge the solution at 16,000 × *g* at 4 °C to remove any insoluble contaminants.
9. Transfer the supernatant into a new tube without disturbing the contaminant pellet.
10. Repeat the hot and cold spin until desired purified protein is obtained.
11. Save purified fractions for SDS-PAGE analysis.

3.2.7 SDS-PAGE

1. Prepare the protein samples by adding the loading dye and boiling the sample at >75 °C on a heating block for 10 min.
2. Load on the polyacrylamide gel, refer to the Bio-rad chart (https://www.bio-rad.com/en-us/category/protein-gel-migration-charts?ID=N3F3TIDN) for choosing the proper gel composition for the optimum separation.
3. Perform the gel electrophoresis using an X-cel SureLock (Life Technologies) or similar apparatus at 200 V for approximately 1 h.
4. Stain the gels with Coomassie stain at room temperature overnight. Detain as needed.

5. Image the gels using a Gel Doc XR imager or a similar equipment, and quantify band intensities were quantified using Image Lab (Bio-Rad) or other similar software.

3.3 HaloTag Mediated Protein-DNA Labeling

1. React the specified oligos with 5′ and 3′ amine modifications with the HaloTag succinimidyl (O_4) chlorohexane ligand (Promega P6751) for HaloTag attachment. Mix the chlorohexane ligand with DNA at a molar ratio of 30:1 and incubate at room temperature for 24 h. Remove excess chlorohexane ligand using a 3000 Da ultrafiltration column (Vivaspin 500, Sartorius Stedim Biotech) (*see* **Note 11**).
2. React purified chlorohexane-oligos with the purified HaloTag fusion protein, using a 3× molar excess of chlorohexane-oligo overnight at 4 °C (*see* **Note 12**).
3. Remove excess/unattached oligo by purifying the protein–DNA and washing away any residual DNA through the purifying elastin-like polypeptide (ELP) tag. Add 1 M NaCl buffer to the purified protein samples, incubate the sample at 37 °C for 10 min, and centrifuge for 10 min with 13,000 × *g*. Remove supernatants and suspend pellets in cold PBS buffer to get the purified protein–DNA complex (*see* **Note 13**).
4. Load the protein only, protein plus excess DNA, and purified protein–DNA samples onto a 10% SDS-PAGE and stain with Coomassie blue for analysis of labeling efficiency by densitometry.
5. Incubate purified protein-DNA samples with single-stranded DNA template containing the target sequence for protein immobilization.
6. Perform native PAGE electromobility shift experiments in a 4.5% non-denaturing acrylamide gel. Load the protein only, protein plus DNA linkers and purified protein-DNA linkers onto the gel and run for 45 min at a constant 90 V. Stain the gel with ethidium bromide for 5 min before imaging.

3.4 Zinc Finger Protein Enabled Protein Immobilization on DNA Template

1. Purified zinc finger protein fusions were incubated with DNA samples at 37 °C for 1 h (*see* **Note 14**).
2. The complexes were subjected to a 10% native acrylamide gel and run under 90 V for 30 min to check on the binding.
3. Membrane transferring and chemiluminescence signaling were conducted with protocols suggested by manufacturer of the Lightshift Chemiluminescent EMSA kit (Pierce Chemical Co., Rockford, IL) (*see* **Note 15**).

3.5 Strand Displacement for Enzyme Function Programming

1. Equal concentrations (500 nM) of Protein A–strand A and Protein B–strand D were mixed with the scaffold strand C in DNA hybridization buffer (*see* **Note 16**).
2. The mixture was heated at 37 °C for 10 min.
3. Then 5× detergent buffer (25 mM DTT, 50% glycerol, and 0.5% IGEPAL) was added into the mixture to prevent potential association of components.
4. After 4 °C overnight incubation, the mixture was mixed with 0.1% cellulose buffer.
5. For disassembly, 1 μM of the NOT strand was added. Samples were collected periodically to measure reducing sugar production.
6. All the samples will be loaded on 10% native acrylamide gel and run under 90 V for 30 min to check on the assembly and disassembly (*see* **Note 17**).

4 Notes

1. Common proteins and peptides used in recent studies include dockerin/cohesin pairs, SH3/GBD/PDZ domains and their ligands, leucine zippers, the Z domain and the Fc domain pairs, and SpyCatcher and tag pairs.
2. Glucose solution after autoclaving will turn brownish, indicating a change in chemical structure. Therefore, when preparing yeast media, it is suggested to add the filter sterilized solution of glucose after autoclaving the rest of the media components.
3. The expression of the mini-scaffoldin in this study is under the control of a tryptophan-inducible promoter. Therefore, SGC medium is used as an induction medium. SDC medium may be used when using constitutive promoters.
4. The interaction between dockerin and cohesin is calcium dependent. Without the addition of calcium, the affinity between dockerin and cohesin is significantly reduced. Furthermore, the presence of calcium ion will help the crystallization of cellulases. This helps cellulases avoid the attach from proteases.
5. The choice of secondary antibody depends on what kind of first antibody is used. The conjugated fluorophore can also be changed depending on the need and the microscope filter one may have.
6. β-ME is toxic, unstable, and has a strong odor. Make sure that β-ME is added under the hood and be the last component to be added before use. 4× solution can be made in large amount without β-ME for further use and dilution.

7. The right ratio of protein moiety and nanoparticles is critical to get the correct ligation efficiency on the nanoparticles. It can be optimized by carrying out the reaction with different ratios and figure out the best ratio.
8. Incubation time can also be optimized by incubating the target protein with the nanoparticles at different time points and check. Sortase A reaction is reversible, so incubation for longer time might be problematic.
9. Protein-Spytag and HBV-SpyCatcher fusion were not described but it is necessary for Spytag-SpyCatcher ligation.
10. To achieve desired decoration density on the HBV nanoparticles, the ratio of Spytagged protein and HBV-SpyCatcher need to be optimized. To fully decorate the HBV, the Spytagged protein can be added in extra amount to ensure maximal consumption of the HBV reactant.
11. Protein fusion with HaloTag was not described but it is necessary for target protein-DNA labeling.
12. Extra molar of chlorohexane-oligo can guarantee higher protein-DNA labeling efficiency, which is approximately 100% according to SDS-PAGE gel analysis in our case.
13. ELP purification is special for our experiments. Other protein purification methods to remove the excess/unattached oligo are also feasible.
14. Protein fusions with zinc finger protein were not described but they are necessary for target protein-DNA template binding.
15. The binding affinity of each zinc finger protein is different. To achieve the best usage of this system, binding affinity of each binding protein should be checked before use.
16. HaloTag mediated protein-DNA labeling was described in Subheading 3.3.
17. Reducing sugar assay was not described. It is to measure the enzyme activity in the system. If other enzymes are used in this system, please use the corresponding enzyme activities.

Acknowledgements

This review was supported by grants from NSF (CBE1803008, CBE1911950, MCB1817675, MCB2013991, and DMR1609621).

References

1. Chen Q, Rozovsky S, Chen W (2017) Engineering multi-functional bacterial outer membrane vesicles as modular nanodevices for biosensing and bioimaging. Chem Commun 53(54):7569–7572
2. Chen Q, Sun Q, Molino NM, Wang S-W, Boder ET, Chen W (2015) Sortase A-mediated multi-functionalization of protein nanoparticles. Chem Commun 51(60): 12107–12110
3. Liu F, Banta S, Chen W (2013) Functional assembly of a multi-enzyme methanol oxidation cascade on a surface-displayed trifunctional scaffold for enhanced NADH production. Chem Commun 49(36): 3766–3768
4. Tsai SL, DaSilva NA, Chen W (2013) Functional display of complex cellulosomes on the yeast surface via adaptive assembly. ACS Synth Biol 2(1):14–21
5. Dueber JE, Wu GC, Malmirchegini GR, Moon TS, Petzold CJ, Ullal AV, Prather KL, Keasling JD (2009) Synthetic protein scaffolds provide modular control over metabolic flux. Nat Biotechnol 27(8):753–759
6. Berckman EA, Chen W (2019) Exploiting dCas9 fusion proteins for dynamic assembly of synthetic metabolons. Chem Commun 55(57):8219–8222
7. Zhao EM, Suek N, Wilson MZ, Dine E, Pannucci NL, Gitai Z, Avalos JL, Toettcher JE (2019) Light-based control of metabolic flux through assembly of synthetic organelles. Nat Chem Biol 15(6):589–597
8. Park M, Sun Q, Liu F, DeLisa MP, Chen W (2014) Positional assembly of enzymes on bacterial outer membrane vesicles for cascade reactions. PLoS One 9(5):6
9. Wheeldon I, Minteer SD, Banta S, Barton SC, Atanassov P, Sigman M (2016) Substrate channelling as an approach to cascade reactions. Nat Chem 8(4):299–309
10. Sun Q, Tsai S-L, Chen W (2019) Artificial scaffolds for enhanced biocatalysis. In: Schmidt-Dannert C, Quin MB (eds) Methods in enzymology, vol 617. Academic Press, Cambridge, pp 363–383
11. Price JV, Chen L, Whitaker WB, Papoutsakis E, Chen W (2016) Scaffoldless engineered enzyme assembly for enhanced methanol utilization. Proc Natl Acad Sci 113(45): 12691–12696
12. Sun Q, Chen W (2016) HaloTag mediated artificial cellulosome assembly on a rolling circle amplification DNA template for efficient cellulose hydrolysis. Chem Commun 52(40): 6701–6704
13. Chen Q, Yu S, Myung N, Chen W (2017) DNA-guided assembly of a five-component enzyme cascade for enhanced conversion of cellulose to gluconic acid and H2O2. J Biotechnol 263:30–35
14. Lee H, DeLoache WC, Dueber JE (2012) Spatial organization of enzymes for metabolic engineering. Metab Eng 14(3):242–251
15. Gao X, Yang S, Zhao C, Ren Y, Wei D (2014) Artificial multienzyme supramolecular device: highly ordered self-assembly of oligomeric enzymes in vitro and in vivo. Angew Chem Int Ed Engl 53(51):14027–14030
16. Qu J, Cao S, Wei Q, Zhang H, Wang R, Kang W, Ma T, Zhang L, Liu T, Wing-Ngor Au S, Sun F, Xia J (2019) Synthetic multienzyme complexes, catalytic nanomachineries for cascade biosynthesis in vivo. ACS Nano 13(9): 9895–9906
17. Garbett D, Bretscher A (2014) The surprising dynamics of scaffolding proteins. Mol Biol Cell 25(16):2315–2319
18. Sun Q, Madan B, Tsai SL, DeLisa MP, Chen W (2014) Creation of artificial cellulosomes on DNA scaffolds by zinc finger protein-guided assembly for efficient cellulose hydrolysis. Chem Commun 50(12):1423–1425
19. Tsai SL, Oh J, Singh S, Chen RZ, Chen W (2009) Functional assembly of minicellulosomes on the saccharomyces cerevisiae cell surface for cellulose hydrolysis and ethanol production. Appl Environ Microbiol 75(19): 6087–6093
20. Tsai SL, Goyal G, Chen W (2010) Surface display of a functional minicellulosome by intracellular complementation using a synthetic yeast consortium and its application to cellulose hydrolysis and ethanol production. Appl Environ Microbiol 76(22):7514–7520
21. Goyal G, Tsai SL, Madan B, DaSilva NA, Chen W (2011) Simultaneous cell growth and ethanol production from cellulose by an engineered yeast consortium displaying a functional minicellulosome. Microb Cell Factories 10:8
22. Gilbert HJ (2007) Cellulosomes: microbial nanomachines that display plasticity in quaternary structure. Mol Microbiol 63(6): 1568–1576
23. Bayer EA, Belaich JP, Shoham Y, Lamed R (2004) The cellulosomes: multienzyme machines for degradation of plant cell wall polysaccharides. Annu Rev Microbiol 58:521–554

24. Fierobe H-P, Mingardon F, Mechaly A, Bélaïch A, Rincon MT, Pagès S, Lamed R, Tardif C, Bélaïch J-P, Bayer EA (2005) Action of designer cellulosomes on homogeneous versus complex substrates: controlled incorporation of three distinct enzymes into a defined trifunctional scaffoldin. J Biol Chem 280(16): 16325–16334
25. You C, Myung S, Zhang YHP (2012) Facilitated substrate channeling in a self-assembled trifunctional enzyme complex. Angew Chem Int Ed 51(35):8787–8790
26. You C, Zhang YHP (2013) Self-assembly of synthetic metabolons through synthetic protein scaffolds: one-step purification, co-immobilization, and substrate channeling. ACS Synth Biol 2(2):102–110
27. Lee EJ, Lee NK, Kim I-S (2016) Bioengineered protein-based nanocage for drug delivery. Adv Drug Deliv Rev 106:157–171
28. Zdanowicz M, Chroboczek J (2016) Virus-like particles as drug delivery vectors. Acta Biochim Pol 63(3):469–473
29. Hartzell EJ, Lieser RM, Sullivan MO, Chen W (2020) Modular hepatitis B virus-like particle platform for biosensing and drug delivery. ACS Nano 14(10):12642–12651
30. MacKay JA, Chen M, McDaniel JR, Liu W, Simnick AJ, Chilkoti A (2009) Self-assembling chimeric polypeptide-doxorubicin conjugate nanoparticles that abolish tumours after a single injection. Nat Mater 8(12):993–999
31. Estrada LH, Chu S, Champion JA (2014) Protein nanoparticles for intracellular delivery of therapeutic enzymes. J Pharm Sci 103(6): 1863–1871
32. Lieser RM, Yur D, Sullivan MO, Chen W (2020) Site-specific bioconjugation approaches for enhanced delivery of protein therapeutics and protein drug carriers. Bioconjug Chem 31(10):2272–2282
33. Axup JY, Bajjuri KM, Ritland M, Hutchins BM, Kim CH, Kazane SA, Halder R, Forsyth JS, Santidrian AF, Stafin K, Lu Y, Tran H, Seller AJ, Biroc SL, Szydlik A, Pinkstaff JK, Tian F, Sinha SC, Felding-Habermann B, Smider VV, Schultz PG (2012) Synthesis of site-specific antibody-drug conjugates using unnatural amino acids. Proc Natl Acad Sci 109(40): 16101–16106
34. Lyu Z, Kang L, Buuh ZY, Jiang D, McGuth JC, Du J, Wissler HL, Cai W, Wang RE (2018) A switchable site-specific antibody conjugate. ACS Chem Biol 13(4):958–964
35. Berckman EA, Hartzell EJ, Mitkas AA, Sun Q, Chen W (2020) Biological assembly of modular protein building blocks as sensing, delivery, and therapeutic agents. Annu Rev Chem Biomol Eng 11(1):35–62
36. Khoshnejad M, Greineder CF, Pulsipher KW, Villa CH, Altun B, Pan DC, Tsourkas A, Dmochowski IJ, Muzykantov VR (2018) Ferritin nanocages with biologically orthogonal conjugation for vascular targeting and imaging. Bioconjug Chem 29(4):1209–1218
37. ElSohly AM, Netirojjanakul C, Aanei IL, Jager A, Bendall SC, Farkas ME, Nolan GP, Francis MB (2015) Synthetically modified viral capsids as versatile carriers for use in antibody-based cell targeting. Bioconjug Chem 26(8):1590–1596
38. Raeeszadeh-Sarmazdeh M, Hartzell E, Price JV, Chen W (2016) Protein nanoparticles as multifunctional biocatalysts and health assessment sensors. Curr Opin Chem Eng 13:109–118
39. Raeeszadeh-Sarmazdeh M, Parthasarathy R, Boder ET (2015) Site-specific immobilization of protein layers on gold surfaces via orthogonal sortases. Colloids Surf B: Biointerfaces 128: 457–463
40. Raeeszadeh-Sarmazdeh M, Parthasarathy R, Boder ET (2017) Fine-tuning sortase-mediated immobilization of protein layers on surfaces using sequential deprotection and coupling. Biotechnol Prog 33(3):824–831
41. Zakeri B, Fierer JO, Celik E, Chittock EC, Schwarz-Linek U, Moy VT, Howarth M (2012) Peptide tag forming a rapid covalent bond to a protein, through engineering a bacterial adhesin. Proc Natl Acad Sci 109(12): E690–E697
42. Keeble AH, Howarth M (2020) Power to the protein: enhancing and combining activities using the Spy toolbox. Chem Sci 11(28): 7281–7291
43. Berckman EA, Chen W (2020) A modular approach for dCas9-mediated enzyme cascading via orthogonal bioconjugation. Chem Commun 56:11426–11428
44. Swartz AR, Sun Q, Chen W (2017) Ligand-induced cross-linking of Z-elastin-like polypeptide-functionalized E2 protein nanoparticles for enhanced affinity precipitation of antibodies. Biomacromolecules 18(5): 1654–1659
45. Liu F, Tsai SL, Madan B, Chen W (2012) Engineering a high-affinity scaffold for non-chromatographic protein purification via intein-mediated cleavage. Biotechnol Bioeng 109(11):2829–2835
46. Madan B, Chaudhary G, Cramer SM, Chen W (2013) ELP-z and ELP-zz capturing scaffolds for the purification of immunoglobulins by

affinity precipitation. J Biotechnol 163(1): 10–16

47. Kim H, Siu KH, Raeeszadeh-Sarmazdeh M, Sun Q, Chen Q, Chen W (2015) Bioengineering strategies to generate artificial protein complexes. Biotechnol Bioeng 112(8):1495–1505
48. Levary DA, Parthasarathy R, Boder ET, Ackerman ME (2011) Protein-protein fusion catalyzed by sortase A. PLoS One 6(4):e18342
49. Parthasarathy R, Subramanian S, Boder ET (2007) Sortase A as a novel molecular “stapler” for sequence-specific protein conjugation. Bioconjug Chem 18(2):469–476
50. Wei Q, He S, Qu J, Xia J (2020) Synthetic multienzyme complexes assembled on virus-like particles for cascade biosynthesis in cellulo. Bioconjug Chem 31(10):2413–2420
51. Pinheiro AV, Han D, Shih WM, Yan H (2011) Challenges and opportunities for structural DNA nanotechnology. Nat Nanotechnol 6(12):763–772
52. Blackstock D, Chen W (2014) Halo-tag mediated self-labeling of fluorescent proteins to molecular beacons for nucleic acid detection. Chem Commun 50(89):13735–13738
53. Blackstock D, Sun Q, Chen W (2014) Fluorescent protein-based molecular beacons by zinc finger protein-guided assembly. Biotechnol Bioeng. 112(2):236–241
54. Chen RP, Blackstock D, Sun Q, Chen W (2018) Dynamic protein assembly by programmable DNA strand displacement. Nat Chem 10(4):474–481
55. Siu KH, Chen W (2019) Riboregulated toehold-gated gRNA for programmable CRISPR-Cas9 function. Nat Chem Biol 15(3):217–220
56. Los GV, Encell LP, McDougall MG, Hartzell DD, Karassina N, Zimprich C, Wood MG, Learish R, Ohana RF, Urh M, Simpson D, Mendez J, Zimmerman K, Otto P, Vidugiris G, Zhu J, Darzins A, Klaubert DH, Bulleit RF, Wood KV (2008) HaloTag: A novel protein labeling technology for cell imaging and protein analysis. ACS Chem Biol 3(6): 373–382
57. Conrado RJ, Wu GC, Boock JT, Xu H, Chen SY, Lebar T, Turnšek J, Tomšič N, Avbelj M, Gaber R, Koprivnjak T, Mori J, Glavnik V, Vovk I, Benčina M, Hodnik V, Anderluh G, Dueber JE, Jerala R, DeLisa MP (2012) DNA-guided assembly of biosynthetic pathways promotes improved catalytic efficiency. Nucleic Acids Res 40(4):1879–1889
58. Choo Y, Isalan M (2000) Advances in zinc finger engineering. Curr Opin Struct Biol 10(4):411–416
59. Dhanasekaran M, Negi S, Sugiura Y (2006) Designer zinc finger proteins: tools for creating artificial DNA-binding functional proteins. Acc Chem Res 39(1):45–52
60. Siu KH, Chen RP, Sun Q, Chen L, Tsai SL, Chen W (2015) Synthetic scaffolds for pathway enhancement. Curr Opin Biotechnol 36:98–106

Chapter 8

Random and Positional Immobilization of Multi-enzyme Systems

Hassan Maleki, Kamyar Khoshnevisan, and Hadi Baharifar

Abstract

In recent years, three key techniques including random co-immobilization, positional co-immobilization, and compartmentalization for multi-enzyme immobilization were extensively considered. Herein, we investigate random co-immobilization and positional co-immobilization techniques for multi-enzyme systems in detail. We describe randomly co-immobilized glucose oxidase (GOx) and horseradish peroxidase (HRP) on reduced graphene oxide (rGO) as the most used methods. Materials and methods are presented in terms of preparation of GO and rGO as well as enzyme immobilization procedure. Moreover, the principles of positional co-immobilization have been reviewed, and the relevant methods based on microfluidic systems and DNA structure considering HRP and GOx enzymes have been individually studied. It is believed that the benefits of using the methods associated with random and specifically positional immobilized multi-enzyme systems include not only enhanced cascade enzymatic activity via manipulated surface such as microfluidic systems (including porous materials) and DNA structure but also improved enzyme stability and ease of recovery for recycle.

Key words Random co-immobilization, Positional co-immobilization, Multi-enzyme systems, Glucose oxidase, Horseradish peroxidase, Microfluidic systems, DNA structure

1 Introduction

Methods for enzyme immobilization can be divided into three categories, comprising binding to a support, cross-linking, and encapsulation or entrapment [1–4]. (1) Binding to a support denotes enzyme immobilization on suitable carriers by hydrophobic reaction, ion interaction, hydrogen bonding, specific affinity, covalent bonding, etc. (2) Cross-linking is a support-free immobilization method, which can link enzymes together to form a network by a cross-linker. For instance, the Schiff base bond can be formed between the double aldehyde group of glutaraldehyde (a typical cross-linking agent) and the amino group of the enzyme molecules, which can efficiently connect enzymes. (3) Encapsulation means the entrapment of enzymes in a network or a hollow

Haralambos Stamatis (ed.), *Multienzymatic Assemblies: Methods and Protocols*,
Methods in Molecular Biology, vol. 2487, https://doi.org/10.1007/978-1-0716-2269-8_8,
© The Author(s), under exclusive license to Springer Science+Business Media, LLC, part of Springer Nature 2022

fiber or a microcapsule or membranes (complex systems) during material preparation. In recent years, the mentioned categories have been exclusively applied for immobilization of two or more than two enzymes which is called multi-enzyme immobilization.

Multi-enzyme immobilization is a technique or method that merges two or more than two enzymes on appropriate supports/carriers or integrates enzymes by a linker without supports [5, 6]. Enzymes immobilization will take place while they will nearly locate to one another and the mass transfer drawback can be decreased through co-immobilization, which has been verified to raise the enzymatic activity and enhance the stability and reusability. Support selection has played a key role in the enzyme immobilization process owing to the significant impact of support materials on the properties of enzymes. To the present time, various materials, such as graphene-based nanomaterials (e.g., graphene oxide (GO), reduced graphene oxide (rGO)), carbon nanotubes (CNTs), metal-organic frameworks (MOFs), DNA nanostructures, silica porous materials, organic materials including polymers have been widely employed for multi-enzyme immobilization, which can beneficially keep enzymes safe from heavy metals, high temperatures, and other biologically challenging conditions [5, 7].

There are three main techniques for multi-enzyme immobilization, including random co-immobilization, positional co-immobilization, and compartmentalization (Fig. 1). As the basic interaction between enzymes and supports is critical to understand the co-immobilized techniques, the basic types of enzyme immobilization will be discussed briefly.

1.1 Random Co-immobilization

Random co-immobilization is the easiest strategy to develop a multi-enzyme system, in which multiple enzymes can accidentally link on the surface or endowed inside the carrier/support via adsorption, encapsulation, covalent attachment, cross-linking, etc. (Fig. 2). To create co-immobilized multi-enzymatic systems, different enzymes can be first blended and then immobilized onto a support/substrate or directly linked to the support individually. Based on their chemical composition, various organic and/or inorganic support materials have been utilized for construction of co-localized multiple enzyme systems, including graphene-based matrices, silica-based supports, carbon nanotubes, metal-organic frameworks, DNA nanostructures, synthetic polymers, and biopolymers [8, 9].

Organic materials include synthetic polymers and natural polymers (biopolymers) that have broadly been applied to construct supports intended for co-immobilization of various enzymes under different immobilization protocols. Materials such as collagen [9], chitosan [10], cellulose [11], and agarose [12] as well as polyaniline [13], polyvinyl alcohol [14], and polyurethane [15] are examples of polymers utilized for enzyme immobilization. These materials

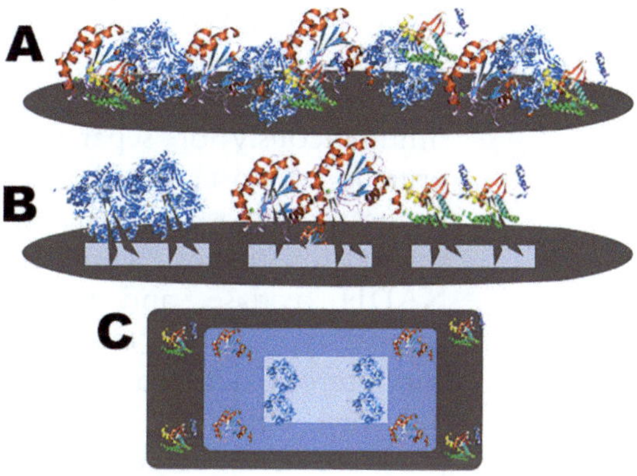

Fig. 1 Methods for multi-enzyme immobilization. (**a**) Random co-immobilization; (**b**) positional co-immobilization; and (**c**) compartmentalization

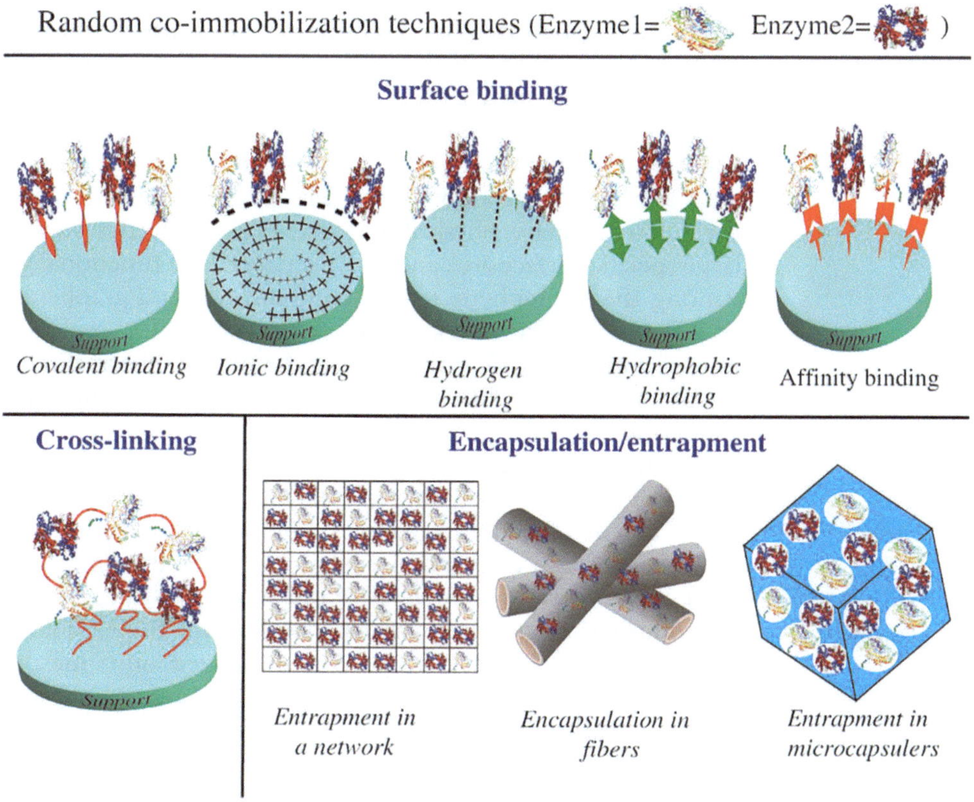

Fig. 2 Different techniques of random co-immobilization to construct a multi-enzyme system

possess excellent characteristics for multi-enzyme immobilization, such as good biocompatibility and biodegradability, affinity to biomolecules, and possibilities to create various geometrical shapes, etc., which has led to this organic support material being of great interest for multifarious applications. Organic supports can be prepared in diverse forms and shapes, including polymersomes, polymer nanoparticles, polymer hydrogels, nanofibers, etc. [16–19]. In numerous studies, two, three, and even more enzymes simultaneously or separately have been immobilized on/into organic supports by different techniques for various purposes, including glucose and catalase co-immobilized on chitosan microspheres for various commercial uses [20], formate dehydrogenase, NADH oxidase, and peroxidase enzymes co-immobilized on agarose-type carriers to oxidize phenols [21], cholesterol esterase, cholesterol oxidase, and peroxidase enzymes co-immobilized on a polyvinyl chloride sheet for serum cholesterol determination [22]. However, usage of the polymers for multi-enzyme immobilization systems encounters several limitations/problems, including insufficient mechanical strength and chemical resistance under various operational conditions as well as the possibility of microbial contamination. On the other hand, inorganic materials possess important standards for construction of support materials in terms of insolubility in water, reactivity towards any functionalizing agents, excellent mechanical strength and chemical resistance as well as excellent thermal, mechanical, and microbial resistance [23–25]. Graphene and its derivatives consist the most used materials/substrates for random multi-enzyme immobilization owing to their extraordinary physical and chemical properties such as the large specific surface area and the diversity of the functional groups. Among them, graphene oxide (GO) and reduced graphene oxide (rGO) are the most common-used supports for multi-enzyme immobilization. rGO is obtained by chemical or thermal reduction of GO to acquire different mechanical, electrical, and chemical properties.

Glucose oxidase (GOx), horseradish peroxidase (HRP), lactate dehydrogenase, urease, and alcohol dehydrogenase are among the most extensively used enzymes for immobilization purposes [26]. Furthermore, co-immobilization of enzymes such as GOx, HRP, glucoamylase, and cellulase onto GO substrates via random co-immobilization strategy has been well studied [27–31]. Besides, the same enzymes have been co-immobilized onto mesoporous silica nanoparticles [32], functionalized magnetic nanoparticles [33], and carbon nanotubes [34] to fabricate bioelectrodes.

The loading and activity of enzyme as well as morphology, performance, and efficiency of fabricated biocatalytic systems can be characterized by various methods such as UV/Vis spectroscopy, Fourier-transform infrared spectroscopy, atomic force microscopy, transmission electron microscopy, and scanning electron microscopy.

Herein, we describe an experimental procedure for random co-immobilization of GOx and HRP enzymes onto rGO substrate via non-covalent bonds intended for one-pot production of gluconic acid from starch/glucose. In such a glucose biosensor, GOx catalyzes the oxidation of glucose to hydrogen peroxide and subsequently reduced by HRP as an electron relay. The entire cycle could be described by the following reaction schemes:

$$\text{Glucose} + O_2 \rightarrow \text{Gluconic acid} + H_2O_2$$

$$H_2O_2 + \text{HRP(red)} \rightarrow H_2O + \text{HRP(ox)}$$

1.2 Positional Co-immobilization

Positional co-immobilization has been confirmed as a potent strategy for well-organized multi-enzyme immobilization. This method is also a good strategy for accumulating a consecutive order of enzymes, controlled by the reaction rate or pathway and the enzyme loading [8, 35, 36]. Consequently, the positional method leads to providing multi-enzyme systems in a different method than compartmentalization method (Fig. 3). Furthermore, it can regulate and enhance the cascade enzymatic reaction rates via changing the immobilized sequence. Polymers and DNA nanostructures are the common-applied carriers for positional co-immobilization to manage the relative positions of enzymes via certain interactions.

Among these, microfluidic channels are a common model of the positional co-immobilization in multi-enzyme systems. Logan et al. fabricated photopatterned porous polymer monoliths within microfluidic systems and applied them to obtain divided multi-enzymatic reactions [37]. The pore surface is improved by attaching PEG, using surface photoactivation and enzyme immobilization in the presence of a nonionic surfactant, to diminish nonspecific adsorption in enzyme complex. A common three-enzyme cascade reaction has been carried out by using invertase, GOx, and HRP. Several immobilization sequences were considered, however, the main product formation is observed only when the enzymes embedded in the exact sequential order. Luckarift et al. provided silica-immobilized enzymes in multi-enzyme systems within microfluidic tools [38]. Inorganic materials including metallic zinc, silica-immobilized soybean peroxidase, and silica-

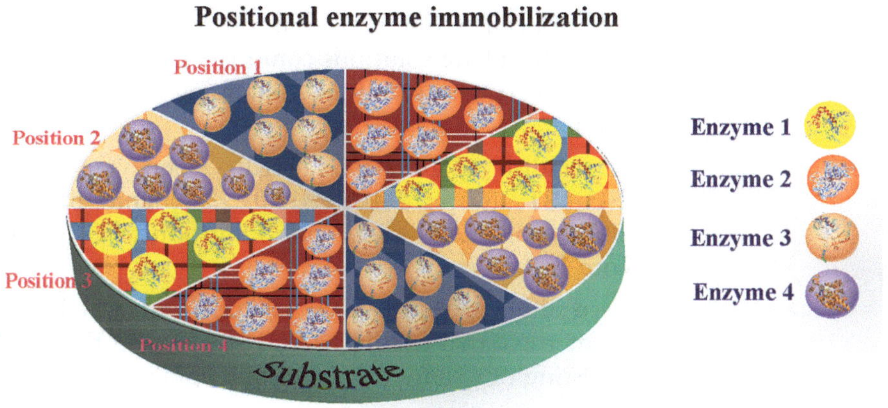

Fig. 3 Illustration of positional co-immobilization in multi-enzyme systems

immobilized hydroxylamino-benzene mutase are linked in sequence inside of distinct microfluidic chips to achieve cascade enzymatic combination. Ono et al. planned and fabricated a microfluidic chip with three reaction chambers to investigate three progressive glycosyltransfer reactions [39].

As alluded to above, microfluidic channels/chips and DNA structure are the most common surfaces in co-immobilization of multi-enzyme systems. Therefore, herein, we investigated the methods based on microfluidic systems and DNA structure with focus on HRP and GOx enzymes. Besides, the relevant methods of positional co-immobilization in multi-enzyme systems have been exclusively studied.

2 Materials

2.1 Random Co-immobilization

1. Graphite powder should be supplied with high purity (D50 < 600 nm, 99.95% purity).
2. Lyophilized powder of glucose oxidase (GOx, EC 1.1.3.4) from *Aspergillus niger.*
3. Lyophilized powder of horseradish peroxidase (HRP, EC 1.11.1.7) from horseradish.
4. Sulfuric acid (H_2SO_4), hydrochloric acid (HCl), potassium permanganate ($KMnO_4$), sodium nitrate ($NaNO_3$), and L-ascorbic acid (L-AA) in analytical grade.
5. Working solutions of hydrogen peroxide (H_2O_2) should be prepared fresh daily from 30% (v/v) H_2O_2.
6. Phosphate-buffered saline (PBS pH 7.0) and deionized (DI) water.

2.2 Positional Co-immobilization

1. All trials should be carried out with a 10 mM sodium phosphate solution arranged with ultrapure Milli-Q water. The NaCl concentration in all solutions is 0.15 M. The pH value of the phosphate solutions comprising GOx is almost 5.0; the pH value of the solutions containing HRP is 7.2 (enzyme assay).
2. Horseradish peroxidase (HRP isoenzyme C, product PEO-131, RZ = A403/A280 >3.1) and glucose oxidase (GOx from *Aspergillus niger*, product GLO-2022) (*see* **Note 1**).
3. Fused silica capillaries (ID = 100 μm) with a Teflon AF fluoropolymer external coating, TSU100375, and the DNA strands consisting of twenty-one bases in lyophilized form.
4. B-de-PG2 solution: 0.05 mg/mL in phosphate solution, pH 7.2.

5. 15 avidin: 0.1 mg/mL in the phosphate solution.
6. B-HRP solution: 0.1 mg/mL HRP in phosphate solution, pH 7.2, 0.15 M NaCl.
7. B-GOx solution: 0.06 mg/mL GOx in phosphate solution, pH 5.0, 1 M NaCl.
8. Activity of immobilized GOx: 0.5 mL of an o-phenylenediamine (OPD)/D-glucose/HRP solution and flow cell (3.14 mM OPD, 3.45 mM D-glucose, 1.98 nM HRP).
9. Activity of dissolved GOx: 1. 5 mL glass vial containing 3.14 mM OPD, 3.45 mM D-glucose, and 1.98 nM HRP (VTOT ¼ 2.92 mL), 80 mL of a 15 nM solution of GOx or B-GOx.
10. Activity of immobilized GOx following.
    (a) Qualitative in situ activity test: TInAS flow cell, 0.5 mL of an OPD/D-glucose/HRP solution into the flow cell (3.14 mM OPD, 3.45 mM D-glucose, 1.98 nM HRP).
    (b) Quantitative ex-situ activity measurements: 3.14 mM OPD, 3.45 mM D-glucose, and 1.98 nM HRP.
11. Lipase activity measurements: 80 nM enzyme in PBS, 5% isopropanol, and 0.1% (w/v) Triton X-100 with variable concentrations of 0.05–1.75 mM pNPB.
12. Enzyme immobilization, removal, and re-hybridization: 0.2 mM saline sodium citrate buffer (SSC) with 0.1% sodium dodecyl sulfate (SDS), 0.1 M NaOH, 0.2 mM SSC + 0.1% Tween-20.
13. Three-enzyme cascade reaction: 2 mM Gluc-Ac, 5 mM ABTS, and 100 nM GOx.
14. *Candida antarctica* lipase B and its azide-functionalized counterpart (AHA-CalB) and the azide-functionalized HRP (azido-HRP) and the substrate glucose mono-acetate (1-O-acetyl-D-gluco-pyranose, Gluc-Ac).
15. Tris ((1-((Oethyl) carboxymethyl)-(1,2,3-triazol-4-yl)) methyl) amine, additional denoted as the tris-triazole ligand, applied for the copper catalyzed acetylene azide cycloaddition (CuAAC) [40, 41]. 3 and 0.2 mM SSC, 3 mM sodium citrate and 30 mM NaCl.
16. Phosphate-buffered saline (50 mM NaH_2PO_4, 150 mM NaCl, calibrated with NaOH to pH 7).
17. Oligo sequences: DNA-A: H_2N-(CH_2)6-50-CCA CGG ACT ACT TCA AAA CTA-30 (complementary strand for DNA-CalB) and DNA-B: H_2N-(CH_2)6-50-GTC AAT ACT TAG GTC AAT ACT-30 (complementary strand for DNA-HRP). DNA-AC: hexynyl-50-TAG TTT TGA AGT

AGT CCG TGG-30 (complementary strand of DNA-A, to be used for conjugation with CalB) and DNA-BC: hexynyl-50-AGT ATT GAC CTA AGT ATT GAC-30 (complementary strand of DNA-B, to be used for conjugation with HRP) (*see* **Note 2**).

3 Methods

3.1 Random Co-immobilization

3.1.1 Preparation of GO

Graphene oxide is synthesized through the oxidation of graphite powder using a modified Hummers method, so that the functional group –OH of GO is activated by a nucleophilic substitution reaction.

1. One gram of graphite powder dispersed into a mixture of sodium nitrate ($NaNO_3$) (1 g) and concentrated H_2SO_4 (46 mL) in an ice bath with constant stirring to cool the mixture around 0 °C.
2. 6.13 g of $KMnO_4$ added slowly in the reaction flask to keep the reaction temperature below 10 °C and left for 2 h, obtaining a dark mixture.
3. The mixture stirred at 0 °C for 1 h, stored at room temperature (2 h), and subsequently heated to 35 °C for 3 h.
4. After slowly instilling 300 mL deionized water, 30% H_2O_2 (20 mL) should be added dropwise to the mixture and result in color changing in the reaction solution to a bright yellow.
5. The mixture centrifuged and the isolated materials washed with 10% HCl three times, followed by washing with distilled water until neutral pH is attained.
6. Finally, the product sonicated (250 W, 40 kHz) in buffer solution (10 mM Sodium phosphate at pH 7.0) for 40 min and then centrifuged (1252 × *g* for 1 h) to separate any unexfoliated graphite.
7. The supernatant, which is the fully exfoliated GO, collected, and freeze-dried at 50 °C under vacuum for 24 h to afford a dark brown solid.

3.1.2 Preparation of rGO

Several approaches have been developed for converting GO to rGO, among which the reduction of GO with L-ascorbic acid (L-AA) is a simple approach and environmentally friendly route to water-soluble graphene. It is a facile method for reducing GO sheets using L-AA as a reductant in an aqueous solution at room temperature (~ 25 °C). L-AA and its oxidized products are safe materials to the environment or biological systems and can act as a reductant as well as a capping reagent to stabilize as-reduced GO sheets.

1. First, 50 mg of the freeze-dried GO powder added in 100 mL of DI water and sonicated (37 Hz) for 2 h to obtain a 0.5 mg/mL colloidal solution.
2. After that, 500 mg of L-AA added to 100 mL of the prepared GO aqueous dispersion under vigorous stirring (*see* **Note 3**).
3. Then, the resultant suspension centrifuged at 11,269 × *g* for 30 min to separate the reduction product and washed three times with DI water to remove redundant L-AA, resulting in the formation of aqueous dispersions of reduced GO (*see* **Note 4**).

3.1.3 Enzyme Immobilization

GOx and HRP are separately co-immobilized on rGO to obtain a multi-enzyme system [27, 29].

1. The prepared rGO is first dispersed in 10 mM PBS buffer at pH 7 with a concentration of 1 mg/mL.
2. GOx enzyme added to the prepared rGO solution with a concentration of 0.5 mg/mL.
3. Incubate the mixture on ice for 2 h to perform the immobilization.
4. Centrifuge the mixture at 3835 × *g* for 10 min and then wash three times the GOx-bound rGO with the same buffer at pH 7 to remove non-adsorbed enzymes.
5. HRP enzyme added to the solution of GOx-bound rGO with the concentration of 1 mg/mL.
6. Incubate the mixture on ice for 2 h to perform the immobilization.
7. Centrifuge the mixture at 7000 rpm for 10 min and then wash three times the GOx/HRP-bound rGO with the same buffer at pH 7 to remove unbound enzymes.
8. After centrifugation, collect the supernatant to measure the enzyme loading (*see* **Note 5**).
9. The amount of the immobilized enzymes can be measured by the enzyme concentration change of the solution before and after loading.
10. The catalytic activity of GOx and HRP enzymes on GOx/HRP-rGO is measured by the chromogenic 3,3′,5,5′-Tetramethylbenzidine (TMB) assay operated on the UV–Vis spectra. The GOx activity is determined by incubating GOx/HRP-rGO with TMB (0.25 mM in PBS) and different concentrations of glucose. Also, the enzymatic activity of HRP is determined by incubating GOx/HRP-rGO with TMB (0.25 mM in PBS) and different concentrations of H_2O_2. First, GOx/HRP-rGO solution is centrifuged at a speed of 12,228 × *g* for 30 min and the resulting pellet is resuspended in 2 mL 0.1 M PBS

(pH 7.2) before the measurements. Then, in separate reactions 100 μL TMB solution, different concentrations of glucose (0.5–50 mM) and H_2O_2 (0.1–10 mM) are added to the prepared suspension for measuring the activity of GOx and HRP, respectively. Afterward, using UV–vis spectroscopy, the absorbance of oxidized TMB is recorded for 1200 s at wavelength of 652 nm. Finally, the obtained data are fitted to the Lineweaver–Burk plot to determine apparent values of the Michaelis–Menten parameters (Vmax and Km). The Michaelis constant (Km), defined as the concentration of substrate ([S]) providing half of enzyme maximal activity (Vmax), is a reflection of the enzyme substrate affinity and enzyme activity that calculated from the slope of Lineweaver–Burk plots according to the Michaelis–Menten equation [$V = (V\text{max}[S])/(\text{Km} + [S])$].

Schematic illustration of co-immobilize multi-enzyme (GOx@HRP) onto a 2D GO-based material is shown, to fabricate an artificial bio-microsystem intended for the one-pot conversion of glucose to gluconic acid (Fig. 4).

3.2 Positional Co-immobilization

3.2.1 Methods Used on Inorganic Structure

Biotinylation

1. HRP biotinylation should be done with Sulfo-NHS-LC-Biotin [42]. Solutions of biotinylated HRP should be kept at 5 °C (pH 7.2, 5 mM enzyme).
2. GOx (27.5 mg) should be dissolved in 6.875 mL phosphate solution. To separate the enzyme from the stabilizers, a volume of 1.5 mL of the solution should be added onto a Sephadex G75 column.
3. Ultrafiltration should be applied to concentrate the fractions comprising GOx (Millipore 50 kDa, 20 min, 2000 rpm).
4. If B-GOx and native GOx are stored at 5 °C in phosphate solutions (pH 5.0), their activity is remained for at least 2 months (5 mM enzyme).

De-PG2-mediated GOx Immobilization onto Sensor Chips

1. Measurements should be done with a flow cell connected to a syringe pump under continuous flow (20 mL/min).
2. Bath sonication should be applied to clean the sensor chips (Sonorex Super 10 P, Bandelin, Germany) for 30 min in pure ethanol followed by 30 min UV/ ozone cleaning (UV clean 135500, Boekel Industry Inc., USA).
3. The cleaned sensor chips should be stored in phosphate solutions (pH 7.2) until the chips are located in the transmission interferometric adsorption sensor (TInAS) unit.
4. The protocol for the immobilization of B-GOx is according to the protocol established for the immobilization of B-HRP [42].

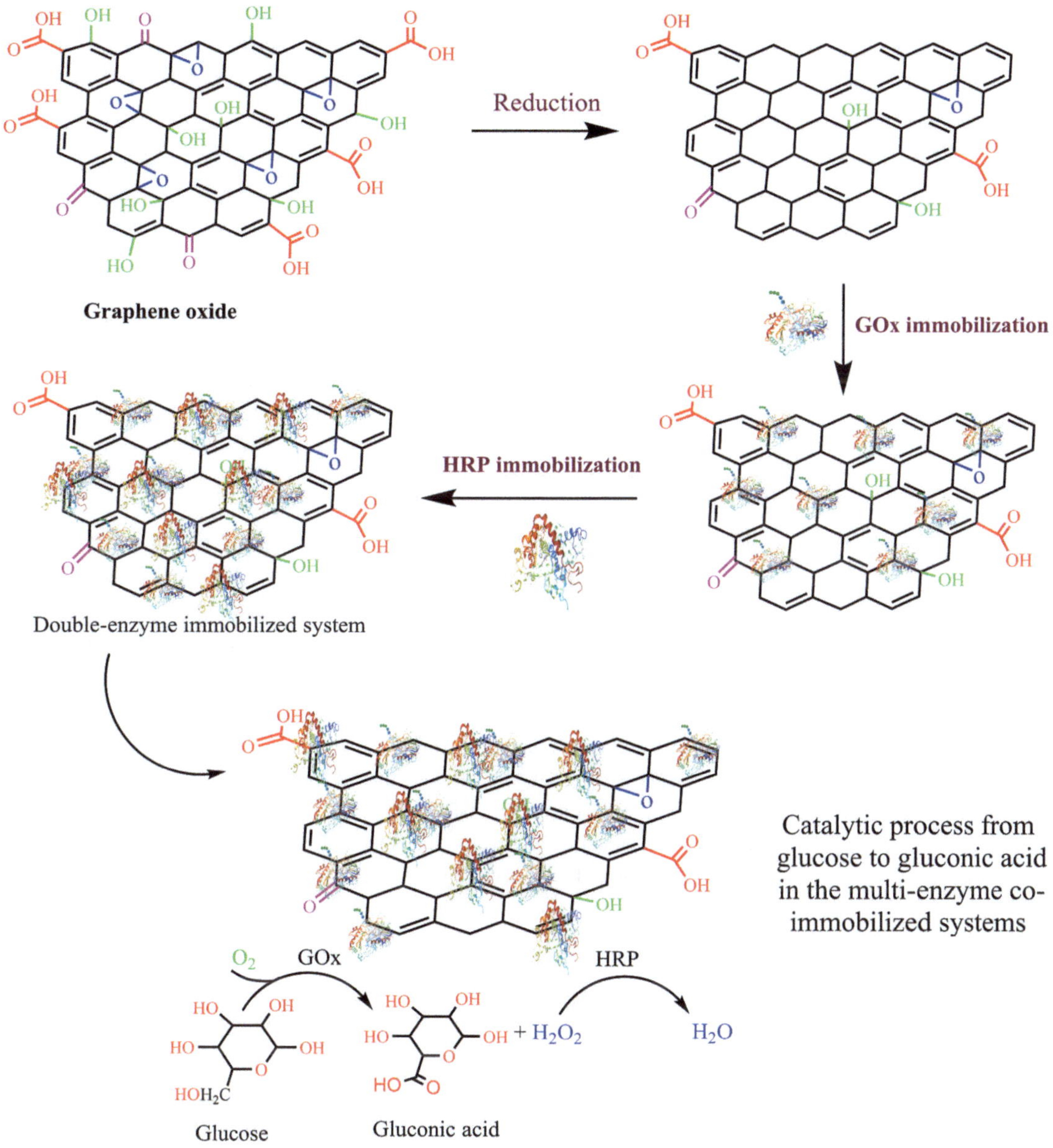

Fig. 4 A schematic scheme of preparation of rGO substrate, GOx@HRP immobilization, and the catalytic process of gluconic acid production from glucose

De-PG2-mediated Enzyme Immobilization Inside Glass Tubes

1. First, the tubes should be filled with a B-de-PG2 solution (0.05 mg/mL in phosphate solution, pH 7.2).
2. After washing with the buffer solution (5 mL), 15 avidin is adsorbed (0.1 mg/mL in the phosphate solution, incubation time 1 h).
3. Lastly, after a rinsing step with the corresponding phosphate solution, the glass tubes should be filled either with a B-HRP

solution (0.1 mg/mL in phosphate solution, pH 7.2, 0.15 M NaCl, incubation time 30 min) or with a B-GOx solution (0.06 mg/mL in phosphate solution, pH 5.0, 1 M NaCl, incubation time 1 h) (*see* **Note 6**).

HRP and GOx Activity Measurements

1. The activity of HRP should be measured by OPD/H_2O_2 as substrates. The formed product, DAP, provides a maximum absorbance at 417 nm.
2. The activity of GOx should be also measured using D-glucose as substrate and the OPD/HRP system for quantifying the rate of H_2O_2 formation upon oxidation of D-glucose [43].
3. All GOx activity measurements should be carried out in the corresponding phosphate solution. Under these conditions, the formed reaction product DAP possesses a maximum absorbance at 427 nm.

Activity of Dissolved GOx

1. 5 mL glass vial containing 3.14 mM OPD, 3.45 mM D-glucose, and 1.98 nM HRP ($V_{TOT} = 2.92$ mL), 80 μL of a 15 nM solution of GOx or B-GOx are added.
2. The UV/Vis spectrum between 325 and 600 nm of the solution is documented (*see* **Note 7**).

Activity of Immobilized GOx

1. Qualitative in situ activity test

   After B-GOx is immobilized onto the sensor chip in the TInAS flow cell, 0.5 mL of an OPD/D-glucose/HRP solution should be injected into the flow cell (3.14 mM OPD, 3.45 mM D-glucose, 1.98 nM HRP).
2. Quantitative ex-situ activity measurements
   (a) The sensor chip should be added into a 50 mL Falcon tube comprising 3.14 mM OPD, 3.45 mM D-glucose, and 1.98 nM HRP (in the corresponding phosphate solution, VTOT ¼ 3 mL).
   (b) After incubation in the dark for 30 min, the sensor chip is discarded and the UV/Vis spectrum of the solution is documented between 325 and 600 nm.

D-Glucose Quantification Via the GOx/HRP Cascade Reaction

The enzymatic cascade reaction exhibited in Fig. 5b is employed for quantifying D-glucose in the corresponding phosphate solution (with 0.15 M NaCl) in three different systems:

(1) both enzymes dissolved in the solution; (2) GOx immobilized on a TInAS chip immersed in the solution containing dissolved HRP; (3) both enzymes immobilized inside glass tubes (*see* **Note 6**)

As shown in Fig. 5a, a glass tube containing immobilized B-GOx (C) is linked to a glass tube containing immobilized B-HRP (D). The linking of the two tubes should be made with a

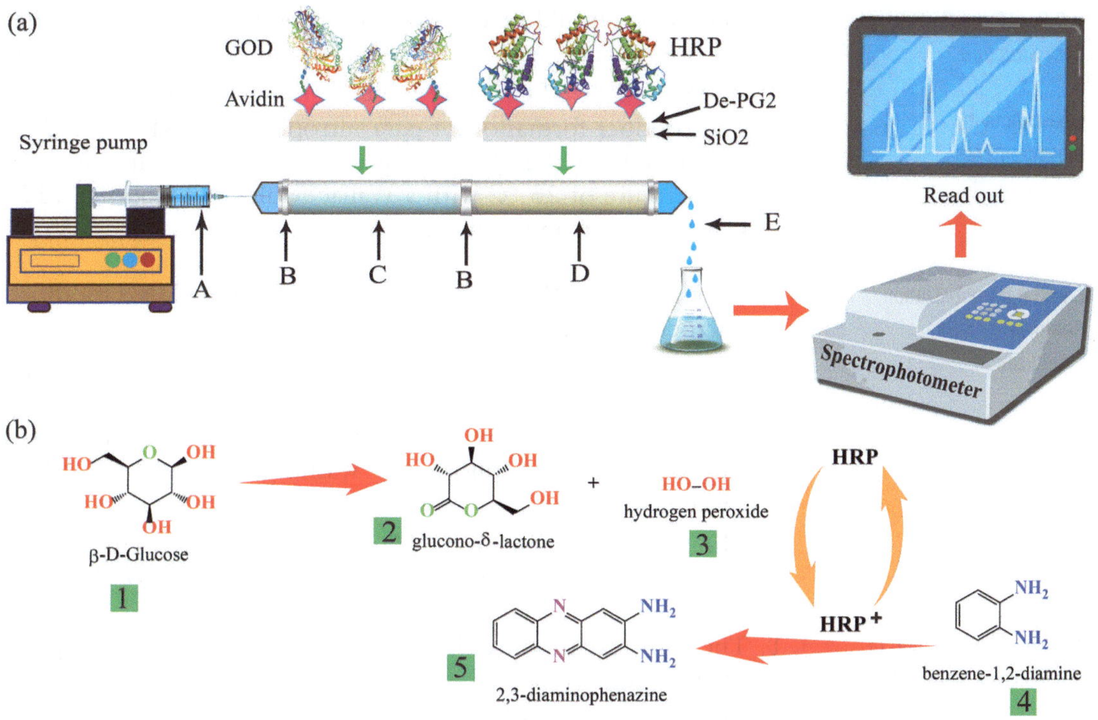

Fig. 5 (**a**) Schematic of the glass tube system for the measurement of D-glucose with immobilized B-GOx and immobilized B-HRP. (A) Syringe comprising of an aqueous solution of the substrates D-glucose (0–540 mM) and OPD (3.14 mM), 10 mM phosphate buffer, 0.15 M NaCl, pH 5.0; a syringe pump organized the constant flow of the solution from the syringe into the tubes, 70 mL/min; (B) silicone tubing for joining the syringe and the two glass tubes; (C) glass tube inside which B-GOx is immobilized; (D) glass tube inside which B-HRP is immobilized; (E) droplets eluting from the second tube (D) are collected (one fraction per 2 min) and then analyzed by UV/Vis spectroscopy for identifying the amount of DAP formed (via absorption measurements at 427 nm). Each tube has a length of 140 mm and an internal diameter of 1.6 mm. (**b**) Schematic representation of the enzymatic cascade reaction for the determination of D-glucose. β-D-Glucose (1) is oxidized by GOx to glucono-δ-lactone (2) and H_2O_2 (3). H_2O_2 oxidizes HRP to the active form HRP* (so-called compound I) which then oxidizes the colorless o-phenylenediamine (OPD, 4) to produce the yellow-orange reaction product DAP, 5

flexible silicone tubing (B) into which the two glass tubes should be injected. In this way, the glass tubes are in direct contact with each other without any visible gap between them. Then a phosphate solution containing OPD (3.14 mM) and D-glucose (0–540 mM) should be passed through the two connected tubes (70 μL/min, a tube containing immobilized B-GOx first, A in Fig. 5a).

Specimens of 0.14 mL eluting from the second tube containing immobilized B-HRP should be collected after 10 min and the UV/Vis spectrum in the 600–325 nm range is measured instantly after collection (Fig. 5a). After completion of the assessment, the two tubes with immobilized B-GOx or B-HRP should be separated, washed with phosphate solution kept for further use [43].

3.2.2 Methods Used on DNA Structure

Lipase Activity Measurements

1. The assays should be carried out with 80 nM enzyme in PBS, 5% isopropanol, and 0.1% (w/v) Triton X-100 with variable concentrations of pNPB ranging from 0.05–1.75 mM.
2. The product formation of para-nitrophenol (pNP) should be monitored via the absorbance at 405 nm overtime at a set temperature of 25 °C with a Wallac Multilabel Counter 1420 UV-VIS spectrophotometer.

Enzyme Immobilization, Removal, and Re-hybridization

1. A solution of DNA-CalB is flushed through the DNA modified microchannel (*see* **Note 7**).
2. The microchannel is then washed with 0.2 SSC with 0.1% sodium dodecyl sulfate (SDS). After activity determination, the DNA-CalB can be separated by cleaning the system 30 s with 0.1 M NaOH followed by washing with 0.2 SSC + 0.1% Tween-20.
3. The system is regenerated by flushing with 3 SSC for at least 2 h before re-hybridization with a fresh batch of DNA-CalB.

Activity Determination of Immobilized DNA-CalB

1. The substrate mixture should be stored identical to the quantities performed in bulk solution.
2. The capillary is equilibrated with PBS and 0.1% Triton X-100 prior to use and before each measurement, the flow is set to 10 μL/min for 5 min before adjusting flow to the desired speed at 0.5 mL/min. The production of pNP should be monitored over time by measuring the absorbance at 405 nm (Fig. 6).

Three-Enzyme Cascade Reaction

1. The cascade reaction, as shown in Fig. 7 should be performed by connecting two modified capillaries with zero-dead volume PEEK microtight unions P-720.
2. The patches of enzymes are positioned 10 or 50 cm apart from each other. The oligo sequences used for the modified capillaries are H_2N-$(CH_2)6$-50-CCA CGG ACT ACT TCA AAA CTA-30 for attachment of DNA-CalB and H_2N-$(CH_2)6$-50-GTC AAT ACT TAG GTC AAT ACT-30 for the attachment of DNA-HRP following the procedure described above.
3. The reaction should be performed in PBS, with 2 mM Gluc-Ac, 5 mM ABTS, and 100 nM GOx. Product formation (ABTS +) should be monitored by measuring the absorbance at 640 nm. Similar to the measurement with DNA-CalB, the flow is equilibrated at 10 mL/min, before setting the desired lower flow speed.

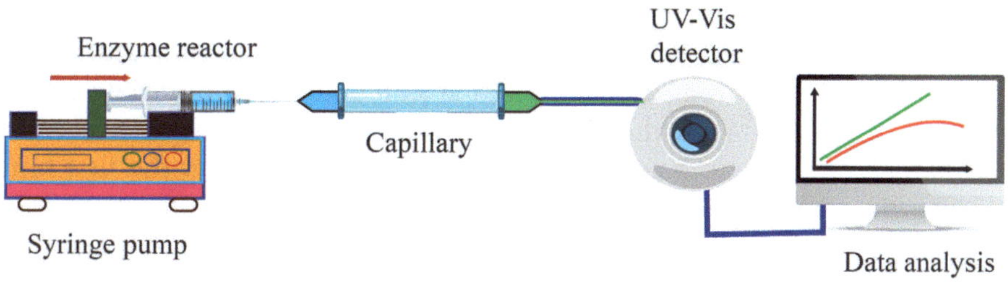

Fig. 6 Schematic outline of the set-up applied for detecting substrate conversion of the immobilized enzyme in a capillary system

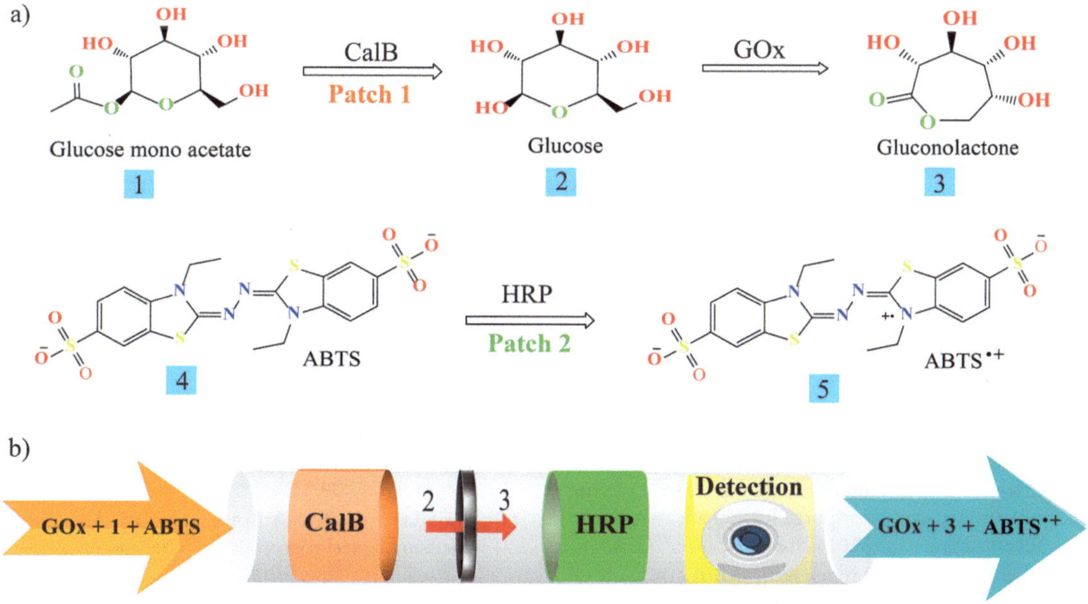

Fig. 7 (**a**) Schematic representation of the 3-enzyme cascade reaction. Glucose mono-acetate (1-O-acetyl-b-D-glucopyranose) 1 is hydrolyzed by CalB to produce glucose 2, which is subsequently oxidized by GOx to gluconolactone 3 generating H_2O_2 as a side product. HRP uses H_2O_2 to transform ABTS (4) into $ABTS^+$ (5). (**b**) Schematic representation of the microfluidic set-up used for performing a cascade reaction. The patches display the immobilized enzymes CalB and HRP (GOx is carried along by the mobile phase; see text) and the block signifies the zero dead-volume PEEK connectors

4 Notes

1. All buffer and aqueous solutions should be prepared with Milli-Q water at room temperature and the pH should be controlled with a pH-meter. All reactions should be carried out at room temperature ($T = 25$ °C) unless noted otherwise.
2. Some DNA strands are synthesized and some should be purchased as well as all DNA strands applied to provide the DNA-enzyme conjugates.

3. The reduction progress can be monitored by UV-vis spectroscopy. The UV-vis spectrum of GO demonstrates two absorption peaks centered at 230 and 300 nm that decayed gradually with the increase of reduction time, while a new absorption peak appears around 264 nm, providing the first hint that the GO is reduced.
4. The prepared rGO sheets can be suspended stably in the water at room temperature for several days without adding any capping reagent, if there are signs of instability in the solution, it is better to reconstitute the solution freshly.
5. The amount of enzyme absorbed onto support and enzyme activity can be determined by Lowry protein assay and tetramethylbenzidine assay, respectively.
6. The incubation should be carried out generally within 30 min unless otherwise stated.
7. Generally, the time required for immobilizing an enzyme inside a cleaned glass tube is about 3.5 h

References

1. Khoshnevisan K, Poorakbar E, Baharifar H et al (2019) Recent advances of cellulase immobilization onto magnetic nanoparticles: an update review. Magnetochemistry. 5(2):36
2. Khoshnevisan K, Vakhshiteh F, Barkhi M et al (2017) Immobilization of cellulase enzyme onto magnetic nanoparticles: applications and recent advances. Mol Catal 442:66–73
3. Cacicedo ML, Manzo RM, Municoy S et al (2019) Chapter 7 - Immobilized enzymes and their applications. In: Singh RS, Singhania RR, Pandey A et al (eds) Biomass, biofuels, biochemicals. Elsevier, Amsterdam, pp 169–200
4. Bilal M, Zhao Y, Noreen S et al (2019) Modifying bio-catalytic properties of enzymes for efficient biocatalysis: a review from immobilization strategies viewpoint. Biocatal Biotransformation 37:159–182
5. Ren S, Li C, Jiao X et al (2019) Recent progress in multienzymes co-immobilization and multienzyme system applications. Chem Eng J 373: 1254–1278
6. Sheldon RA, van Pelt S (2013) Enzyme immobilisation in biocatalysis: why, what and how. Chem Soc Rev 42:6223–6235
7. Sheldon RA, Woodley JM (2018) Role of biocatalysis in sustainable chemistry. Chem Rev 118:801–838
8. Xu K, Chen X, Zheng R et al (2020) Immobilization of multi-enzymes on support materials for efficient biocatalysis. Front Bioeng Biotechnol 8:660
9. Wahab RA, Elias N, Abdullah F et al (2020) On the taught new tricks of enzymes immobilization: an all-inclusive overview. React Funct Polym 152:104613
10. Krajewska B (2004) Application of chitin- and chitosan-based materials for enzyme immobilizations: a review. Enzym Microb Technol 35: 126–139
11. Sulaiman S, Mokhtar MN, Naim MN et al (2015) A review: potential usage of cellulose nanofibers (CNF) for enzyme immobilization via covalent interactions. Appl Biochem Biotechnol 175:1817–1842
12. Zucca P, Fernandez-Lafuente R, Sanjust E (2016) Agarose and its derivatives as supports for enzyme immobilization. Molecules 21: 1577
13. Ashly PC, Joseph MJ, Mohanan PV (2011) Activity of diastase α-amylase immobilized on polyanilines (PANIs). Food Chem 127:1808–1813
14. Porto MDA, dos Santos JP, Hackbart H et al (2019) Immobilization of α-amylase in ultrafine polyvinyl alcohol (PVA) fibers via electrospinning and their stability on different substrates. Int J Biol Macromol 126:834–841
15. Thomson T (2018) Polyurethane immobilization of cells and biomolecules: medical and

environmental applications. John Wiley & Sons, Inc., Hoboken, NJ
16. Wei Q, Xu M, Liao C et al (2016) Printable hybrid hydrogel by dual enzymatic polymerization with superactivity. Chem Sci 7:2748–2752
17. Smith S, Goodge K, Delaney M et al (2020) A comprehensive review of the covalent immobilization of biomolecules onto electrospun nanofibers. Nanomaterials (Basel) 10:1–39
18. Zdarta J, Meyer AS, Jesionowski T et al (2018) A general overview of support materials for enzyme immobilization: characteristics, properties, practical utility. Catalysts. 8:92
19. Klermund L, Poschenrieder ST, Castiglione K (2017) Biocatalysis in polymersomes: improving multienzyme cascades with incompatible reaction steps by compartmentalization. ACS Catal 7:3900–3904
20. Zhang J, Zhou X, Wang D et al (2013) Studies on the co-immobilized GOD/CAT on cross-linked chitosan microsphere modified by lysine. J Mol Catal B Enzym 97:80–86
21. Rocha-Martin J, Velasco-Lozano S, Guisán JM et al (2014) Oxidation of phenolic compounds catalyzed by immobilized multi-enzyme systems with integrated hydrogen peroxide production. Green Chem 16:303–311
22. Chauhan N, Pundir CS (2011) Co-immobilization of cholesterol esterase, cholesterol oxidase and peroxidase on PVC strip for serum cholesterol determination. Anal Methods 3:1360–1365
23. Hartmann M, Kostrov X (2013) Immobilization of enzymes on porous silicas – benefits and challenges. Chem Soc Rev 42:6277–6289
24. Zucca P, Sanjust E (2014) Inorganic materials as supports for covalent enzyme immobilization: methods and mechanisms. Molecules. 19:14139–14194
25. Amirnejat S, Movahedi F, Masrouri H et al (2013) Silica nanoparticles immobilized benzoylthiourea ferrous complex as an efficient and reusable catalyst for one-pot synthesis of benzopyranopyrimidines. J Mol Catal A Chem 378:135–141
26. Homaei AA, Sariri R, Vianello F et al (2013) Enzyme immobilization: an update. J Chem Biol 6:185–205
27. Zhao F, Li H, Jiang Y et al (2014) Co-immobilization of multi-enzyme on control-reduced graphene oxide by non-covalent bonds: an artificial biocatalytic system for the one-pot production of gluconic acid from starch. Green Chem 16:2558–2565
28. Zore OV, Pattammattel A, Gnanaguru S et al (2015) Bienzyme-polymer-graphene oxide quaternary hybrid biocatalysts: efficient substrate channeling under chemically and thermally denaturing conditions. ACS Catal 5: 4979–4988
29. Mathesh M, Liu J, Barrow CJ et al (2017) Graphene-oxide-based enzyme nanoarchitectonics for substrate channeling. Chem A Eur J 23:304–311
30. Li F, Ma W, Liu J et al (2018) Luminol, horseradish peroxidase, and glucose oxidase ternary functionalized graphene oxide for ultrasensitive glucose sensing. Anal Bioanal Chem 410: 543–552
31. Zhang H, Hua SF, Zhang L (2020) Co-immobilization of cellulase and glucose oxidase on graphene oxide by covalent bonds: a biocatalytic system for one-pot conversion of gluconic acid from carboxymethyl cellulose. J Chem Technol Biotechnol 95:1116–1125
32. Gustafsson H, Küchler A, Holmberg K et al (2015) Co-immobilization of enzymes with the help of a dendronized polymer and mesoporous silica nanoparticles. J Mater Chem B 3: 6174–6184
33. Zhuo Y, Yuan PX, Yuan R et al (2009) Bienzyme functionalized three-layer composite magnetic nanoparticles for electrochemical immunosensors. Biomaterials 30:2284–2290
34. Yang H, Gong C, Miao L et al (2017) A glucose biosensor based on horseradish peroxidase and glucose oxidase co-entrapped in carbon nanotubes modified electrode. Int J Electrochem Sci 12:4958–4969
35. Hwang ET, Lee S (2019) Multienzymatic cascade reactions via enzyme complex by immobilization. ACS Catal 9:4402–4425
36. Wu R, Song H, Wang Y et al (2020) Multienzyme co-immobilization-based bioelectrode: design of principles and bioelectrochemical applications. Chin J Chem Eng 28:2037–2050
37. Logan TC, Clark DS, Stachowiak TB et al (2007) Photopatterning enzymes on polymer monoliths in microfluidic devices for steady-state kinetic analysis and spatially separated multi-enzyme reactions. Anal Chem 79:6592–6598
38. Luckarift HR, Ku BS, Dordick JS et al (2007) Silica-immobilized enzymes for multi-step synthesis in microfluidic devices. Biotechnol Bioeng 98:701–705
39. Ono Y, Kitajima M, Daikoku S et al (2008) Sequential enzymatic glycosyltransfer reactions on a microfluidic device: synthesis of a glycosaminoglycan linkage region tetrasaccharide. Lab Chip 8:2168–2173
40. Chan TR, Hilgraf R, Sharpless KB et al (2004) Polytriazoles as copper(I)-stabilizing ligands in catalysis. Org Lett 6:2853–2855

41. van Kasteren SI, Kramer HB, Gamblin DP et al (2007) Site-selective glycosylation of proteins: creating synthetic glycoproteins. Nat Protoc 2: 3185–3194
42. Fornera S, Balmer TE, Zhang B et al (2011) Immobilization of peroxidase on SiO_2 surfaces with the help of a dendronized polymer and the avidin-biotin system. Macromol Biosci 11: 1052–1067
43. Fornera S, Bauer T, Schlüter AD et al (2012) Simple enzyme immobilization inside glass tubes for enzymatic cascade reactions. J Mater Chem 22:502–511

Chapter 9

Compartmentalized Immobilization of Multi-enzyme Systems

Hadi Baharifar, Kamyar Khoshnevisan, and Hassan Maleki

Abstract

The methods of compartmentalized immobilization in multi-enzyme systems containing inorganic complexes and organic scaffolds (i.e. nucleic acid (RNA and DNA), protein and lipid) have been thoroughly investigated. Compartmentalization mostly focuses on dividing individual enzyme(s) into specific location or orientation of the enzymes cooperating in cascade reaction. Organic scaffolds are preferred because of their capability for simultaneous synthesis in biological systems. Besides, the most required methods of horseradish peroxidase (HRP) and glucose oxidase (GOD) enzymes including enzyme activity measurement, enzyme immobilization, removal, and re-hybridization, and enzyme attaching have been provided because they have been extensively applied in multi-enzyme systems. Organic scaffolds have a wide range and properties. Therefore, two methods including dockerin–cohesin linker and nucleotides interaction have been demonstrated for immobilization of enzyme on protein and DNA scaffold, respectively.

Key words Compartmentalized immobilization, Multi-enzyme systems, Glucose oxidase, Horseradish peroxidase, Organic scaffolds

1 Introduction

As a mimic of natural enzyme groups in environments, compartmentalization can spatially divide enzymes with distinct patterns and ratios (Fig. 1), protecting enzymes against proteolysis, microbial degradation, or other harmful conditions [1]. The simplest method for orientation of the enzymes is achieved by using metal ions. Metal ions commonly have a positive charge and could easily interact with a negative charge of amino acids and proteins. Therefore, arranging metal ions on the substrate results in arranging enzymes on the substrate. An inorganic nanocrystal complex could be prepared using a wide range of proteins and metals and could have different shapes and sizes [2]. Copper II was used in different studies as an inorganic component for protein complex preparation [3]. For example, Horseradish peroxidase (HRP) is blended with copper II to form the HRP incorporated complexes,

Haralambos Stamatis (ed.), *Multienzymatic Assemblies: Methods and Protocols*,
Methods in Molecular Biology, vol. 2487, https://doi.org/10.1007/978-1-0716-2269-8_9,
© The Author(s), under exclusive license to Springer Science+Business Media, LLC, part of Springer Nature 2022

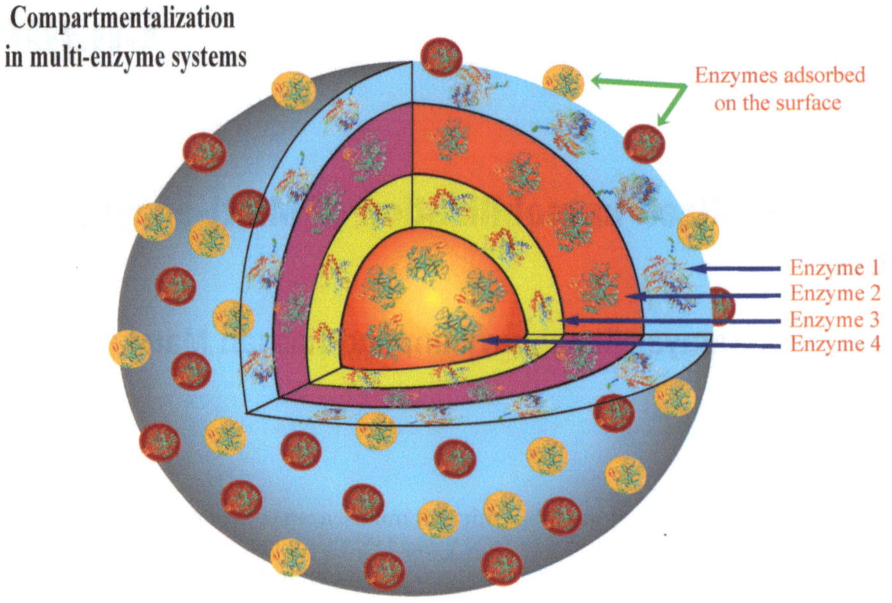

Fig. 1 Compartmentalization method in multi-enzyme systems

then glucose oxidase (GOx) is adsorbed on the surface of the complexes by the coordination interaction between Cu^{2+} and amino acids of the protein [4]. Resulted complex (GOx@HRP) displayed higher catalytic performance and activity compared with free and random enzyme complexes (GOx-HRP).

Other nano and micro-particles such as polymersomes and colloidosomes have been proposed as the scaffold for compartmentalized multi-enzyme immobilization [5, 6]. Polymersomes and colloidosomes form hollow capsules/spheres in solution providing a large area for enzyme immobilization. Hydrophilic enzymes are usually attached inside polymersomes and colloidosomes while hydrophobic ones interact with their outer surface. Immobilization of enzymes using polymersomes showed that final product yield increased 2.2 fold in comparison to noncompartmentalized reaction [7].

Using biological interactions for immobilization of enzymes in a specific site has been discussed in numerous studies [5–8]. Different organic substrates such as proteins, nucleic acids (DNA and RNA), and lipids have been used for compartmentalized enzyme immobilization. For fixing the enzyme at specific sites, enzymes are attached to different ligands or linkers interacting to the desired site of scaffold. Some studies focus on simultaneous preparation of scaffold and enzymes using genetic engineering and biotechnological methods. Numerous enzyme and organic scaffolds have been used for compartmentalization of the enzymes [8, 9]. Due to differences in nucleic acid, protein, and lipid structure and monomers, it is hard to categorize general protocols in this area.

For protein scaffold, cohesin–dockerin [10] based interaction has been exploited more than other ligands. Chun You and his colleague have used a specific protein scaffold for immobilization of cascade enzymes [11]. The interaction between enzymes and their scaffold is based on cohesin and dockerin existing in natural extracellular complexed cellulose systems (i.e., cellulosomes). Dockerin containing triosephosphate isomerase (TIM, EC 5.3.1.1), aldolase (ALD, EC 4.1.2.13), and fructose 1,6-bisphosphatase (FBP, EC3.1.3.11) as cascade enzymes in the glycolysis and gluconeogenesis pathways are attached to synthetic trifunctional scaffoldings carrying a family 3 cellulose-binding module (CBM3) at the N terminus followed by three different types of cohesins. Each TIM, ALD, FBP sequence is integrated by specific dockerin from certain bacteria at their C terminals attached to a specific sequence of cohesins at the scaffolding. Compartmentalization of the mentioned enzymes in protein scaffolds using specific attaching sequence for each enzyme resulted in having better control on concentration and location of the desired enzymes. This enzyme complex increased Kcat from 3.90 to 36.9 (min^{-1}) and decreased Km from 1.63 to 0.46 mM in comparison to the enzyme mixture.

Nucleic acid scaffolds are another favorite structure for fixing cascade enzymes. Both DNA [12] and RNA [13] (Fig. 2) based patterns have been used for immobilization of different enzymes.

Attachment of the enzyme could be utilized by specific nucleic acid motif and protein domain interactions or common ligand-protein such as streptavidin-biotin. Plasmid is used as one of the simple scaffolds for the arrangement and regulation of enzyme concentrations. Enzyme genes could be altered to have certain sequences (i.e., zinc finger) bonded to specific parts of plasmids. In a study, three enzymes including methylglyoxal synthase (MgsA), 2,5-diketo-D-gluconic acid reductase (DkgA), and glycerol dehydrogenase (GldA) that produced 1,2-PD from dihydroxyacetone phosphate (DHAP) have different zinc finger motifs attached to the specific sequence of plasmid scaffold [14]. The scaffold was capable to interact with the majority of expressed enzymes that resulted in the formation of substrate channeling on plasmid. Obtained results demonstrated that scaffold attached enzymes improve product concentrations up to 4.5 fold in comparison to the mixture of unattached enzymes. Notably, the efficiency of the scaffold-driven enzyme compartmentalization is affected by enzyme to enzyme and enzyme to scaffold ratio.

There is a variety of methods for enzyme compartmentalization. In the following sections, using copper ions as the simplest way for immobilization of HRP and GOx [4], designing protein scaffold based on cohesin–dockerin interactions [8] and

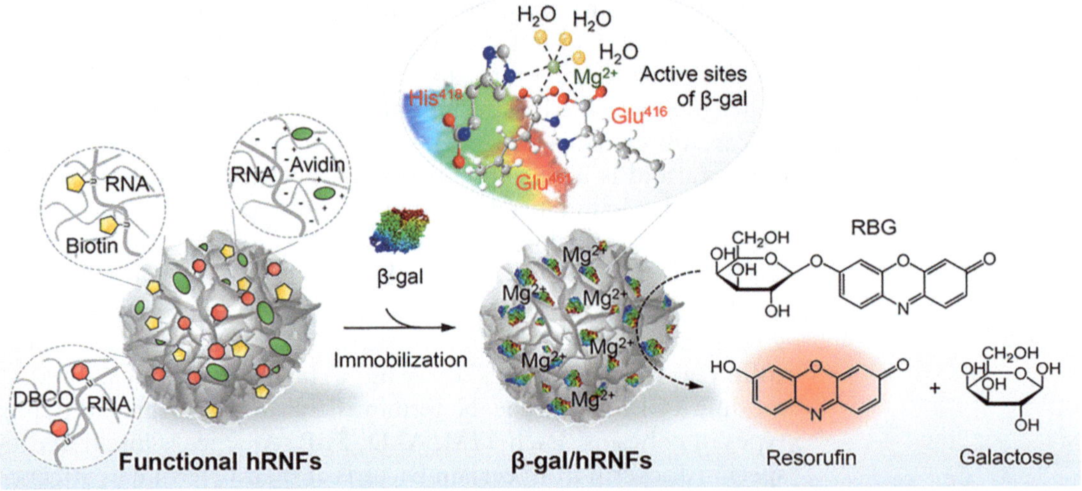

Fig. 2 Schematic illustration of progressing circle transcription-amplified hierarchically designed organic–inorganic hybrid RNA flowers for multi-enzyme systems. Reproduced from [Wang, Ye, et al. 2019] with permission from [American Chemical Society, https:/pubs.acs.org/doi/10.1021/acsami.9b04663]. Ref. 13

construction of a plasmid as DNA scaffold having zinc finger sequence [11] are described in detail.

2 Materials

All chemicals that used in the following studies are of analytical grade or higher and prepared with deionized water.

1. $CuSO_4$ solution: 200 mM. Add 15.96 g of the $CuSO_4$ powder to 100 mL deionized water in a glass beaker. Then stir it for 20 min at RT. After that filter the solution using 0.22 μm syringe filter and kept it at 4 °C.
2. Phosphate buffer saline solution (PBS): 10 mM, pH 7. Add 10 mL of commercial 1X PBS solution to 90 mL of DW in a glass beaker and mix it well. Then adjust its pH using 1 M HCL and/or 1 M NaOH.
3. GOx solution: 0.25 mg/mL. Add 12.5 mg of GOx (Sigma Aldrich) to 50 mL of PBS (10 mM, pH 7) at 4 °C and mix it carefully.
4. HRP solution: 0.25 mg/mL. Add 12.5 mg of HRP (Sigma Aldrich) to 50 mL of PBS (10 mM, pH 7) at 4 °C and mix it carefully.
5. GOx-HRP solution: 0.25 mg/mL. Add 12.5 mg of HRP and 12.5 mg of GOx to 50 mL of PBS (10 mM, pH 7) at 4 °C.

6. Glucose solution: 20 mM. Add 0.36 g of glucose powder to 100 mL of 1× commercial PBS in a glass *Erlenmeyer flask* and mix it well using a magnetic stirrer for 20 min.
7. 2,2′-azino-bis(3-ethylbenzothiazoline-6-sulfonic acid (*ABTS)* solution: 1 mM. Add 0.051 g of ABTS to 100 mL of 1× commercial PBS in a glass *Erlenmeyer flask* and mix it well using a magnetic stirrer for 20 min.
8. Lysogeny *broth* (*LB*) medium with 1.2% glycerol: add 12 mL glycerol to a glass *beaker and mix it with 988 mL of LB.*
9. IPTG solution: 100 mM. Add 0.002 g of IPTG to 100 mL of DIW and mix it well.
10. HEPES buffer: 100 mM, pH 7.4. Add 2.3 g HEPES powder to 100 mL DIW in a glass beaker and stir it using a magnetic stirrer for 30 min. Then adjust its pH using NaOH (1 M).
11. HEPES buffer containing $CaCl_2$: 100 mM HEPES and 1 mM $CaCl_2$. Add 0.11 g $CaCl_2$ to 100 mL of 100 mM HEPES and stir it for 30 min.
12. Amplified dockerin from the *C. thermocellum CelS by* IF-CTDoc and IRCTDoc primers (*see* **Note 1** for primers sequences), dockerin from *C. cellulovorans EngE using* IF-CC-Doc/IR-CCDoc, and dockerin from *R. flavefaciens ScaA using* IF-RFDoc/IR-RFDoc. (GenBank Accession number: L06942, AAD39739.1, CAC34384.3 respectively).
13. Cohesins from *C. thermocellum* (ATCC 27405 CipA, *C. cellulovorans* ATCC 35296 CbpA, and *R. flavefaciens ScaB* (cohesins CTCoh, CCCoh, and RFCoh) (Piscataway, NJ).
14. Plasmids and Hosts.
    (a) Alter pET20b plasmid (Novagen, Madison, WI) to flowing ones:
        - A plasmid with one CBM3 module and three cohesin modules from *C. thermocellum*, *C. cellulovorans,* and *R. flavefaciens* in tandem (pET20b-mini-scaf:) (Piscataway, NJ, USA)).
        - A plasmid with the gene of TIM from *T. thermophilus* HB27 and a dockerin module from *C. thermocellum* CelS (pET20b-tim-ctdoc).
        - A plasmid with the gene of ALD from *T. maritima* and a dockerin module from endoglucanase EngE of *C. cellulovorans* (pET20bald- ccdoc).
        - A plasmid with the gene of fructose 1,6-bisphosphatase from *T. maritima* and a dockerin domain from the *R. flavefaciens* ScaA (pET20b-fbp-rfdoc).
        - A plasmid with *tim* gene (pET20b-tim).

- A plasmid with *ald* gene (pET28a-ald) (provided by Dr. J.J. Zhong of East China University of Science and Technology (ECUST)).
- *Escherichia coli Top10 (F- mcrA Δ(mrr-hsdRMS-mcrBC) φ80lacZΔM15 ΔlacX74 nupG recA1 araD139 Δ(ara-leu)7697 galE15 galK16 rpsL(StrR) endA1 λ)* (Invitrogen, Carlsbad, CA).
- *E. coli* BL21 Star (F-*ompT* *hsd*SB (rB-mB-) *gal dcm rne131* (DE3)) and E. coli Rosetta (DE3), W3110, DP10 (Invitrogen, Carlsbad, CA).

15. HPLC (Thermo Finnigan) equipped with a stainless-steel column Hypersil ODS C18 (150 × 4.6 mm I.D., particle size: 5 mm) for resveratrol separation.
16. Agilent Technologies chiral cyclosil-B column to determine extracted mevalonate.

3 Methods

3.1 Compartmentalized Multi-enzyme Immobilization on Inorganic Complexes

1. Prepare HRP incorporated complex by adding 1.5 mL of the $CuSO_4$ solution to 50 mL of the HRP solution. Then incubate the final solution at 4 °C for 24 h.
2. Incubate HRP incorporated complex in the GOx solution at 4 °C 24 h (*see* **Note 2**).
3. Centrifuge obtained solution and wash with DIW to separate the GOx@HRP complexes (GOx@HRP) from unattached or loosely bonded enzymes.
4. Add 1.5 mL of the $CuSO_4$ solution to 50 mL of the GOx solution and keep it at 4 °C overnight.
5. Incubate obtained GOx incorporated complex to 50 mL of HRP solution and incubate it at 4 °C for 24 h.
6. Centrifuge and wash the solution to obtain HRP@GOx complex (HRP@GOx).
7. Add GOx-HRP solution to 1.5 mL of the $CuSO_4$ solution and incubate it for 24 h at 4 °C. Then centrifuge and wash it to make GOx-HRP randomly enzyme complex as control (*see* **Note 3**).
8. Use electron microscopies for investigation of enzyme complexes formation and shapes and high-resolution transmission electron microscopy for visualization of crystal structure.
9. Calculate enzyme concentration by a standard Bicinchoninic Acid (BCA) method [15].

10. Check enzyme activities in all three complexes (i.e., GOx@HRP, HRP@GOx, and GOx–HRP) using glucose and ABTS as the substrates.
11. Add glucose (20 mM) and ABTS (1 mM) in PBS for enzyme activities. Then add the same concentrations of GOx@HRP, HRP@GOx, and GOx–HRP complexes to substrate solution. Incubate enzymes complexes and substrate for 5 min in RT. After that read out absorbance of the solutions at 450 nm using microplate reader.

3.2 Protein Scaffold for Enzyme Compartmentalization

1. Use two pairs of primer (IF-scaf and IR-scaf) and (VF-scaf and VR-scaf) to amplify scaffold sequence and the pET20b vector backbone, respectively, to construct pET20b-mini-scaf. Then assemble two DNA templates using overlap extension PCR.
2. Amplify the *tim* gene by a primer pair of IF-TIM and pET20b vector backbone by a primer pair of VF-TIM and VR-TIM and merge it by other fragment using simple cloning to construct t pET20b-tim-ctdoc.
3. Amplify the *ald* gene, the dockerin domain, and the vector backbone using a primer pair of IF-ALD/IR-ALD, IF-CC-Doc/IR-CCDoc and VF-ALD and VR-ALD, respectively, to construct pET20b-ald-ccdoc and merge all three fragments by simple cloning.
4. Amplify the FBP, the dockerin domain and the vector backbone using a primer pair of IF-FBP/IR-FBP, IF-RFDoc/IR-RFDoc, and VF-FBP/VR-FBP, respectively, and merge all three fragments by simple cloning to construct pET20b-fbp-rfdoc.
5. Amplify the sequence of TIM by a primer pair of IF-TIM and TIM-XhoI-R, then digest the PCR product with NdeI /XhoI and ligate it into NdeI/XhoI-digested pET20b to construct pET20b-tim.
6. *Culture E. coli (*BL21 Star (DE3)) containing the protein expression plasmids in the LB medium with glycerol at 37 °C for 24 h.
7. When A600 of the culture medium reaches to 0.75, add IPTG solution as inducer and drop the medium temperature to 16 °C for 16 h.
8. Centrifuge the culture and suspend obtained pellets in a 100 mM HEPES buffer containing 1 mM $CaCl_2$ and lyses the cells by ultrasonication.
9. Use 10 μL of the cell extracts in 12% SDS-PAGE to investigate expression of all proteins.
10. Purify His-tag containing protein by Ni-NTA resin.

11. Add 20 mL of the cell lysate supernatant containing TIMCT-DoC, 50 mL of the cell lysate supernatant of ALD-CCDoc and 10 mL of the cell lysate supernatant of FBP-RFDoc to 6 mL of the cell lysate supernatant of mini scaffoldin.
12. Add 100 mg RAC (regenerated amorphous cellulose) to enzyme and scaffoldin mixture. Centrifuge the solution at 1957 × *g* for 10 min and wash the RAC pellet three times in 20 mL of 100 mM HEPES (pH 7.5) containing 1 mM $CaCl_2$.
13. Resuspend 2000 μg of pellet (i.e., immobilized enzyme on RAC) in 40 μL of SDS loading buffer and load it into 12% SDS-PAGE to investigate the interaction of mini-scaffoldin and dockerin containing enzymes.
14. Protein concentration in scaffoldin can be calculated by Bradford method (Bio-Rad, Hercules, CA).
15. Check the activity of the TIM, ALD, and FBP using standard methods as published previously [14].
16. Examine the purity of protein samples by 12% SDS-PAGE ((Bio-Rad, Hercules, CA).

3.3 Nucleic Acid Scaffold for Enzyme Compartmentalization

3.3.1 Pathway Plasmid Construction

1. Link the genes encoding 4CL (from *A. thaliana*) and STS (from *V. vinifera*) to the 3′-end of the genes encoding Zif268 and PBSII, respectively (GeneArt).
2. Use GGSGGGSGGS polypeptide linker to the enzyme separation from the ZF domain.
3. Use overlap extension PCR to fuse enzymes and zinc fingers.
4. Digest Zif268-4CL PCR product using XbaI /ApaI and PBSII-STS PCR product using ApaI/BamHI.
5. Construct pET-Res-ZF-Enz plasmid by XbaI/BamHI digested pET19b vector and previously produced fragments.
6. Construct pET28a plasmid that contains 4CL-STS fusion protein using previously described method [16].
7. Extract and amplify the genes encoding MgsA, DkgA, and GldA from E. coli MG1655 and clone it in pBAD18 [6] (*see* **Note 4**).
8. Attach OZ052 (ZFa), OZ300 (ZFb), and OZ076 (ZFc) to the 3′-end of mgsA, dkgA, and gldA respectively [17] and append each ZF with a C-terminal HA epitope tag (pBAD-PD-ZF-Enz plasmid).
9. Fuse ZFa to AtoB, ZFb to HMGS, and ZFc f to HMGR, with an HA epitope tag on the C-terminus of all proteins (pTet-Mev-ZF-Enz plasmid).
10. Anneal primers containing Zif268 or PBSII binding sites with 2, 4, and 8 bp spacer and standard Biobrick restriction sites with 10 min incubation at 95 °C and slow cooling to RT.

11. Assemble DNA scaffold by standard Biobrick assembly [18] and clone it into the high copy pSB1K3 vector (http://partsregistry.org/Part:pSB1K3). The DNA scaffold composed of pUC19 for ZFa, ZFb, and ZFc binding sites (*see* **Note 5**).
12. Digest backbone of the plasmid by XbaI and then ligate it by SpeI/XbaI-digested insert at the 3′-end for construction of scaffold plasmid (have the SpeI site at the 5′-end and XbaI site at the 3′-end for future ligations).
13. Use 4- or 12-bp spacer for separation of binding sites.
14. Subclone the obtained scaffolds into pBAD18 (between the b-lactamase and PBAD promoters) and use the scaffold directly from the pUC19 plasmid to produce mevalonate.
15. Culture host (*see* **Note 6**) in 2xYT medium overnight and dilute the culture to reach OD of 0.2 at 600 nm and keep in shake flasks at 30 °C and 160 rpm.
16. When A600 of the culture reaches 0.8, add IPTG to induce gene expression and coumaric acid (0.3 mM) as the substrate to the culture medium.
17. Analyze ZF-enzyme chimeras induction after 6 h of incubation.
18. Setup 1,2-PD anaerobic fermentation [19] to optimize (E. coli W3110 harboring the pBAD-PD-ZF-Enz plasmid encoding the ZF-enzyme chimeras and corresponding DNA scaffolds) growth and at the time of inoculation, add L-Arabinose (0.2% w/v) to medium to induce gene expression.
19. Collect samples after 9 h and 24 of induction, and analyze them via western blot and calculate fermentation yield.
20. For the production of Mevalonate (E. coli DP10 cells harboring the pTet-Mev-ZF-Enz plasmid and a pUC19-based DNA scaffold) use anhydrotetracycline (250 nM) as an inducer [20] (*see* **Note** 7).
21. Separate cells from the supernatant by centrifugation at 13,226 × *g* for 20 min. Then, acidify 1 mL of the supernatant by HCl (1 M) and extract resveratrol using ethyl acetate [21].
22. Use a vacuum evaporator to remove ethyl acetate and suspend the remaining pellet in methanol.
23. Check the identity of resveratrol by MS analysis.
24. Use immunoblot technique for study of the compartmentalized enzyme.
25. For immunoblotting study, harvest E. coli Rosetta (DE3) cells that expressed the 4CL-STS fusion after 6 h, W3110 cells expressing the 1,2-PD ZF-enzyme chimeras after 9 h and DP10 cells expressing the mevalonate ZF-enzyme chimeras after 25 h by centrifugation (13,000 rpm for 20 min). Then

resuspend cells in 1× PBS and lyse cells by sonication and centrifuge the solution at 20,035 × *g* for 10 min. After that, use anti-His antibodies (Sigma) to detect Zif268-4CL, PBSII-STS, or 4CL-STS chimeras and use anti-HA antibodies (Sigma) for detection of 1,2-PDor mevalonate-related chimeras in soluble lysates.

4 Notes

1. Primers sequences

VF-Scaf	TACTGATGGCGCTATCATTGTTAAGC TCGAGCACCACCACCACCACCACT
VR-Scaf	CCTTCAAATTGCCTGATACCGGCATATGTATATCTCC TTCTTAAAGTTAA
IF- Scaf	TTAACTTTAAGAAGGAGATATACATATGCCGGTA TCAGGCAATTTGAAGG
IR-Scaf	AGTGGTGGTGGTGGTGGTGCTCGAGCTTAACAATGA TAGCGCCATCAGTA
VF-TIM	AATAGATACATTGCCGTACAAGAACTAGGGATCCGAA TTCGAGCTCCGTC
VR-TIM	AGTTGCCCGCCACCAAGACGCGCCGCATATGTATATC TCCTTCTTAAAGT
IF-TIM	ACTTTAAGAAGGAGATATACATATGCGGCGCGTC TTGGTGGCGGGCAACT
IR-TIM	AAGCAGTTGCCTTGTTCCAATCAACACCCGCGATCC TGAGGAGGGCGAGG
IF-CTDoc	CCTCGCCCTCCTCAGGATCGCGGGTGTTGA TTGGAACAAGGCAACTGCTT
IR-CTDoc	GACGGAGCTCGAATTCGGATCCCTAGTTCTTG TACGGCAATGTATCTATT
TIM-XhoI-R	GTCCACTCGAGACCCGCGATCCTGAGGAGGGCGAGG
VF-ALD	TCTTGCATTCTTAAAAAAAGCAATATGAG TCGACAAGCTTGCGGCCGCAC
VR-ALD	TCTTTACATAAGGCATTGTCACCATATGTATATCTCC TTCTTAAAGTTAA
IF-ALD	TTAACTTTAAGAAGGAGATATACATATGGTGACAA TGCCTTATGTAAAGA
IR-ALD	CTTGGATTCCTGGTAATACCTTACCCGCCTTTCCGC TGGATCCGAATATT

(continued)

IF-CCDoc	AATATTCGGATCCAGCGGAAAGGCGGGTAAGGTATTACCAGGAATCCAAG
IR-CCDoc	GTGCGGCCGCAAGCTTGTCGACTCATATTGCTTTTTTTAAGAATGCAAGA
VF-FBP	CGTTGAACTCATCACACTTCCTCAGCTCGAGCACCACCACCACCACCACT
VR-FBP	TAGAAAAGTCCAGTCTGTCCAACATATGTATATCTCCTTCTTAAAGTTAA
IF-FBP	TTAACTTTAAGAAGGAGATATACATATGTTGGACAGACTGGACTTTTCTA
IR-FBP	GGAACGAGCTTTGTGCCGGGCTTTCCTCCTATTTCTTCTA
IF-RFDoc	TAGAAGAAATAGGAGGAAAGCCCGGCACAAAGCTCGTTCC
IR-RFDoc	AGTGGTGGTGGTGGTGGTGCTCGAGCTGAGGAAGTGTGATGAGTTCAACG

2. Presence of Cu^{2+} in the solution cause adsorption of GOx on HRP-incorporated complex.
3. All three complexes including GOx@HRP, HRP@GOx, and GOx-HRP should be prepared in the same conditions and have an equal concentration of enzymes. It is necessary to keep all enzymes containing solutions at 4 °C.
4. The mgsA, dkgA, and gldA genes are placed between NheI and XbaI, XbaI and SphI, and SphI and HindIII, respectively.
5. pUC19 is completely digested with AatII and PvuII and replaced with a polylinker containing AatII–SacI–SpeI–XbaI–SphI–ClaI–PvuII sites. Then the scaffolds are assembled using cohesive ends of SpeI and XbaI.
6. Resveratrol is produced in E. coli Rosetta (DE3) pLysS harboring pET-Res-ZF-Enz encoding the Zif268-4CL and PBSII-STS chimeras or pET-Res-Enz encoding 4CL and STS without the ZF domains in the presence of a DNA scaffold in plasmid pSB1K3.
7. Add different antibiotics including ampicillin, 100 mg/mL; chloramphenicol, 25 mg/mL; and kanamycin, 50 mg/mL to the medium and after 25 and 50 h of induction, samples should be collected and analyzed via western blot.

References

1. Marguet M, Bonduelle C, Lecommandoux S (2013) Multicompartmentalized polymeric systems: towards biomimetic cellular structure and function. Chem Soc Rev 42:512–529
2. Wang L-B, Wang Y-C, He R et al (2013) A new nanobiocatalytic system based on allosteric effect with dramatically enhanced enzymatic performance. J Am Chem Soc 135:1272–1275

3. Ge J, Lei J, Zare RN (2012) Protein–inorganic hybrid nanoflowers. Nat Nanotechnol 7:428–432
4. Li Z, Zhang Y, Su Y et al (2014) Spatial co-localization of multi-enzymes by inorganic nanocrystal–protein complexes. Chem Commun 50:12465–12468
5. Schmitt C, Lippert AH, Bonakdar N et al (2016) Compartmentalization and transport in synthetic vesicles. Front Bioeng Biotechnol 4:19
6. Liu Z, Wang B, Jin S et al (2018) Bioinspired dual-enzyme colloidosome reactors for high-performance biphasic catalysis. ACS Appl Mater Interfaces 10:41504–41511
7. Mason AF, Yewdall NA, Welzen PLW et al (2019) Mimicking cellular compartmentalization in a hierarchical protocell through spontaneous spatial organization. ACS Cent Sci 5: 1360–1365
8. Schmid-Dannert C, López-Gallego F (2019) Advances and opportunities for the design of self-sufficient and spatially organized cell-free biocatalytic systems. Curr Opin Chem Biol 49: 97–104
9. Chen AH, Silver PA (2012) Designing biological compartmentalization. Trends Cell Biol 22:662–670
10. Li M, Yue Y, Zhang Z-J et al (2016) Site-specific and high-loading immobilization of proteins by using cohesin–dockerin and CBM–cellulose interactions. Bioconjug Chem 27:1579–1583
11. You C, Myung S, Zhang Y-HP (2012) Facilitated substrate channeling in a self-assembled trifunctional enzyme complex. Angew Chem Int Ed 51:8787–8790
12. Wilner OI, Weizmann Y, Gill R et al (2009) Enzyme cascades activated on topologically programmed DNA scaffolds. Nat Nanotechnol 4:249–254
13. Wang Y, Kim E, Lin Y et al (2019) Rolling circle transcription-amplified hierarchically structured organic–inorganic hybrid RNA flowers for enzyme immobilization. ACS Appl Mater Interfaces 11:22932–22940
14. Conrado RJ, Wu GC, Boock JT et al (2012) DNA-guided assembly of biosynthetic pathways promotes improved catalytic efficiency. Nucleic Acids Res 40:1879–1889
15. Arnal, G., Attia, M. A., Asohan, J., & Brumer, H. (2017). A low-volume, parallel copper-bicinchoninic acid (BCA) assay for glycoside hydrolases. In Protein-Carbohydrate Interactions (pp. 3–14). Humana Press, New York, NY
16. Zhang Y, Li S-Z, Li J et al (2006) Using unnatural protein fusions to engineer resveratrol biosynthesis in yeast and mammalian cells. J Am Chem Soc 128:13030–13031
17. Maeder ML, Thibodeau-Beganny S, Osiak A et al (2008) Rapid "open-source" engineering of customized zinc-finger nucleases for highly efficient gene modification. Mol Cell 31:294–301
18. Shetty RP, Endy D, Knight TF (2008) Engineering BioBrick vectors from BioBrick parts. J Biol Eng 2:5
19. Altaras NE, Cameron DC (1999) Metabolic engineering of a 1,2-propanediol pathway in Escherichia coli. Appl Environ Microbiol 65: 1180–1185
20. Dueber JE, Wu GC, Malmirchegini GR et al (2009) Synthetic protein scaffolds provide modular control over metabolic flux. Nat Biotechnol 27:753–759
21. Beekwilder J, Wolswinkel R, Jonker H et al (2006) Production of resveratrol in recombinant microorganisms. Appl Environ Microbiol 72:5670–5672

Chapter 10

3D Printed Polylactic Acid Well-Plate for Multi-enzyme Immobilization

Elena Gkantzou, Anastasia Skonta, Andreas-Georgios Vasios, and Haralambos Stamatis

Abstract

3D printing is lately utilized in biological sciences under the scope to develop customized scaffolds that will host biomolecules, either whole cells or parts of them, like enzymes. In the present work, we present a protocol to modify the surface of 3D printed polylactic acid (PLA) well-plates with the aim to co-immobilize multiple enzymes that will perform cascade reactions. Detailed steps to design and print the final models are described. The developed protocol for surface modification is based on coating with chitosan biopolymer and covalent immobilization of the enzymes β-glucosidase, glucose oxidase, and peroxidase via glutaraldehyde cross-linking. Enzymatic activity measurements indicative of the catalytic performance of the system are also presented.

Key words 3D design, 3D printing, Polylactic acid, PLA well-plate, Surface modification, Chitosan, Multi-enzyme immobilization

1 Introduction

3D printing is a layer-by-layer manufacturing process based on computer data. 3D printing techniques present high reproducibility and control, facile modifications of the product and convenient, cost-effective manufacture [1]. Biological sciences have welcomed 3D printing technologies in several fields such as regenerative medicine, diagnosis, and implants [2]. Polylactic acid (PLA) is one of the most versatile materials used in bio-related applications of 3D printing since it is 100% biocompatible and biodegradable [3]. Surface modification methods are recently under investigation in order to make PLA hydrophilic and hence receptive to biomolecules immobilization [4]. There have been already published works with cells immobilization on the surface of 3D printed PLA and fewer studies with enzymes immobilization on modified PLA [5, 6].

Haralambos Stamatis (ed.), *Multienzymatic Assemblies: Methods and Protocols*,
Methods in Molecular Biology, vol. 2487, https://doi.org/10.1007/978-1-0716-2269-8_10,
© The Author(s), under exclusive license to Springer Science+Business Media, LLC, part of Springer Nature 2022

PLA modification with chitosan for cells immobilization was first published in 2018 by Wang et al. via an entrapment method. Based on this concept, we took advantage of the functional amino groups on the surface of chitosan to covalently immobilize enzymes on PLA-chitosan scaffolds via the bi-functional glutaraldehyde cross-linker. We also introduced an initial step in the protocol to treat the surface of PLA with alkali solution, a practice that has been shown to greatly enhance the hydrophilicity of 3D printed PLA objects [7, 8].

Concerning the model we designed, we named it PLA well-plate mimicking the concept of a typical microplate used for analysis. In this chapter, we describe the design of a 9-well model, but the user can easily add the desired number of wells once he gets familiar with the designing rationale. Each well serves as a scaffold to immobilize enzymes with a final application for biocatalytic cascade reactions. The concept of well-plates can be used for easy and fast screening of several parameters affecting the immobilization or biocatalytic behavior of the enzymes.

2 Materials

2.1 Equipment

1. Fused Deposition modeling (FDM) 3D printer (*see* **Note 1**).
2. Analytical balance.
3. pH meter.
4. Hotplate magnetic stirrer.
5. Vortex mixer.
6. Incubator for 25 °C, 30 °C, and 40 °C.
7. Elisa plate reader.
8. Vacuum pump and silica-containing vessel.
9. Eppendorf tubes of 1 mL and 2 mL capacity.
10. Micropipettes of 200 μL and 1000 μL.

2.2 Three-Dimensional (3D) Design and Printing of Polylactic Acid (PLA) Well-Plates

1. 3D design software: Autodesk Fusion 360 (*see* **Note 2**).
2. Slicing software: Ultimaker Cura (*see* **Note 3**).
3. Printing material: polylactic acid (PLA) filament for FDM 3D printers (*see* **Note 4**).

2.3 Surface Modification and Enzyme Immobilization on 3D Printed PLA Well-Plates

All stock solutions should be used at room temperature and stored at 4 °C until next use. Prepare all solutions using double-distilled water.

1. 1 M sodium hydroxide (NaOH) aqueous solution (10 mL stock solution): Dissolve 0.4 g NaOH in 10 mL water using a hotplate magnetic stirrer.

2. 0.2% w/v chitosan solution in 1% acetic acid aqueous solution (10 mL stock solution): Add 20 mg of powdered chitosan in 9.9 mL water. Use a hotplate magnetic stirrer to gently mix the solution while adding 100 μL acetic acid. Adjust the temperature around 50 °C and leave under stirring for 30 min or complete dissolution.
3. 0.2 M solution of monobasic sodium phosphate (NaH_2PO_4) (1000 mL stock solution): Dissolve 27.8 g of NaH_2PO_4 monohydrate in 1000 mL water. Put under gentle stirring and around 50 °C until complete dissolution.
4. 0.2 M solution of dibasic sodium phosphate (Na_2HPO_4) (1000 mL stock solution): Dissolve 35.4 g of Na_2HPO_4 dihydrate in 1000 mL water. Put under gentle stirring at a hotplate magnetic stirrer around 50 °C until complete dissolution.
5. 50 mM sodium phosphate buffer pH 7 (100 mL stock solution): Dilute 9.75 mL of 0.2 M NaH_2PO_4 solution and 15.25 mL of 0.2 M Na_2HPO_4 solution in a total of 100 mL aqueous solution. Gently mix in a magnetic stirrer at room temperature and adjust the pH of the solution at 7 using a pH meter and alkaline (NaOH) or acidic (HCl) solutions, respectively (*see* **Note 5**).
6. 10% glutaraldehyde (GA) solution (2 mL stock solution) (*see* **Note 6**): Add 800 μL of GA reagent 25% in 1.2 mL sodium phosphate buffer pH 7. Mix well in a vortex mixer.
7. Enzyme commercial preparations: *β*-glucosidase from almonds, glucose oxidase from *Aspergillus niger*, peroxidase from horseradish.
8. 50 mM HEPES buffer pH 6 (stock solution 100 mL): Dissolve 1.2 g HEPES (4-(2-hydroxyethyl)-1-piperazineethanesulfonic acid) in 100 mL water. Gently mix in a magnetic stirrer at room temperature and adjust the pH of the solution at 6 using a pH meter and alkaline (NaOH) or acidic (HCl) solutions, respectively.
9. Multi-enzymes solution (2 mL stock solution): Weigh 4 mg of *β*-glucosidase, 2 mg of glucose oxidase, and 6 mg of peroxidase and mix the amounts by dissolving them in a total of 2 mL 50 mM HEPES buffer pH 6 (*see* **Note 7**).

2.4 Enzymatic Activity Measurement in the PLA Well-Plates

1. Substrate reagents: *p*-NPG (4-Nitrophenyl β-D-glucopyranoside) and ABTS (2,2′-azino-bis(3-ethylbenzothiazoline-6-sulfonic acid)).
2. 50 mM sodium phosphate buffer pH 6 (100 mL stock solution): Follow the same procedure described at Subheading 2.2, **item 5** and adjust the pH of the solution at 6 instead of 7.

3. 20 mM *p*-NPG stock solution (1 mL stock solution): Dissolve 6 mg of *p*-NPG reagent in 1 mL sodium phosphate buffer pH 6. Mix well in a vortex mixer.
4. 20 mM ABTS stock solution (1 mL stock solution): Dissolve 11 mg of ABTS reagent in 1 mL sodium phosphate buffer pH 6. Mix well in a vortex mixer. Keep the solution under dark during experiments.
5. Substrates solution (2 mL solution): Add 200 μL *p*-NPG 20 mM and 300 μL ABTS 20 mM in an Eppendorf tube (final concentrations: 2 mM *p*-NPG, 3 mM ABTS). Add 1500 μL sodium phosphate buffer pH 6 and mix well in a vortex mixer. Keep the solution under dark.

3 Methods

3.1 3D Design of a PLA Well-Plate

3.1.1 2D Sketch Design

When working with a 3D design software it is advisable to create all your basic structures in the 2D sketch palette. This way you will have a sketch to guide you when working in the 3D design space.

1. Hit the *Create sketch* button and choose XY plane (red-green).
2. On the sketch surface set grid settings (bottom bar) on the fixed mode with a major grid spacing 10 mm. Hit *Close* on the grid settings panel once you finish setting parameters.
3. Hit *2-point rectangle* from the CREATE panel and set the dimensions to 30 mm for both sides of the rectangle. Place each rectangle corner in a way to include 9 squares of the sketch surface.
4. Hit the *Line* button from the CREATE panel and create 2 vertical and 2 horizontal lines inside your original rectangle in a way to divide your sketch in 9 equal (10 mm × 10 mm) squares. Right click on each line and choose *Normal/Construction*. Now the lines should appear in a dotted form (*see* **Note 8**).
5. Hit the *2-point rectangle* button from the CREATE panel and create 9 equal (10 mm × 10 mm) squares in the interior of your original (30 mm × 30 mm) square. Construction lines that you created in the previous step will guide you to complete this step.
6. Hit the *Point* button from the CREATE panel and create a point in the center of each sub-square you designed in the previous step. Starting from the corner of each square the software will guide you to find the center of each square.
7. Hit *Finish Sketch* on the sketch palette (on the right of the sketch surface).

3.1.2 3D Model Design

Once you finish the sketch design you will be transferred to the 3D design workspace. Here you will find all the necessary tools to transform your sketch into a 3D model.

1. Select the whole surface of your sketch by left-clicking your mouse. If you have selected all the elements of your sketch, they must appear in bold blue lines. With the surface selected, hit the *Extrude* button from the CREATE panel. A blue arrow appears that shows you the direction of the extrusion. Next to the arrow, there is a dimensions box where you can insert the value of the extrusion you desire in mm. Insert value 12 and hit enter (*see* **Note 9**).
2. Choose the back side of your model by hitting on the guidance cube on the upper right corner of your workspace (*see* **Note 10**). Now you should have a planar view of your model with your original sketch. Hit the *Hole* button from the CREATE panel. Choose *At point (single hole)* in the placement status from your parameters panel on the right and move your mouse to the central sketch point of one sub-square on your model. Once you left-click on the sketch point a red-colored hole will appear on your model (the red area indicates where the model will be cut or intersected). Refer again to the parameters panel and choose *simple* for Hole Type, *simple* for Hole Tap Type, and *Flat* for Drill Point. Two boxes for inserting dimensions must have now appeared in the parameters panel. Insert 11.4 for length and 5.2 for diameter size. If this step has been executed properly, *Body 1* must be automatically selected in the section Objects to cut. Hit *OK* in the parameters panel.
3. Repeat the above step for each central point of your sub-squares. Once you complete this **step 9** wells must have been created in the interior of your main model.
4. Hit the *Create Sketch* button from the CREATE panel. Choose the front side of your model (the opposite side from the one you created the holes). The directions box in the upper right corner of the workspace will help you find your way in the different sides of the model. Once you choose the desired plane for your sketch, you will be transferred to the sketching workspace. Hit the *Line* button and draw a line starting from the upper left corner of the original rectangle. Two boxes for inserting dimensions values must have appeared. Insert 2 mm and 135°, respectively, using the Tab button on your keyboard. Repeat this step and draw a line on the lower right corner of your model, with 2 mm and 45° values. Hit the *2-point rectangle* button and draw a rectangle starting from the upper left line and finishing at the lower right line you draw previously on this step. Now you must have drawn a rectangle concentric with

your original rectangle that will be used as a base to support your model. Hit *Finish sketch* on your sketch palette.

5. On your 3D design workspace, hit the *Extrude* button and choose the rectangle you designed in the previous step. Insert 2 mm in the dimensions box and make sure that the chosen features in the parameters panel on your right are the following: Type *Extrude*, Start *Profile Plane*, Direction *One Side*, Extent Type *Distance*, Distance *2 mm* (this is the value you inserted in the dimensions box), Taper Angle *0.0°*, Operation *Join* (*see* **Notes 11** and **12**). After finishing this step you must have created a support base for your well-plate.
6. On your upper left corner of the design workspace choose File → 3D print. A parameters panel will appear on your right. Left-click anywhere on your model in order to select it for printing (the selected model will turn blue). On the parameters panel, Send to 3D print Utility box must be selected. For the Print Utility choose *Custom* and hit on the folder next to Application in order to find the slicing software application on your computer files. Once you find the path for the application you chose, the name of the software will appear next to Application parameter. Hit *OK* on the parameters panel.

3.2 3D Printing

To this point, they were described all the necessary steps to design a well-plate 3D model. The 3D design software redirects the user to a slicing software. This software is used to provide all the necessary instructions for the printer. The user can define several features of the printing process concerning both the model morphology and the printer's function, like extrusion temperature and build plate temperature. The morphology of the model lies on features like infill density, support structures for floating parts of the model and additional rafts that will ensure the adhesion of the model on the printer's platform.

1. By selecting the 3D print option in the design software (Fusion 360) you will be redirected to a new window of the slicing software (Cura). Probably the model will appear reversed (you constructed the base on the front side of the model). Once you left-click on the model surface different options of the bar on your left will be enabled. Choose the rotation tool. Three options will appear, choose the third one (*Select face to align to the build surface*) and left-click on the base of your model. Now the model must appear reversed from the original view (you will be able to see the holes). Alternatively, you can use the colored arrows that appear around the model and move it the way you desire to be printed (*see* **Note 13**).
2. On the print settings panel (*see* Fig. 1) choose *Super quality* as a Profile, *80%* as an Infill Density, *210 °C* as Printing

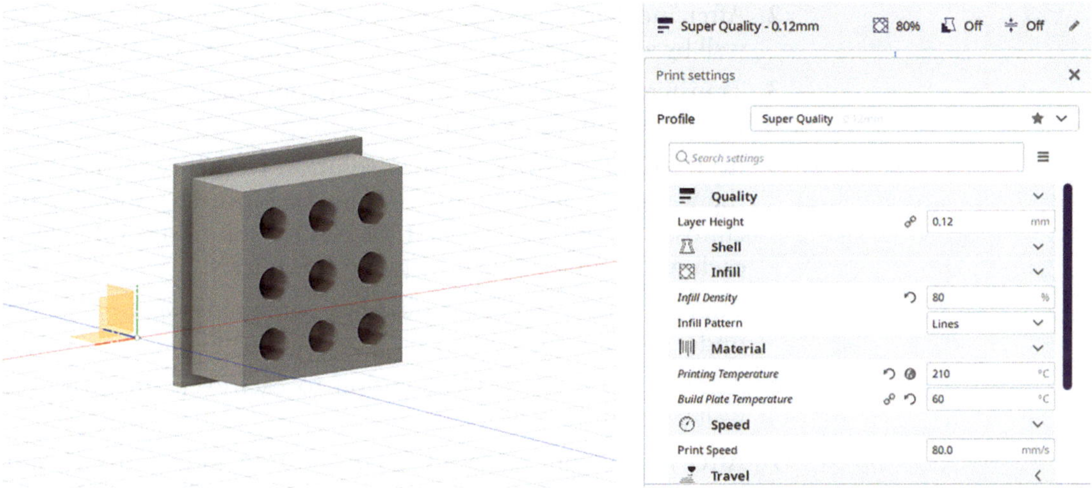

Fig. 1 Final model illustrated at the design software (left) and print settings panel at the slicing software (right)

Temperature, *60 °C* as Build Plate Temperature, *80 mm/s* as a Print Speed and *Skirt* as Build Plate Adhesion Type (*see* **Note 14**).

3. Hit the *Slice* button (lower right corner). Once the slicing process is completed (it takes some seconds depending on the complexity of each model) it will appear an information box that depicts the estimated time for the printing process and the amount of material that will be consumed (*see* **Note 15**). In this box, you have the option to save the slicing file (STL) in an external device in order to transfer it to a 3D printer or to save it on your computer to use it later. You can also rename the file by clicking on the panel on the lower left of the window, where you will find the list of the objects that you have sliced.
4. Once you have saved the STL file in an external device (or you have a direct connection of the printer with your computer) you can transfer the file to the 3D printer. There you will choose the model you desire to print from the control panel of the printer and the printing process will automatically start since all the necessary parameters were recorded in the STL file. You only need to feed the printer with filament and check that the printing platform is properly leveled.

3.3 Surface Modification and Enzyme Immobilization on 3D Printed PLA Well-Plates

One of the advantages of FDM 3D printing is that the models need no post-printing treatment and thus they are ready to use. Once printing is complete, leave the model to cool down to room temperature, remove it from the printer and transfer it to the laboratory bench in order to proceed with the experimental part.

1. Add 200 μL of 1 M NaOH solution to each well. Incubate the well-plate at 25 °C for 2 h.

2. After incubation, remove the NaOH solution and wash each well by adding 200 μL water and pipetting quickly. Wash twice.
3. Transfer the well-plate to a silica-containing vessel and dry under vacuum for 30 min.
4. Add 200 μL of 0.2% chitosan solution to each well. Incubate the well-plate at 25 °C for 10 min (*see* **Note 16**). Remove the chitosan solution and add 1 M NaOH solution for 1 min in each well, and remove it by pipetting quickly.
5. Transfer the well-plate to a silica-containing vessel and dry under vacuum for 30 min.
6. Add 200 μL of 10% GA solution to each well. Incubate the well-plate at 30 °C for 1 h.
7. After incubation, remove the GA solution and wash each well by adding 200 μL 50 mM sodium phosphate buffer pH 7 and pipetting quickly. Wash thrice.
8. Transfer the well-plate to a silica-containing vessel and dry under vacuum for 30 min.
9. Add 200 μL of multi-enzymes solution in each well, except for one which is going to be the blank measurement. Incubate the well-plate at 30 °C for 2 h.
10. After incubation, remove the enzymatic solution and wash each well by adding 200 μL 50 mM HEPES buffer pH 6 and pipetting quickly. Wash thrice.
11. Transfer the well-plate in a silica-containing vessel and dry under vacuum for 30 min.

3.4 Enzymatic Activity Measurement in the PLA Well-Plates

By completing the steps in the previous paragraph, you will have immobilized β-glucosidase, glucose oxidase, and peroxidase on the inner surface of each well of the 3D printed PLA well-plate. In this paragraph, it will be described how to monitor the activity of the 3-enzyme immobilized system at different time intervals, until the system yields no more product.

1. Mark each well for the different time intervals (*see* Fig. 2). Use a timer and set it at 120 min. Put the well-plate inside an incubator which is set at 40 °C. Fill the 120 min well and the blank well (*see* **Note 17**) with the substrates solution and start the timer. Fill each well when the timer reaches the specified minute. Notice that the samples need to be under dark during the reaction.
2. When the time is up, transfer the content of each well in an ELISA well-plate. Measure the absorbance of the different samples in an ELISA plate reader at 730 nm (*see* **Note 18**). The obtained values from the different reaction time intervals can then be plotted in a graph (Fig. 2) indicative of the enzymes catalytic behavior.

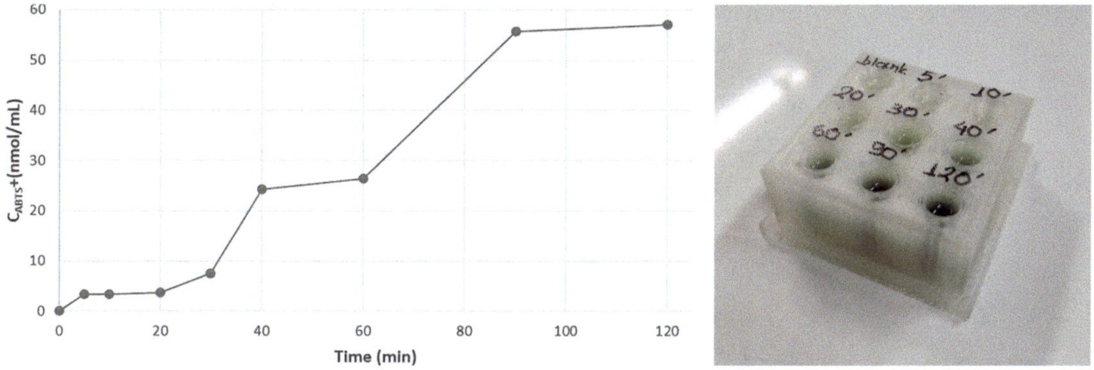

Fig. 2 (left) Absorbance measurements for the cascade reaction performed at different time intervals in each well of the well-plate. Each well bares the same amount of immobilized enzymes and has been modified under the same conditions. Concentration of $ABTS^+$ was calculated based on Beer–Lambert law with absorbance coefficient $\varepsilon = 36.8$/mM/cm. (right) 3D printed well-plate for cascade reactions at different time intervals

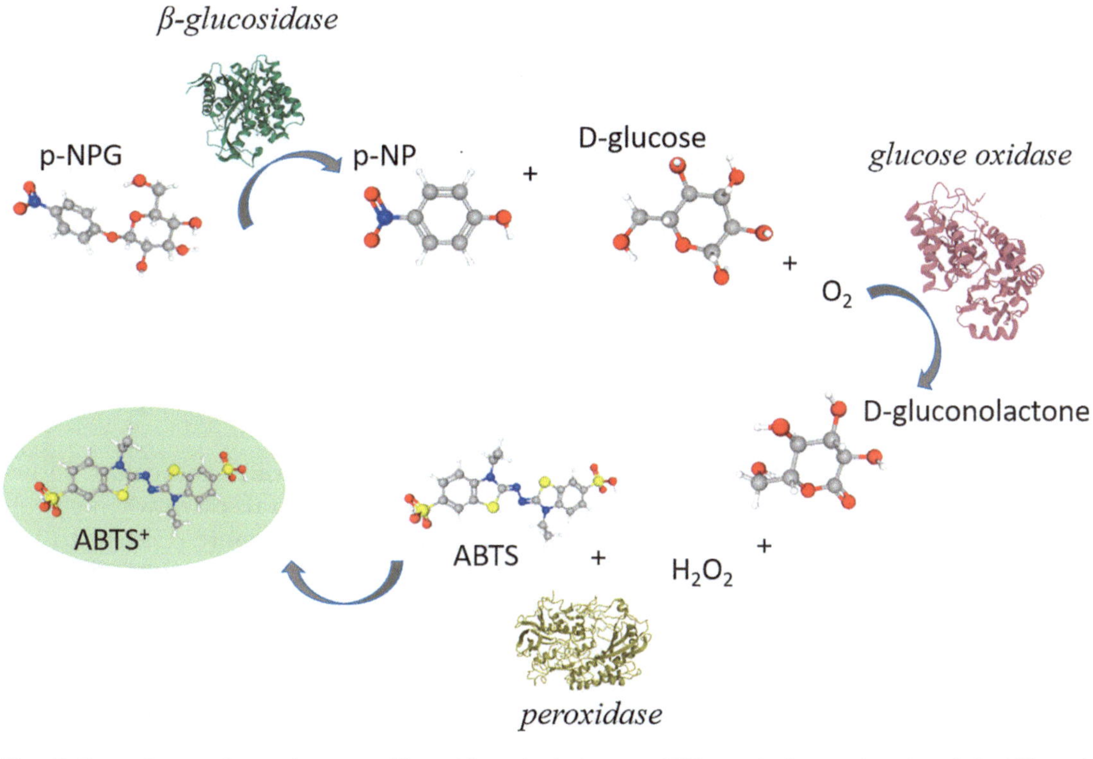

Fig. 3 Cascade reaction scheme. β-Glucosidase hydrolyzes *p*-NPG producing *p*-nitrophenol (*p*-NP) and D-glucose. Glucose oxidase uses atmospherical oxygen to oxidize D-glucose to D-gluconolactone. Peroxidase uses H_2O_2 produced from the action of glucose oxidase to oxidize ABTS. The final product, oxidized ABTS ($ABTS^+$) exhibits a characteristic green color

Figure 3 illustrates the reaction scheme for the multi-enzymatic system used in this chapter as a model reaction system (*see* **Note 19**). The described protocols offer the potential to design and 3D print well-plates with the desired number of positions to immobilize a variety of different enzymatic systems in order to perform a variety of screening measurements.

4 Notes

1. PLA is printable in almost every FDM 3D printer on the market. However, some printers appear more suitable for printing PLA in a way that they will guarantee minimum technical problems when printing consecutively for a long period of time. In the present work, Ender 5 from Creality was used.
2. There is plenty of 3D design software that can be used to come up with the same final model. Fusion 360 from Autodesk is available for free with a student license for 1 year, while Autodesk offers free-of-charge courses for beginners (see useful links).
3. There is also a variety of slicing software that will produce the same STL files for printing no matter the design software you will choose to work. However, choosing a slicer needs attention as to the compatibility with the printer that will be used. Usually, companies that produce 3D printers produce also slicing software that recognizes several 3D printers other than their own models. There are, however, some companies that produce printers that need their specialized slicer from the same company. So, it is advisable that, along with choosing a 3D printer, the appropriate slicer should also be taken under consideration. Cura from Ultimaker is available online free-of-charge and is compatible with a great variety of 3D printers (see Useful links).
4. PLA filaments come in a variety of textures and colors. However, natural PLA is usually preferred in research when chemicals are going to be employed. It is also more resistant to UV treatment and more appropriate for composting after use, compared to filaments with additives.
5. You will need a strong acid solution and/or a strong base solution in order to adjust the pH of the buffer solutions. We use 2 M aqueous HCl solution and 2 M aqueous NaOH solution.
6. Glutaraldehyde acts as a bi-functional cross-linker, reacting with both the amino groups of the enzymes and the support material. Hence, the concentration of the linker affects critically the enzyme loading, activity recovery, and operational stability of the biocatalytic system [9]. 10% was chosen as the appropriate concentration in this protocol based on previous studies from our group [10].

7. Enzymes mass ratio during random co-immobilization has been shown to have a crucial effect on the activity recovery of the co-immobilized system [10, 11]. In the present study, we use the 2:1:3 ratio for β-gluc, GOx, and HRP, respectively.
8. It is advisable to use *Construction* option when you draw lines that will serve only as a guide for you to draw normal lines, calculate dimensions, etc. Construction lines do not interfere with your main shape and are not recognizable from the program as part of the main model (so the model is not divided by these lines).
9. This is the first extrude that you will perform during your model design. This extrude will create a new body in your workspace and every action made from now on will be referenced on *Body 1*. The size of the extrusion (12 mm) as well as the size of the holes made later in the process depends on the volume you want to create for the wells. With the parameters set in this work, each well will hold 200 μL working volume.
10. The guidance cube helps you understand the 3D working environment. There is also an *orbit* button on the down central toolbar that you can hit and pivot your model in order to observe changes you have made.
11. Operation *join* integrates the new component in the original body so that the slicing software recognizes the whole model as one body.
12. The base created in this step is not obligatory for the model to be functional, however, it helps in the integrity and the looks of the final model. You can choose to make it thicker or patterned, you should however keep in mind that the more material is going to be consumed by the printer the more printing time that will be needed.
13. When preparing a model for printing in the slicing software, it is advisable to rotate it accordingly in order to achieve the most model surface possible to have adhered to the build surface. This will reassure that the model will be printed with a stable structure.
14. These settings have been optimized for the printer we used. The reader should keep in mind that every printer may need optimization as to the appropriate characteristics that have to be set at the slicing software. Temperatures (printing temperature and build plate temperature) depend on the type of filament. PLA is printed in a range of 180–210 °C. The bed might not need to be heated at all or heated at 50–60 °C to ensure better adhesion.
15. Printing time depends on the infill density of the model and the print speed chosen. Infill density should be adjusted in a way to

reassure a rigid structure (100% infill density means no spaces in the interface). Print speed should be adjusted according to the capability of the printer. Low print speed means more precise control of the printing process (slower motion of the extruder).

16. Chitosan solution should not exceed 10 min residence in the-wells. It is advisable to use a timer that will be started when the first well is filled and, when the time is up, start removing the solution from the well you filled first at the beginning of the countdown.
17. The blank measurement well is considered a well modified with the same process as the other wells except incubating with the enzymes solution. This measurement will be subtracted from the obtained measurements from the other wells. This way, a possible effect on the substrates from the modified support material will be eliminated.
18. The maximum absorbance wavelength for $ABTS^+$ is at 405 nm. However, we measure the absorbance at a second peak of the absorbance spectrum of $ABTS^+$ since *pNP* produced from the action of β-glucosidase also absorbs at 405 nm.
19. The protocol described in this work is simultaneously applied in every well of the well-plate, so there is a need to run some tests before proceeding with activity measurements at different time intervals. Every well was checked for its activity at a fixed time point in order to reassure that all wells will have the same biocatalytic capacity. We also checked the remaining activity of the wells after incubation at 40 °C for 1 h (thermal stability). All tests were run with the protocol described in Subheading 3.3 for 30 min residence time.

Useful links

Fusion 360: https://www.autodesk.com/products/fusion-360/overview?term=1-YEAR

Online Fusion 360 courses: https://help.autodesk.com/view/fusion360/ENU/courses/AP-GET-STARTED-OVERVIEW

Ultimaker Cura slicing software: https://ultimaker.com/software/ultimaker-cura

Acknowledgments

This research has been co-financed by the European Regional Development Fund of the European Union and Greek national funds through the Operational Program Competitiveness, Entrepreneurship, and Innovation, under the call RESEARCH– CREATE–INNOVATE (project code: T2EDK- 01410). E.G. is thankful for the financial support to the Operational Program

≪Human Resources Development, Education and Lifelong Learning≫ in the context of the project "Strengthening Human Resources Research Potential via Doctorate Research" (MIS-5000432), implemented by the State Scholarships Foundation (IKY), which is co-financed by Greece and the European Union (European Social Fund-ESF).

References

1. Ghilan A, Chiriac AP, Nita LE et al (2020) Trends in 3D printing processes for biomedical field: opportunities and challenges. J Polym Environ 28:1345–1367
2. Kong YL, Gupta MK, Johnson BN et al (2016) 3D printed bionic nanodevices. Nano Today 11:330–350
3. Lupuleasa D, Drăgănescu D, Hîncu L et al (2018) Biocompatible polymers for 3D printing. Farmacia 66:737–746
4. Baran EH, Yildirim Erbil H (2019) Surface modification of 3D printed PLA objects by fused deposition modeling: a review. Colloids Interfaces. 3(2):43
5. Maia-Pinto MOC, Brochado ACB, Teixeira BN et al (2021) Biomimetic mineralization on 3D printed PLA scaffolds: on the response of human primary osteoblasts spheroids and in vivo implantation. Polymers (Basel) 13:1–26
6. Ye J, Chu T, Chu J et al (2019) A versatile approach for enzyme immobilization using chemically modified 3D-printed scaffolds. ACS Sustain Chem Eng 7:18048–18054
7. Chen W, Nichols L, Brinkley F et al (2021) Alkali treatment facilitates functional nano-hydroxyapatite coating of 3D printed polylactic acid scaffolds. Mater Sci Eng C 120:111686
8. Schneider M, Fritzsche N, Puciul-Malinowska A et al (2020) Surface etching of 3D printed poly(lactic acid) with NaOH: a systematic approach. Polymers (Basel) 12:1711
9. Barbosa O, Ortiz C, Berenguer-Murcia Á et al (2014) Glutaraldehyde in bio-catalysts design: A useful crosslinker and a versatile tool in enzyme immobilization, https://pubs.rsc.org/en/content/articlehtml/2014/ra/c3ra45991h
10. Giannakopoulou A, Patila M, Spyrou K et al (2019) Development of a four-enzyme magnetic nanobiocatalyst for multi-step cascade reactions. Catalysts. 9:995
11. Yang Y, Zhang R, Zhou B et al (2017) High activity and convenient ratio control: DNA-directed coimmobilization of multiple enzymes on multifunctionalized magnetic nanoparticles. ACS Appl Mater Interfaces 9: 37254–37263

Chapter 11

Rational Design of Self-Assembling Supramolecular Protein Nanostructures Utilizing the Cucurbit[8]Uril Macrocyclic Host

Elisavet Ioannou and Nikolaos E. Labrou

Abstract

Self-assembly is a phenomenon that governs molecular structural organization in nature, therefore raising research interest for the fabrication of novel nanomaterials with diverse applications in biocatalysis, biomedicine, material templating, and biosensor development. In this chapter we provide protocols for the development of supramolecular protein complexes based on host–guest interactions in the presence of the macrocyclic host, cucurbit[8]uril (CB[8]). CB[8] is reported to exhibit high binding affinity towards the tripeptide Phe-Gly-Gly (FGG), therefore it can be utilized as a selective adhesive of protein molecules, after fusion of FGG to an accessible protein surface.

Key words Protein self-assembly, Supramolecular nanostructures, Host–guest interactions, Cucurbit [8]uril, FITC labeling of proteins, Aggregate formation

1 Introduction

Self-assembly is a sophisticated reoccurring phenomenon in nature. By definition molecular self-assembly is the spontaneous organization of molecules into ordered and stable arrangements due to non-covalent interactions [1]. Inspired by native protein nanostructures that function as biocatalysts, regulators of biochemical reactions and form structural matrixes or partake in signaling [2], recent research advances have been focused on the manipulation of protein self-assembly for the construction of novel nanomaterials with desired functions and diverse applications [3]. Proteins are an inherently fertile ground for intermolecular and intramolecular interactions; therefore they are rendered as an attractive multifunctional building block for manipulation [4]. Native protein self-assemblies form symmetric polyhedral structures [5], inspiring the fabrication of self-assembled fused oligomers in order to form nanohedral materials [6]. Non-covalent interactions such as metal

Haralambos Stamatis (ed.), *Multienzymatic Assemblies: Methods and Protocols*,
Methods in Molecular Biology, vol. 2487, https://doi.org/10.1007/978-1-0716-2269-8_11,
© The Author(s), under exclusive license to Springer Science+Business Media, LLC, part of Springer Nature 2022

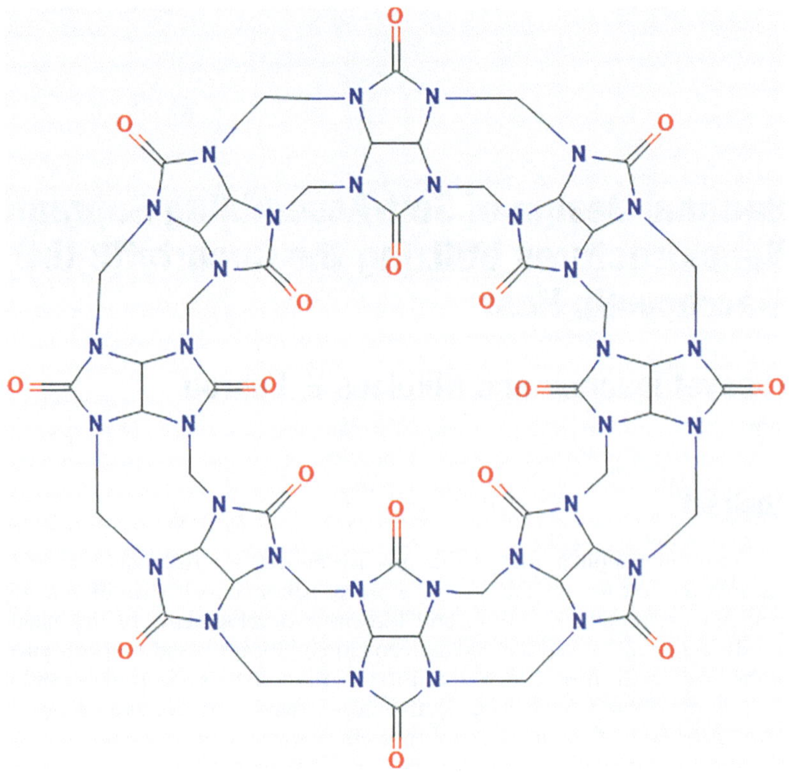

Fig. 1 The structure of cucurbit[8]uril

coordination, host–guest and electrostatic interactions, implemented in modified protein sites have been widely equipped for the construction of protein nanostructures [7, 8].

CB[8] (Fig. 1) is a synthetic macrocyclic host of eight glycoluril units, with a hydrophobic cavity that is able to bind one or two guest molecules, resulting in the formation of homo- or heteroternary complexes. Generally, potential guests for CB[8] may contain (1) an aromatic group that binds into the cavity, therefore releasing trapped high energy water molecules, and (2) a positively charged side group in order to form ion dipole interactions with its carbonyl rims [9]. It has been established that CB[8] is able to bind and induce dimerization of N-terminal aromatic tripeptides Trp-Gly-Gly (WGG) and Phe-Gly-Gly (FGG) with high binding affinity (K_{ter} 10^9–$10^{11}/M^2$), although when Phe and Trp are not at the end of the tripeptide, the estimated binding constant corresponded to a single binding event [10]. On the other hand, Sozini et al. [11], showed that 2:1 binding is possible even with a middle Phe residue located in a chain of amino acids. Due to the high binding affinity and selectivity of this macrocyclic host towards the recognition motif FGG, it can be utilized as an adhesive of protein molecules via host–guest interactions [12]. Depending on the application, incorporation of a suitable recognition motif can be directly attached to the protein building block or conjugated before the formation of the complex [4]. Previous research has repeatedly utilized CB[8] for the construction of supramolecular protein complexes and nanostructures, either by incorporation of an FGG tail at the N-terminal site of a protein [12–14] or at the protein's external surface [15]. The morphology of the derived complex depends on the position of recognition motif on the protein building block along with its overall structural arrangement.

2 Materials

All solutions are prepared with double-distilled water and analytical grade reagents. All buffers are prepared at room temperature and stored at 4 °C.

2.1 Addition of Phenylalanine-glycine-glycine (FGG-) to the N-terminal Site of Selected Enzyme Forms

1. *E. coli* plasmid containing the gene of interest.
2. High fidelity DNA polymerase and its corresponding reaction buffer (e.g., KAPA HiFi DNA Polymerase, KAPA Biosystems, Roche).
3. dNTPs mixed solution (10 mM of each dATP, dTTP, dCTP, dGTP).
4. Phosphorylated oligonucleotide primers designed to amplify a full vector containing the enzyme of interest with the addition of FGG to its N-terminal site (Subheading 3.1).
5. T4 DNA ligase and its corresponding reaction buffer (e.g., TAKARA T4 DNA Ligase, TAKARA BIO Inc.).
6. DNP1 restriction enzyme and its corresponding reaction buffer [e.g., Dpnl (10 U/μL), Thermo Fisher Scientific].
7. *E. coli* competent cells and appropriate growth media, e.g., Luria-Bertani Broth (LB): 1% w/v tryptone/peptone, 1% w/v NaCl, 0.5% yeast extract. Autoclave at 121 °C for 20 min and let to cool before use.
8. Appropriate antibiotics for plasmid selection.
9. Plasmid mini prep kit (e.g., NucleoSpin Plasmid (NoLid), Mini kit for plasmid DNA purification, Macherey-Nagel).

2.2 Labeling of the Purified Enzyme with Fluorescein Isothiocyanate (FITC)

1. Purified enzyme with FGG tag.
2. Dialysis cellulose membrane tubes, with appropriate molecular weight cutoff (e.g., 12 kDa) and secure clips.
3. Appropriate dialysis buffer (e.g., 20 mM KH_2PO_4, pH 7.0) stored at 4 °C. Add 5.44 g KH_2PO_4 to 1.9 L ddH_2O in a beaker, mix and adjust pH at 7 with KOH. Fill with ddH_2O up to 2 L.

4. Labeling buffer I: 0.1 M Sodium carbonate buffer, pH 9. In 90 mL ddH_2O dissolve 0.798 g $NaHCO_3$ and 0.053 g Na_2CO_3. Mix and fill with ddH_2O until 100 mL.
5. Labeling buffer II: 0.1 M KH_2PO_4 buffer, pH 6.5. Add 1.3609 g KH_2PO_4 to 90 mL ddH_2O. Mix and adjust pH at 6.5 with KOH. Fill with ddH_2O up to 100 mL.
6. FITC (1 mg/mL) freshly dissolved in anhydrous DMSO.
7. FITC quenching buffer: 0.5 M NH_4Cl. Add 1.337 g of NH_4Cl to 45 mL ddH_2O, mix and fill with ddH_2O until 50 mL.

2.3 Addition of CB[8] to the FITC-Enzyme at Different Molar Ratios and Examination of Aggregate Formation Induced by CB[8]

1. Solid cucurbit[8]uril (CB[8]).
2. FITC labeled enzyme with FGG tag.
3. CB[8] dissolution buffer I: 10 mM Sodium Phosphate buffer (PB), pH 7.4. Add 1.343 g Na_2HPO_4 (Na_2HPO_4:12-H_2O, MW:358.14 g/mol) and 0.195 g NaH_2PO_4 (NaH_2.PO_4:2H_2O, MW:156.01 g/mol) in 45 mL ddH_2O. Mix and adjust pH with NaOH as needed. Fill with ddH_2O until 50 mL.
4. CB[8] dissolution buffer II: 3 M HCl. Slowly add 2.5 mL of 12 M HCl stock solution in 6 mL of ddH_2O. Fill up to 10 mL with ddH_2O.

3 Methods

The following methodology describes a protocol for construction and recognition of self-assembled protein nanostructures induced by selective binding of the FGG-tagged enzymes with CB[8] at 1:1 or 2:1 enzyme to host stoichiometry.

3.1 Addition of Phenylalanine-glycine-glycine (FGG-) Tag at the N-terminal Site of Selected Enzyme Forms

The procedure for the addition of FGG- at the N-terminus of the target enzyme consisted of the following steps: (1) amplification of the enzyme's gene with suitably designed primers that contain the nucleotide sequence that encodes for the FGG sequence, (2) reaction with T4 DNA ligase, (3) removal of vectors carrying the parental genes by DPN1I, and d) transformation of *E. coli* cells for enzyme expression.

1. Amplify the target gene with a high fidelity polymerase following the provided manufacturer's protocol at 25 μL final volume. Forward primer contains the nucleotide sequence 5′-ATG-TTT-GGA-GGA-3′ (corresponding to Met-Phe-Gly-Gly) and the first 18–20 nucleotides of the gene of interest, lacking the start codon (ATG). Reverse primer corresponds to

the inverse and complementary sequence of the plasmid upward the gene's 5′ ATG. The primers are phosphorylated. The final PCR product is a linear plasmid vector with the FGG-flagged gene.

2. Add 2 μL of the PCR product, 1 μL of 10× ligase reaction solution, 6 μL of sterile ddH_2O, and 1 μL of T4 DNA ligase (350 U/μL). Incubate at 16 °C for 16 h. Proceed to enzyme inactivation at 65 °C for 15 min.
3. Add 1.5 μL of 10× DNP1 reaction buffer and 0.5 μL DNP1 (10 U/μL) to ligation products. Add sterile ddH_2O to 15 μL final volume. Incubate for 3.5 h at 37 °C. Stop the reaction at 80 °C for 20 min (*see* **Note 1**).
4. The final reaction product can be used to transform *E. coli* cells according to the manufacturer's protocol. Proceed with DNA sequencing to confirm the addition of FGG.

3.2 Labeling of the Purified Enzyme with Fluorescein Isothiocyanate (FITC)

Expression and purification protocols are dependent on the particular target enzyme; therefore a specified protocol will not be included in the methodology. Labeling is based on Hermanson (2013) protocol [16].

1. Dialyze overnight the purified enzyme in appropriate buffer (e.g., 20 mM KH_2PO_4, pH 7.0) at 4 °C, to remove excess salts that may have been used during purification. Dialysis buffer should provide and maintain the functional stability of the enzyme.
2. Add dialyzed enzyme at 1.5–2 mg/mL, final concentration, to 0.1 M sodium carbonate buffer, pH 9, for selective FITC binding to the N-terminal and the ε-amino group of lysine residues (Fig. 2). For FITC binding to cysteine's thiol group, perform the reaction at 0.1 M KH_2PO_4, pH 6.5 (*see* **Note 2**).
3. Add 50 μL of 1 mg/mL FITC stock solution (freshly dissolved in anhydrous DMSO) for each mL of protein solution. Incubate the reaction overnight at 4 °C in the dark.
4. If the labeling reaction takes place at pH 9, add ammonium chloride (Stock solution: 0.5 M NH_4Cl) as needed at 50 mM final concentration to quench remaining isothiocyanate groups. Incubate the reaction for 2 h at 4 °C in the dark.
5. Centrifuge the labeled protein sample at 9,500 × *g* for 10 min (*see* **Note 3**).
6. Dialyze the mixture overnight in appropriate buffer (20 mM KH_2PO_4, pH 7.0) at 4 °C, in order to remove free FITC molecules.
7. Store the FITC labeled enzymes in the dark and 4 °C.

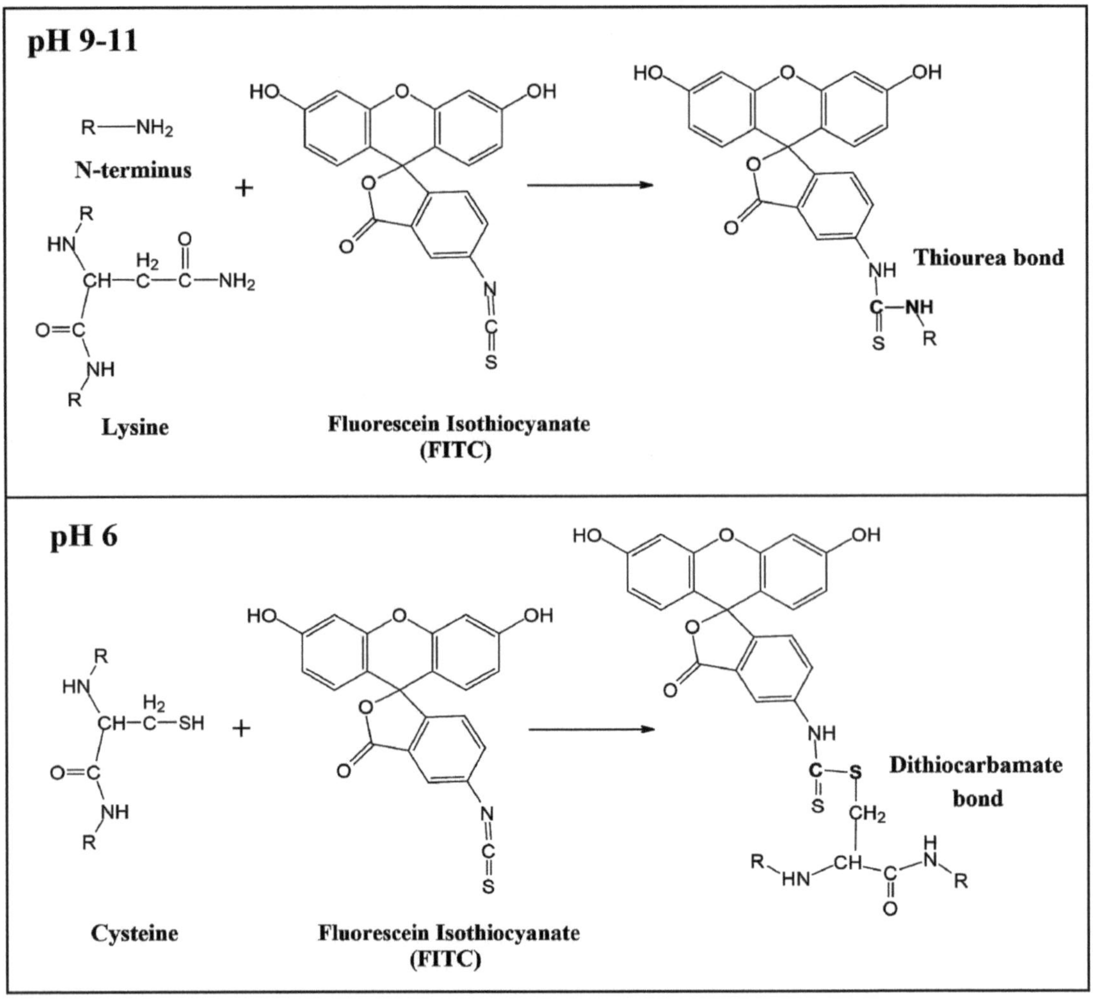

Fig. 2 Nucleophilic attack of the electrophilic central carbon of FITC isothiocyanate group that favors the formation of a thiourea bond at pH 6-8 and a dithiocarbamate bond at pH 9-11

8. Use the following formulas for the molar ratio determination of fluorescent dye per molecule of protein (F/P):

$$\text{Protein concentration}(M) = \frac{A280 - (A494 \times \text{CF})}{\varepsilon} \times df$$

$$\text{mol of FITC per mol of protein}\left(\frac{F}{P}\right) = \frac{\text{A494}}{\varepsilon' \times (M)} \times df$$

where.

A_{280}: the absorbance of the protein sample at 280 nm.

A_{494}: the absorbance of the protein sample at 494 nm.

CF: the absorption correction factor of the fluorescent probe at 280 nm. $CF_{(FITC)} = 0.3$.

ε: extinction coefficient of the protein (obtain by Expasy ProtParam tool, https://web.expasy.org/protparam/).

ε': extinction coefficient of FITC (68,000/M/cm).

df: the sample dilution factor.

3.3 Addition of CB[8] to the FITC-Enzyme at Different Molar Ratios and Examination of Aggregate Formation Induced by CB[8]

1. Weigh the suitable amount of CB[8] for the preparation of a stock solution.
2. Add CB[8] to 10 mM PB buffer, pH 7.4, at a final concentration of 70–100 μM. For stock solutions of higher concentration (up to 1.5 mM), add CB[8] to 3 M HCl solution (*see* **Notes 4** and **5**).
3. Solubilize CB[8] in a bath sonicator for 2 h. For higher concentration solutions, let CB[8] stir at 280 rpm and 60 °C, overnight.
4. Add CB[8] to a known concentration enzyme solution at 1:1 and 2:1 stoichiometry (i.e., 15 μM of enzyme/15 μM of CB [8]; 15 μM of enzyme/7.5 μM of CB[8]). Use neutral reaction buffer (e.g., 50 mM KH_2PO_4, pH 7.0) to maintain enzyme's structural integrity and stability (*see* **Note 6**).
5. Keep the reaction of the FITC labeled enzyme with CB[8] in the dark and 4 °C.
6. Monitor changes in the fluorescence intensity of the sample (Fig. 3). Use excitation wavelength 494 nm. Emission wavelength is expected at approximately 525 nm. Add a standard amount of the enzyme sample (i.e., 5 μL) in final volume of 2 mL (20 mM KH_2PO_4, pH 7.0), and scan for changes in the fluorescence intensity due to the formation of aggregates over the course of a few days. The measurements can be taken at 25 °C (*see* **Note 7**).
7. Quantify the percentage of fluorescence intensity reduction against a "control" reaction of the labeled enzyme and the solvent of CB[8]. The order and morphology of the assembled structures can be examined with further experiments (*see* **Note 8**).

4 Notes

1. Sufficient digestion of the ligation product by DPN1 is crucial in order to optimize the outcome of the procedure, since the template for the initial PCR reaction is a methylated circular plasmid that contains the gene of interest.

2. During FITC labeling of molecules, the isothiocyanate group of the fluorescent dye reacts with amines, sulfhydryls, or the phenolic side group of tyrosine [17]. Isothiocyanate primarily reacts with either non-protonated aliphatic amino groups of the N-terminus and lysine residues (Lys), or with cysteine's thiol group (Cys) [18, 19]. The selectivity of labeling reaction towards the amino acids Lys and Cys is mainly determined by the pH of the reaction and it is affected by the ionization state of their side groups. Lysine amino groups are protonated at lower pH, therefore labeling requires higher pH (9 to 11) for optimal coupling [20]. Accordingly, modification of cysteine's thiol group is achieved at lower pH values [21]. Therefore, protein's nucleophilic group reacts with the electrophilic central carbon of the isothiocyanate group, forming a thiourea bond in the case of lysine, and a dithiocarbamate bond in the case of cysteine (Fig. 1).
3. Filter or centrifuge the FITC labeled enzyme in order to remove the aggregates formed by the reaction.
4. CB[8] is practically insoluble in water and in most common solvents. Solubility increases in the presence of cations such as Na^+ and K^+ due to formation of ion-dipole bonds with CB[8]'s carbonyl rims [22, 23]. Moreover, HCl can be used, since H^+ bonds to CB[8]'s portals in acidic environment [22]. For example, a stock solution of CB[8] can be prepared in 0.1 M HCl in a bath sonicator for 2 h.
5. CB[8]'s low solubility in aqueous solutions, as well as its tendency to form monomer conjugates, complicate the formation of high degree of supramolecular polymers [24]. Improved solubility, while retaining its dimerization functionality, has been accomplished by other synthesized derivatives [25–27].
6. The pH of the reaction between the enzyme and CB[8] should be measured in the case that CB[8] is solubilized in HCl. This can be accomplished either by raising the buffer's concentration or its initial pH value to 7.5.
7. Fluorescence quenching due to aggregate formation may occur with the formation of oligomers induced by host–guest interactions in the presence of CB[8] [15]. Lower emission or even non-emissive rates are recorded when most of the common fluorophores form aggregates although they are highly emissive in dilute solutions [28, 29]. Fluorescence reduction can either be observed instantly after the addition of CB[8], or gradually over the course of a few days. In the latter case, it is crucial to measure a "control" reaction of the FITC labeled enzyme (including appropriate amount of CB[8]'s solvent, especially if it's HCl) in order to estimate potential fluorescence losses that are not attributed to the hosts addition (Fig. 3).

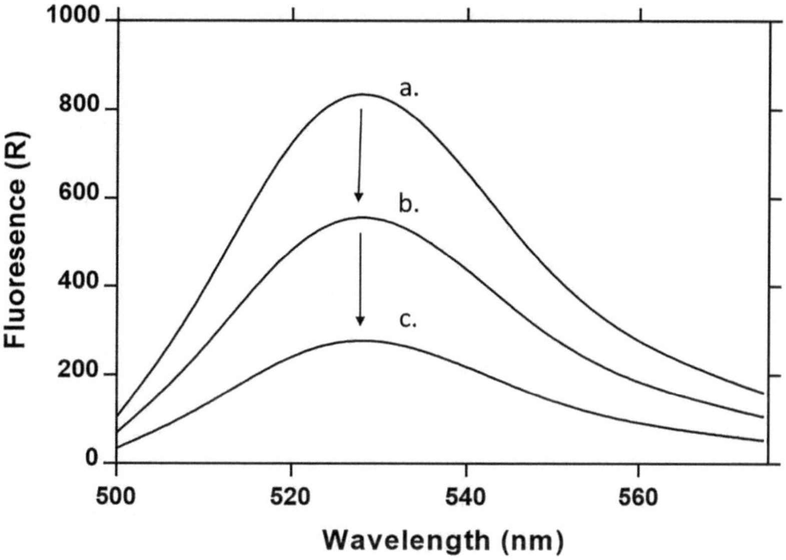

Fig. 3 Gradual fluorescence reduction (a.→b.→c.) due to the formation of protein aggregate, following the addition of CB[8] to the FGG tagged protein

8. A quick and reproductive method for the size estimation of protein aggregates (dimers or higher order structures) is size exclusion chromatography (SEC-HPLC) [30]. Moreover, determination of morphological characteristics of FGG-enzyme: CB[8] self-assemblies can be achieved by tapping mode atomic force microscopy (AFM) and transmission electron microscopy (TEM).

Acknowledgements

This research is co-financed by Greece and the European Union (European Social Fund- ESF) through the Operational Programme «Human Resources Development, Education and Lifelong Learning» in the context of the project "Strengthening Human Resources Research Potential via Doctorate Research" (MIS-5000432), implemented by the State Scholarships Foundation (IKY).

References

1. Whitesides GM, Boncheva M (2002) Beyond molecules: self-assembly of mesoscopic and macroscopic components. Proc Natl Acad Sci U S A 99:4769–4774
2. Li F, Wang D, Zhou J et al (2020) Design and biosynthesis of functional protein nanostructures. Sci China Life Sci 63:1142–1158
3. Chronopoulou EG, Ioannou E, Perperopoulou F et al (2020) Protein nanostructures with purpose-designed properties in biotechnology and medicine. In: Shukla P (ed) Microbial enzymes and biotechniques. Springer, Singapore. https://doi.org/10.1007/978-981-15-6895-4_5

4. Kuan SL, Bergamini FRG, Weil T (2018) Functional protein nanostructures: a chemical toolbox. Chem Soc Rev 47(24):9069–9105. https://doi.org/10.1039/c8cs00590g
5. King NP, Sheffler W, Sawaya MR et al (2012) Computational design of self-assembling protein nanomaterials with atomic level accuracy. Science 336(6085):1171–1174. https://doi.org/10.1126/science.1219364
6. Padilla JE, Colovos C, Yeates TO (2001) Nanohedra: Using symmetry to design self-assembling protein cages, layers, crystals, and filaments. Proc Natl Acad Sci U S A 98(5): 2217–2221. https://doi.org/10.1073/pnas.041614998
7. Hamley IW (2019) Protein assemblies: nature-inspired and designed nanostructures. Biomacromolecules 20(5):1829–1848. https://doi.org/10.1021/acs.biomac.9b00228
8. Sun H, Luo Q, Hou C et al (2017) Nanostructures based on protein self-assembly: from hierarchical construction to bioinspired materials. Nano Today 14:16–41., ISSN 1748-0132. https://doi.org/10.1016/j.nantod.2017.04.006
9. Huang Z, Qin K, Deng G et al (2016) Supramolecular chemistry of cucurbiturils: tuning cooperativity with multiple non-covalent interactions from positive to negative. Langmuir 32(47):12352–12360
10. Heitmann LM, Taylor AB, Hart PJ et al (2006) Sequence-specific recognition and cooperative dimerization of N-terminal aromatic peptides in aqueous solution by a synthetic host. J Am Chem Soc 128(38):12574–12581. https://doi.org/10.1021/ja064323s
11. Sonzini S, Ryan STJ, Scherman OA (2013) Supramolecular dimerisation of middle-chain Phe pentapeptides via CB[8] host–guest homoternary complex formation. Chem Commun 2013(49):8779–8781. https://doi.org/10.1039/c3cc45420g
12. Hou C, Li J, Zhao L et al (2013) Construction of protein nanowires through Cucurbit[8]uril-based highly specific host-guest interactions: an approach to the assembly of functional proteins. Angew Chem Int Ed 52(21): 5590–5593. https://doi.org/10.1002/anie.201300692
13. Bosmans RPG, Briels JM, Milroy LG et al (2016) Supramolecular control over split-luciferase complementation. Angew Chem Int Ed 55(31):8899–8903. https://doi.org/10.1002/anie.201602807
14. Biedermann F, Rauwald U, Zayed JM et al (2011) A supramolecular route for reversible protein-polymer conjugation. Chem Sci. 2(2): 279–286. https://doi.org/10.1039/c0sc00435a
15. Li X, Bai Y, Huang Z et al (2017) A highly controllable protein self-assembly system with morphological versatility induced by reengineered host–guest interactions. Nanoscale 9(23):7991–7997. https://doi.org/10.1039/c7nr01612c
16. Hermanson GT (2013) Chapter 10 - Fluorescent probes. In: Hermanson GT (ed) Bioconjugate techniques, 3rd edn. Academic Press, Cambridge, pp 395–463. https://doi.org/10.1016/B978-0-12-382239-0.00010-8
17. Podhradsky D, Drobnica L, Kristian P (1979) Reactions of cysteine, its derivatives, glutathione, coenzyme A, and dihydrolipoic acid with isothiocyanates. Experientia 35:154
18. Karlsson I, Samuelsson K, Ponting DJ et al (2016) Peptide reactivity of isothiocyanates – implications for skin allergy. Sci Rep 6(1): 21203. https://doi.org/10.1038/srep21203
19. Petri L, Szijj PA, Kelemen A et al (2020) Cysteine specific bioconjugation with benzyl isothiocyanates. RSC Adv 10(25):14928–14936. https://doi.org/10.1039/D0RA02934C
20. Nolting D, Aziz EF, Ottosson N et al (2007) pH-induced protonation of lysine in aqueous solution causes chemical shifts in X-ray photoelectron spectroscopy. J Am Chem Soc 129(45):14068–14073. https://doi.org/10.1021/ja072971l
21. Bulaj G, Kortemme T, Goldenberg DP (1998) Ionization–reactivity relationships for cysteine thiols in polypeptides. Biochemistry 37(25): 8965–8972. https://doi.org/10.1021/bi973101r
22. Wheate NJ, Limantoro C (2016) Cucurbit[n]urils as excipients in pharmaceutical dosage forms. Supramol Chem 28(9–10):849–856. https://doi.org/10.1080/10610278.2016.1178746
23. Buschmann HJ, Cleve E, Schollmeyer E (1992) Cucurbituril as a ligand for the complexation of cations in aqueous solutions. Inorg Chim Acta 193(1):93–97., ISSN 0020-1693. https://doi.org/10.1016/S0020-1693(00)83800-1
24. Yin JF, Hu Y, Wang DG et al (2017) Cucurbit[8]uril-based water-soluble supramolecular dendronized polymer: evidence from single polymer chain morphology and force spectroscopy. ACS Macro Lett. 6(2):139–143
25. Vinciguerra B, Zavalij PY, Isaacs L (2015) Synthesis and recognition properties of cucurbit[8]uril derivatives. Org Lett 17(20):

5068–5071. https://doi.org/10.1021/acs.orglett.5b02558

26. Cong H, Ni XL, Xiao X et al (2016) Synthesis and separation of cucurbit[n]urils and derivatives. Org Biomol Chem 14(19):4335–4364. https://doi.org/10.1039/c6ob00268d. PMID: 26991738
27. Wang XX, Tian FY, Liu M et al (2019) A water soluble tetramethyl-substituted cucurbit[8]uril obtained from larger intermediates? Tetrahedron 75(37):130488., ISSN 0040-4020. https://doi.org/10.1016/j.tet.2019.130488
28. Ma X, Sun R, Cheng J et al (2015) Fluorescence aggregation caused quenching versus aggregation-induced emission: a visual teaching technology for undergraduate chemistry students. J Chem Educ 93(2015):345–350
29. Birks JB (1970) Photophysics of aromatic molecules. Wiley- InterScience, London
30. Hong P, Koza S, Bouvier ESP (2012) A review size exclusion chromatography for the analysis of protein biotherapeutics and their aggregates. J Liq Chromatogr Relat Technol 35(20): 2923–2950

Chapter 12

Lectin-Mediated Coimmobilization of Cascade Glycoenzymes

Yifei Zhang, Yunzhenshan Gao, and You Yong

Abstract

As the vast majority of enzymes are glycosylated, lectins can serve as molecular glues to agglutinate multiple glycoenzymes for preparing multienzyme catalysts in an efficient and biocompatible way. Taking glucose oxidase and horseradish peroxidase as a model cascade, we describe in this protocol the coimmobilization of cascade glycoenzymes through lectin-mediated protein agglutination with and without magnetic nanoparticles as carriers.

Key words Lectin, Concanavalin A, Glycoenzyme, ConA-coated magnetic nanoparticles, Multienzyme catalyst

1 Introduction

Multienzyme catalysis that performs consecutive chemical reactions in one pot has shown to be advantageous over traditional step-by-step chemical synthesis in terms of reaction velocity, specificity, and atom economy [1–3]. The aggregation or crowding of cascade enzymes is a key factor for regulating the catalytic activity of multienzyme systems [4, 5]. In living systems, many enzymes are associated together to carry out metabolic reactions through delicate protein–protein interactions. For example, mitochondrial aspartate aminotransferase and α-ketoglutarate dehydrogenase can form a complex in the mitochondrial matrix [6]. Purinosome, a putative six-enzyme complex responsible for purine biosynthesis, was discovered to be dynamically and reversibly formed in response to the environmental changes in purine concentration [7]. Cascade enzymes are also widely used in vitro for the biosynthesis of pharmaceutical compounds. A representative demonstration is the three-step synthesis of an experimental HIV drug islatravir by nine enzymes [8]. Multiple enzymes in practical processes are often coimmobilized, typically on inorganic or polymeric carriers,

Haralambos Stamatis (ed.), *Multienzymatic Assemblies: Methods and Protocols*,
Methods in Molecular Biology, vol. 2487, https://doi.org/10.1007/978-1-0716-2269-8_12,
© The Author(s), under exclusive license to Springer Science+Business Media, LLC, part of Springer Nature 2022

leading to compromised activity and stability [9]. In order to mimic the natural forms of enzyme aggregation, we proposed a new strategy for coimmobilizing cascade enzymes based on lectin-mediated protein agglutination [10].

Lectins are carbohydrate-binding proteins that can specifically agglutinate glycoproteins. More than half of known enzymes, especially those expressed by eukaryotic cells, are glycosylated with different sugar moieties [11]. For example, α-amylase from *Aspergillus oryzae*, α-D-galactosidase from broad bean, peroxidase from horseradish (HRP), glucose oxidase from *Aspergillus niger* (GOx), and chloroperoxidase from *Caldariomyces fumago* are all glycoproteins [12]. These glycoproteins can be recognized by specific lectin proteins through carbohydrate–protein interactions. Concanavalin A (ConA) is a carbohydrate-binding tetrameric lectin from jack beans (*Canavalia ensiformis*) composed of four identical 28 kDa subunits, which can selectively bind to α-D-mannosyl and α-D-glucosyl residues [13]. Taking advantage of this affinity, we have successfully employed ConA as a molecular glue to coimmobilize GOx and HRP as enzyme aggregates or on ConA-coated magnetic nanoparticles [10, 14]. In this protocol, we introduce in detail the preparation of ConA-agglutinated GOx-HRP coimmobilized systems (Fig. 1).

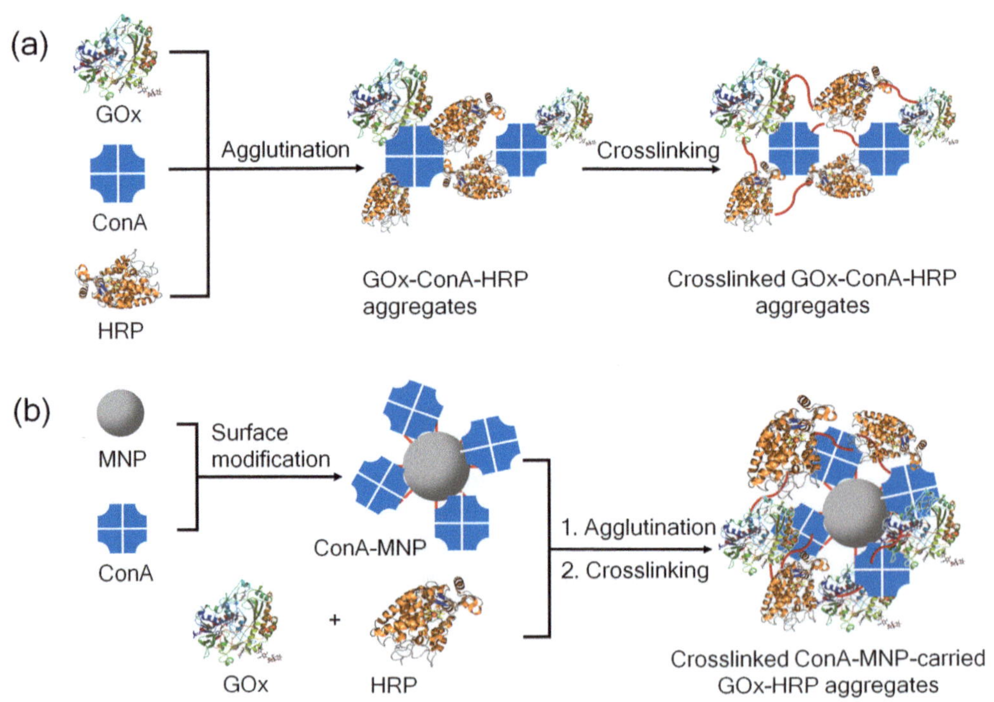

Fig. 1 Schematic of the coimmobilization of GOx and HRP with ConA agglutination (**a**) in the absence and (**b**) presence of magnetic nanoparticles

2 Materials

1. Phosphate buffer: 10 mM sodium phosphate, pH 7.0 at 25 °C.
2. Glucose oxidase solution: 1.0 mg/mL glucose oxidase from *Aspergillus niger* in 10 mM sodium phosphate buffer, pH 7.0.
3. Horseradish peroxidase solution: 1.0 mg/mL horseradish peroxidase in 10 mM sodium phosphate buffer, pH 7.0.
4. Concanavalin A solution: 1.0 mg/mL Concanavalin A (type IV) in 10 mM sodium phosphate buffer, pH 7.0.
5. 5% w/v glutaraldehyde solution: dilute 25% w/v glutaraldehyde solution in phosphate buffer (10 mM, pH 7.0) to produce 5% w/v glutaraldehyde solution.
6. Iron stock solution: 0.5 M $FeCl_3$ + 0.25 M $FeCl_2$ in water.
7. APTES solution: 10% v/v (3-Aminopropyl)triethoxysilane (APTES) in 1:1 water/ethanol mixture.
8. Sodium borohydride solution: 4 mg/mL sodium borohydride ($NaBH_4$) in ddH_2O (freshly prepared).
9. 25% w/w ammonium hydroxide.
10. ABTS solution: 100 mM 2,2′-azino-bis(3-ethylbenzothiazoline-6-sulfonic acid) (ABTS) in ddH_2O.
11. Glucose substrate solution: 50 mM sodium phosphate buffer, pH 7.0, containing 100 mM glucose.
12. Mobile phase: 50 mM phosphate buffer, pH 6.7, containing 100 mM Na_2SO_4 and 0.5‰ w/v NaN_3.
13. An UV-VIS spectrophotometer.
14. An HPLC system equipped with a TSK-GEL G2000SWXL column (5 μm, 7.8 × 300 mm) and an ultraviolet-visible detector (wavelength adjusted to 280 nm).

3 Methods

3.1 Agglutination of GOx and HRP with ConA

1. Mix 20 μL of 1.0 mg/mL GOx solution and 80 μL of 1.0 mg/mL HRP solution (*see* **Note 1**).
2. Add 20 μL of the above solution to 980 μL of 1.0 mg/mL Concanavalin A (ConA) solution. Stir the mixed solution gently at a speed of 250 rpm at room temperature for 1 h to allow the formation of GOx-ConA-HRP aggregate (*see* **Note 2**).
3. Add 40 μL of 5% w/v glutaraldehyde solution to the GOx-ConA-HRP aggregate suspension (final concentration of glutaraldehyde of 0.2% w/v) and stir the mixture at room temperature for 1 h.

4. Add 25 μL of 4 mg/mL freshly prepared $NaBH_4$ solution to the mixture, and gently stir the mixture for 4 h at 4 °C in order to reduce the formed Schiff base.
5. Spin the solution with a quick spin centrifuge at 500 rmp, pull out the supernatant, and add 1.0 mL phosphate buffer (10 mM, pH 7.0) to wash the obtained precipitates, the cross-linked GOx-ConA-HRP aggregates (CLGCHA).
6. Repeat **step 5** two more times.
7. Combine supernatants from **steps 5** and **6** (total volume: V_t), and determine the concentrations of free GOx, HRP, and ConA by size-exclusion chromatography. Typically, 20 μL of sample solution is injected into an HPLC column and eluted by mobile phase (50 mM phosphate buffer, pH 6.7, containing 100 mM Na_2SO_4 and 0.5‰ w/v NaN_3) at a flow rate of 1 mL/min. The protein concentration is determined based on the peak area at 280 nm (*see* **Note 3**).
8. Calculate the agglutination efficiency for GOx and HRP as follows:

$$A\Upsilon(\%) = \left(1 - \frac{[X]_{\text{sup}} \times V_t}{[X]_0 \times V_X}\right) \times 100$$

   where $[X]_{\text{sup}}$ is the concentration of free protein (X = GOx or HRP) in the combined supernatant, $[X]_0$ and V_x are the initial concentration and volume of the protein solution used for the formation of aggregates, respectively.
9. Resuspend the CLGCHA in 1.0 mL of phosphate buffer (10 mM, pH 7.0) by gentle vortexing.
10. Store the suspension of CLGCHA at 4 °C.

3.2 Agglutination of GOx and HRP with ConA-Coated Magnetic Nanoparticles

3.2.1 Synthesis of Magnetic Nanoparticles

1. Cool down 36.4 mL ddH_2O to 4 °C.
2. Add 1.6 mL of iron stock solution to the above ddH_2O at 4 °C.
3. Add 2.0 mL of ice-cooled 25% w/w ammonium hydroxide to the above solution under vigorous magnetic stirring (>1000 rpm). Black precipitates appear immediately upon the addition of ammonium hydroxide.
4. Heat the mixture for 12 h at 70 °C to afford magnetic nanoparticles (MNPs).
5. Collect the MNPs by magnetic decantation, remove the supernatant, wash the MNPs with ddH_2O for 3 times, and suspend the MNPs with 1.0 mL ddH_2O.
6. Disperse 1.0 mL of MNPs suspension in 20 mL of 10% v/v APTES solution.
7. Purge the suspension with nitrogen for 20 min, then reflux the suspension for 5 h at 50 °C under nitrogen.

8. Collect and rinse the MNPs with ddH_2O by magnetic decantation for three times (*see* **Note 4**).

3.2.2 Preparation of ConA-Coated Magnetic Nanoparticles

1. Add 200 μL of 10 mg/mL MNPs suspension to 1.0 mL (V_0) of 1.0 mg/mL ConA solution ($[ConA]_0$), and stir the solution for 30 min for protein adsorption.
2. Add 40 μL of 5% w/v glutaraldehyde solution to the above solution, and stir at room temperature for 2 h.
3. Add 25 μL of freshly prepared 4 mg/mL $NaBH_4$ solution and gently stir the mixture for 2 h.
4. Collect and rinse the ConA-coated MNPs with 1.0 mL phosphate buffer (10 mM, pH 7.0) by magnetic decantation. Repeat three times.
5. Combine the supernatants from each rinse (total volume: V_s).
6. Calculate the immobilization yield of ConA by determining the ConA concentration ($[ConA]_{sup}$) in the supernatant with BCA protein assay (*see* **Note 5**). The immobilization yield was calculated as follows:

$$\Upsilon(\%) = \left(1 - \frac{[\mathrm{ConA}]_{\mathrm{sup}} \times V_s}{[\mathrm{ConA}]_0 \times V_0}\right) \times 100$$

7. Resuspend the ConA-coated MNPs in 1.0 mL of phosphate buffer (10 mM, pH 7.0) by gentle vortexing.

3.2.3 Coimmobilization of GOx and HRP on ConA-Coated MNPs

1. Mix 20 μL of 1.0 mg/mL GOx solution and 80 μL of 1.0 mg/mL HRP solution (*see* **Note 1**).
2. Add 100 μL above solution to 1 mL of ConA-coated MNPs suspension.
3. Stir the solution at room temperature for 1 h for enzyme agglutination.
4. Add 40 μL of 5% w/v glutaraldehyde solution to the above solution, stir the mixture at room temperature for 2 h.
5. Add 25 μL of freshly prepared 4 mg/mL $NaBH_4$ solution to the above solution to reduce the Schiff base for 2 h at 4 °C.
6. Collect and rinse the obtained crosslinked ConA-MNP-carried GOx-HRP aggregates (CLCMGHA) with 1.0 mL of phosphate buffer (10 mM, pH 7.0) by magnetic decantation. Repeat three times.
7. Resuspend the CLCMGHA in 1.0 mL of phosphate buffer (10 mM, pH 7.0) by gentle vortexing.
8. Store the CLCMGHA suspension at 4 °C.

3.2.4 Activity Assay

1. Add 10 μL of 100 mM ABTS solution to 980 μL of glucose substrate solution (50 mM phosphate buffer, pH 7.0, 100 mM glucose) in a quartz cuvette.
2. Add 10 μL of the mixture of GOx and HRP solution (or the suspension of CLGCHA, or the suspension of CLCMGHA) to above solution, and mix the solution immediately by repetitive pipetting.
3. Immediately measure the time-course increase of absorbance at 415 nm for at least 1 min on a UV-Vis spectrophotometer at room temperature.
4. Calculate the activity of the enzyme cascade based on the slope of absorbance versus time curve (*see* **Note 6**).

4 Notes

1. The ratio of GOx to HRP can be varied for the preparation of GOx-ConA-HRP aggregates.
2. The agglutination efficiency is controlled by the concentration ratio of ConA to glycoenzymes. The conditions described here can achieve agglutination efficiencies of 98% for GOx and 90% for HRP. The ConA agglutination is more affinitive to GOx than to HRP.
3. Calibration is done with GOx, HRP, and ConA solutions of known concentrations, respectively.
4. The synthesis of magnetic nanoparticles is modified based on Massart's method [15].
5. Other protein assays may be used besides the BCA protein assay such as absorbance at 280 nm.
6. The extinction coefficient of the final product ABTS radical cation at 415 nm is $\varepsilon_{415} = 3.6 \times 10^4$/M/cm. The activity of GOx-HRP cascade is defined as the production rate of ABTS radical cation.

Acknowledgement

This work is supported by the National Key Research & Development Program of China under Grant No. 2021YFC2104100, and National Natural Science Foundation of China under Grant No. NSFC-22178018.

References

1. Sperl JM, Sieber V (2018) Multienzyme cascade reactions – status and recent advances. ACS Catal 8:2385–2396
2. Zhang Y, Hess H (2017) Toward rational design of high-efficiency enzyme cascades. ACS Catal 7:6018–6027
3. Rabe KS, Müller J, Skoupi M, Niemeyer CM (2017) Cascades in compartments: en route to machine-assisted biotechnology. Angew Chem Int Ed 56:13574–13589
4. Kuzmak A, Carmali S, von Lieres E, Russell AJ, Kondrat S (2019) Can enzyme proximity accelerate cascade reactions? Sci Rep 9:455
5. Küchler A, Yoshimoto M, Luginbühl S, Mavelli F, Walde P (2016) Enzymatic reactions in confined environments. Nat Nanotechnol 11:409–420
6. Fahien LA, Kmiotek EH, MacDonald MJ, Fibich B, Mandic M (1988) Regulation of malate dehydrogenase activity by glutamate, citrate, alpha-ketoglutarate, and multienzyme interaction. J Biol Chem 263:10687–10697
7. An S, Kumar R, Sheets ED, Benkovic SJ (2008) Reversible compartmentalization of de novo purine biosynthetic complexes in living cells. Science 320:103–106
8. Huffman MA et al (2019) Design of an in vitro biocatalytic cascade for the manufacture of islatravir. Science 366:1255–1259
9. Schoffelen S, van Hest JCM (2013) Chemical approaches for the construction of multienzyme reaction systems. Curr Opin Struct Biol 23:613–621
10. Zhang Y, Yong Y, Ge J, Liu Z (2016) Lectin agglutinated multienzyme catalyst with enhanced substrate affinity and activity. ACS Catal 6:3789–3795
11. Apweiler R, Hermjakob H, Sharon N (1999) On the frequency of protein glycosylation, as deduced from analysis of the SWISS-PROT database. Biochim Biophys Acta 1473:4–8
12. Pazur JH, Aronson NN (1972) Glycoenzymes: enzymes of glycoprotein structure. In: Tipson RS, Horton D (eds) Advances in carbohydrate chemistry and biochemistry, vol 27. Academic Press, New York, pp 301–341
13. Gray RD, Glew RH (1973) The kinetics of carbohydrate binding to concanavalin A. J Biol Chem 248:7547–7551
14. Yong Y, Su R, Liu X, Xu W, Zhang Y, Wang R, Ouyang P, Wu J, Ge J, Liu Z (2018) Lectin corona enhances enzymatic catalysis on the surface of magnetic nanoparticles. Biochem Eng J 129:26–32
15. Massart R (1981) Preparation of aqueous magnetic liquids in alkaline and acidic media. IEEE Trans Magn 17:1247–1248

Chapter 13

Self-Assembled Multienzyme Nanostructures for Biocatalysis in Cellulo

Qixin Wei and Jiang Xia

Abstract

Multienzyme complexes naturally exist in cells to catalyze cascade reactions in metabolic pathways. By clustering the enzymes in close proximity, these nanomachineries achieve effective conversion of metabolites. Bioengineers are working on the development of synthetic versions of multienzyme complexes in cells to synergize heterologous biosynthesis. Assembling enzymes on protein scaffolds through protein–protein interactions is a viable and facile way to form synthetic multienzyme complexes. Here, we describe the general methods to construct self-assembled multienzyme nanostructures in *Escherichia coli* for biosynthesis of valuable chemicals.

Key words Multienzyme complexes, Protein scaffold, Protein assembly, Biosynthesis, In cellulo

1 Introduction

In nature, cells often construct multienzyme complexes, distinct from a mixture of enzymes, to catalyze the cascade metabolic reactions [1–4]. The complexes cluster enzymes in sophisticated structures, in which the intermediates directly transport from the active sites of one enzyme to another. Local confinement of the reactants allows for the regulation of the metabolic flux and increases the biosynthetic production yield [5–7].

Scientists are interested in constructing functional protein complexes for a variety of biotechnological applications. Several methods are established, among which protein fusion is the most traditional and direct way. By connecting different protein domains through designed linkers, bifunctional or multifunctional enzyme fusions can be obtained [5, 8]. Forcing the direct linkage of two enzymes in a single peptide however may disrupt the tertiary structures of the enzymes and incur decrease or loss of the enzyme activity [9, 10]. Alternatively, enzymes can form complexes through co-immobilization on solid supports such as polymer

Haralambos Stamatis (ed.), *Multienzymatic Assemblies: Methods and Protocols*,
Methods in Molecular Biology, vol. 2487, https://doi.org/10.1007/978-1-0716-2269-8_13,
© The Author(s), under exclusive license to Springer Science+Business Media, LLC, part of Springer Nature 2022

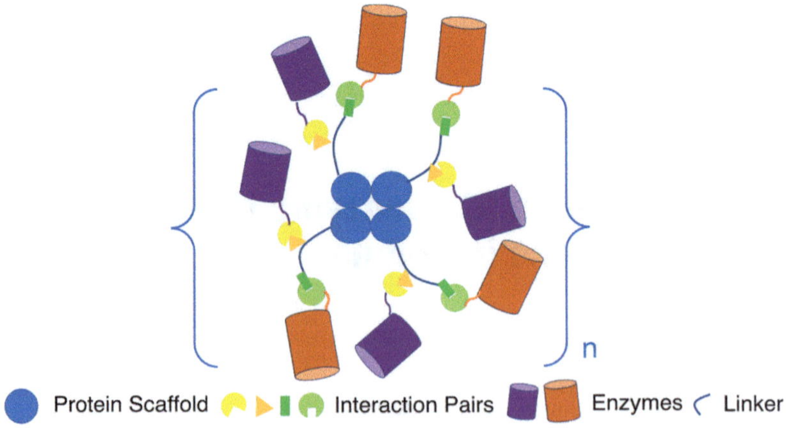

Fig. 1 Enzyme assembly on synthetic protein scaffolds. Biosynthetic enzymes are assembled onto protein scaffold through orthogonal peptide–peptide interaction pairs

particles or quantum dots [7, 11]. These complexes however cannot be constructed inside cells.

Inspired by the structures of cellulosomes, a nature's strategy to assemble enzymes on protein scaffolds [12], researchers have designed synthetic protein scaffolds to dock enzymes that are involved in the biosynthesis of valuable chemicals including lycopene, astaxanthin, amorpha-4,11-diene, catechin, and hydrochione [13–17]. By genetically fusing the interacting pairs with the enzymes and the scaffold, respectively (the interacting pairs are often short peptide tags that do not affect the structure and function of the enzymes), biosynthetic enzymes are assembled onto the protein scaffold to form a multienzyme nanostructure (Fig. 1). Here, we describe the general method to construct such self-assembled multienzyme nanostructures in *Escherichia coli* for biosynthesis of valuable chemicals.

2 Materials

All the solutions are made using deionized water and reagents of analytical grade. Prepare and store all the reagents at room temperature prior to use, unless indicated otherwise.

2.1 Bacterial Media

All the solutions need to be sterilized by autoclaving and be stored at 4–25 °C. Choose the proper media for bacterial growth (*see* **Note 1**). Before inoculating the bacteria, appropriate antibiotics are added at the ratio of 1:1000 (*see* **Note 2**).

1. Lysogeny Broth: 10 g tryptone, 5 g yeast extract, and 10 g NaCl per liter.

2. 2X YT Media: 16 g tryptone, 10 g yeast extract, and 5 g NaCl per liter (adjust pH to 7.0 with 5 N NaOH).
3. Terrific Broth (TB) Medium: 900 mL (a) and 100 mL (b). Prepare and sterile solution (a) and (b) separately (*see* **Note 3**).
   (a) 24 g yeast extract, 20 g tryptone, and 4 mL glycerol.
   (b) Phosphate buffer: 2.31 g KH_2PO_4 and 12.54 g K_2HPO_4 per 100 mL.
4. Agar plates: 5 g NaCl, 5 g tryptone, 2.5 g yeast extract, and 7.5 g agar in 500 mL deionized water. Sterilize, cool down to below 60 °C, and add the appropriate antibiotics (*see* **Note 4**). Pour medium in the plates, use about 20 mL in each 100 mm diameter plate, and allow them to set at room temperature. Seal the plates with parafilm and store at 4 °C.

2.2 Plasmid Construction

1. High-fidelity DNA polymerase for PCR amplification.
2. DNA fragments. DNA fragments after PCR are purified by gel electrophoresis using DNA Electrophoresis Cells & Systems (Bio-Rad) and extracted by Monarch DNA Gel Extraction Kit (New England Biolabs).
3. Gibson Assembly Master Mix (New England Biolabs).
4. Homemade competent *E. coli* strains XL1 Blue or Top 10 for transformation.
5. E.Z.N.A plasmid mini kit I (V-spin) (Omega Biotek) for plasmid extraction.

2.3 Protein Expression

1. Commercial or homemade competent *E. coli* strain BL21 (DE3).
2. Antibiotic stock: 100 mg/mL for ampicillin, 50 mg/mL for kanamycin, and 34 mg/mL for chloramphenicol (*see* **Notes 5** and **6**). All the antibiotics are stored at −20 °C before use.
3. Isopropylthio-β-galactoside (IPTG) for the induction of protein expression (*see* **Note 7**). Usually, a stock of 1 M is prepared by dissolving 2.38 g IPTG in 10 mL sterilized distilled water under room temperature and divided into 1 mL for each tube. Then the stocks are stored at −20 °C.
4. Glycerol as an additional carbon source.
5. UV spectrophotometer for the detection of optical density at 600 nm (OD_{600}) to monitor the cell growth.

2.4 Product Extraction and Quantification

1. Dodecane, extraction solvent.
2. Hexane, wash solvent.
3. Gas chromatography-mass spectrometer for product quantification.

2.5 Transmission Electron Microscopy (TEM)

1. Solvents and chemicals used in fixation, dehydration, and penetration process: glutaraldehyde, agarose, phosphate buffer (pH 7.4), osmic acid, alcohol, acetone, 812 embedding reagent.
2. Embedding plate.
3. Ultra-thin sections for slicing.
4. Uranyl acetate saturated alcohol solution and lead citrate for staining.
5. Transmission electron microscope.

3 Methods

3.1 Plasmid Construction

Preparing the plasmids encoding the genes for multiple proteins and enzymes is the first step to form multienzymes complexes in bacterial cells. Only plasmids with different replication region and antibiotic resistance can be co-transformed to ensure the co-existence of both plasmids in bacterial strain. More than one gene may also be encoded in one plasmid in which a ribosome binding site is needed for each gene. Gibson assembly is a convenient method to construct the plasmid with multiple genes. The general procedure is shown below.

1. Design primers for amplifying the fragments with overlap regions (*see* **Note 8**).
2. Amplify the fragments through polymerase chain reaction (PCR) (*see* **Note 9**).
3. Prepare the vector fragment by restriction digestion or PCR amplification.
4. Mix the DNA fragments with Gibson Assembly Master Mix and incubate at 50 °C for 1 h (*see* **Note 10**).
5. After incubation, add 10 μL reaction mixture into homemade competent cell XL1 Blue or Top 10 for transformation (*see* **Note 11**).
6. Select and inoculate several single colonies in LB for overnight culture to amplify the plasmids. Plasmids can be extracted using the plasmids extraction kit.
7. Sequence the plasmids to determine whether the corresponding fragments have been inserted and detect unwanted mutations.
8. Store the plasmids at −80 °C for further utilization.

3.2 Construction of Multienzyme Complexes in E. coli

1. Construct *E. coli* strains by co-transforming the plasmids that encode the genes for the expression of protein scaffold and enzymes with interaction pairs, into competent cells *E. coli* BL21(DE3) (*see* **Note 12**).
2. Prepare the start culture by picking up the single colonies into LB with antibiotics, followed by overnight incubation in a shaker at 37 °C .
3. Then inoculate the start culture into 2X YT medium with antibiotic and allow it to grow at 37 °C until the OD_{600} reaches 0.6–0.8.
4. Add IPTG to a final concentration of 100 μM to 1.0 mM to induce the protein expression (*see* **Note 13**).
5. Incubate the cell culture at 30 °C for protein expression and product biosynthesis. The recombinant proteins shall self-assemble with the protein scaffold to form the multienzyme nanostructures spontaneously inside the cells.

3.3 Product Extraction and Quantification

For volatile products such as amorpha-4,11-diene and α-farnesene:

1. Overlay dodecane onto the cell culture after IPTG addition.
2. Collect the emulsified layer and centrifuge at 13,000 × *g* for 3 min.
3. Collect the supernatant and subject to GC-MS for product analysis.
4. GC-MS conditions: RTX-5 MS column (inner diameter, 0.25 mm; length, 30.0 m; film thickness, 0.25 μm). Helium is used as the carrier gas. Injector temperature set as 250 °C and the oven temperature starts with holding at 160 °C for 2 min. Then the temperature is increased at the rate of 15 °C/min to 250 °C and let hold for two more minutes [17].

For non-volatile products such as β-carotenoids lycopene and astaxanthin:

1. Obtain the cell pellet by centrifugating the cell cultures.
2. Discard the supernatant and suspend the pellet in DI water to ensure complete removal of the growth medium. Then centrifuge again and collect the pellet.
3. Dry the cell pellet in a SpeedVac concentrator with a vacuum pump and ground the dried pellet into powder.
4. Extract the products by acetone at 55 °C for 15 min.
5. Collect the supernatant after centrifugation, then dilute in methanol and subject it into HPLC for further quantification.
6. Conditions for reverse phase HPLC: Column, analytical C18 reverse phase column. Water and ACN with 0.1% TFA as mobile phases. Products are separated by the gradient running at 1 mL/min.

3.4 Transmission Electron Microscopy (TEM)

If the multienzyme complexes are significantly large, they can be visualized under TEM (*see* **Note 14**). A general procedure is shown below.

1. Prepare the cell cultures and express the enzyme complexes as described in Subheading 3.1.
2. Fixation. Collect the pellet by centrifugation (100 μL cell culture as example). Resuspend the pellet with 500 μL 2.5% glutaraldehyde and incubate at room temperature for 2–4 h. Collect the cell pellet by centrifugation again, followed by wrapping in 1% agarose. Rinse the pellet with 0.1 M phosphate buffer (pH = 7.4) for 3 times.
3. Post-fixation process. Resuspend and incubate the pellet with 0.1 M phosphate buffer (pH = 7.4) with 1% osmic acid at room temperature for 2 h. Rinse with 0.1 M phosphate buffer again.
4. Dehydration. The cells are then sequentially dehydrated with 50% alcohol,70% alcohol, 80% alcohol,90% alcohol,95% alcohol,100% alcohol, 100% alcohol, 100% acetone, 100% acetone, for 15 min each.
5. Permeabilization. Acetone and 812 embedding reagent mixture (1:1 v/v) are used to permeabilize the sample for 2–4 h, then in acetone and 812 embedding reagent mixture (2:1 v/v) overnight, and then in pure 812 embedding reagent overnight at 37 °C.
6. Slicing. Slice the sample into 60–80 nm ultra-thin sections.
7. Staining. The uranium-lead double staining method includes staining the sample with 2% uranyl acetate saturated alcohol solution and lead citrate for 15 min, respectively. Then dry the sample overnight at room temperature.
8. Observe the sample under a transmission electron microscope and collect images for further analysis.

4 Notes

1. LB is the most popular and commonly used medium to culture *E. coli* for cloning and recombinant overexpression. 2X YT medium contains twice the concentration of yeast extract than normal LB medium. It is useful in maintaining the high-density cell culture for longer growth time. Terrific Broth is a nutrient rich medium that supports a rapid growth of bacterial cells. Tryptone and yeast extract are the sources for carbon, nitrogen, vitamins, minerals, and amino acids while glycerol acts as an additional carbon source.

2. The final concentrations of the antibiotics are normally 100 μg/mL for ampicillin, 50 μg/mL for kanamycin, and 34 μg/mL for chloramphenicol.
3. Potassium buffer is added to solution (a) after cooling down to below 60 °C.
4. If the pre-mixed LB agar powder is used, weigh the suggested amount rather than the individual ingredients.
5. Prepare ampicillin and kanamycin with sterilized distilled water and chloramphenicol with absolute ethanol.
6. The final concentration for ampicillin, kanamycin, and chloramphenicol is 10 mg/mL, 5 mg/mL, and 3.4 mg/mL, respectively.
7. IPTG is stored as 1 M stocks .
8. We suggest the overlap region should be 15–25 nt long, with a Tm equal to or higher than 48 °C.
9. Use high-fidelity DNA polymerase here.
10. The mixture normally contains 50–100 ng vector with two–three-fold excess inserts to ensure the high assembly efficiency. The concentration of the DNA fragments can be determined by NanoDrop instrument.
11. Bacteria are selected on antibiotic plate because only the one carrying the correct plasmid is antibiotic-resistant, which can form single colony on the plate.
12. Only plasmids carrying different antibiotic genes can be co-transformed to ensure both recombinant plasmids coexist in *E. coli*.
13. The concentration depends on the strength of induction required, genotype of the cells, and the plasmid type.
14. To better preserve the integrity of cellular ultrastructure, cryo-fixation of biological samples can be an alternative method. Ultra-thin sections can be acquired through high-pressure fast-freezing, freeze-substitution, and then embedded in Epon.

References

1. Pan P, Woehl E, Dunn MF (1997) Protein architecture, dynamics and allostery in tryptophan synthase channeling. Trends Biochem Sci 22:22–27
2. Bayer EA, Belaich JP, Shoham Y, Lamed R (2004) The cellulosomes: multienzyme machines for degradation of plant cell wall polysaccharides. Annu Rev Microbiol 58:521 554
3. Tsuji SY, Cane DE, Khosla C (2001) Selective protein-protein interactions direct channeling of intermediates between polyketide synthase modules. Biochemistry 40:2326–2331
4. Castellana M, Wilson MZ, Xu Y et al (2014) Enzyme clustering accelerates processing of intermediates through metabolic channeling. Nat Biotechnol 32:1011–1018
5. Conrado RJ, Varner JD, DeLisa MP (2008) Engineering the spatial organization of metabolic enzymes: mimicking nature's synergy. Curr Opin Biotechnol 19:492–499

6. Siu KH, Chen RP, Sun Q et al (2015) Synthetic scaffolds for pathway enhancement. Curr Opin Biotechnol 36:98–106
7. Whitaker WR, Dueber JE (2011) Metabolic pathway flux enhancement by synthetic protein scaffolding. Methods Enzymol 497:447–468
8. Ljungcrantz P, Carlsson H, Mansson MO, Buckel P, Mosbach K, Buelow L (1989) Construction of an artificial bifunctional enzyme, beta-galactosidase/galactose dehydrogenase, exhibiting efficient galactose channeling. Biochemistry 28(22):8786–8792
9. Miles EW (1995) Tryptophan synthase: structure, function, and protein engineering. Subcell Biochem 24:207–254
10. Meynial Salles I, Forchhammer N, Croux C et al (2007) Evolution of a *Saccharomyces cerevisiae* metabolic pathway in *Escherichia coli*. Metab Eng 9(2):152–159
11. Betancor L, Betancor L, Luckarift HR, Spain JC (2006) Coimmobilization of a redox enzyme and a cofactor regeneration system. Chem Commun (Camb) 14(34):3640–3642
12. Lior Artzi, Edward A. Bayer, Sarah Moraïs (2017) Cellulosomes: bacterial nanomachines for dismantling plant polysaccharides. Nat Rev Microbiol 15:83–95
13. Kang W, Liu J, Wang J, Nie Y, Guo Z, Xia J (2014) Cascade biocatalysis by multienzyme-nanoparticle assemblies. Bioconjug Chem 25(8):1387–1394
14. Lamed R, Setter E, Bayer E (1983) Characterization of a cellulose-binding, cellulase-containing complex in clostridium thermocellum. J Bacteriol 156(2):828–836
15. Qu J, Cao S, Wei Q, Zhang H, Wang R, Kang W et al (2019) Synthetic multienzyme complexes, catalytic nanomachineries for cascade biosynthesis *in vivo*. ACS Nano 13(9): 9895–9906
16. Wei Q, He S, Qu J, Xia J (2020) Synthetic multienzyme complexes assembled on virus-like particle for cascade biosynthesis *in cellulo*. Bioconjug Chem 31(10):2413–2420
17. Duber JE, Wu GC, Malmirchegini GR et al (2009) Synthetic protein scaffolds provide modular control over metabolic flux. Nat Biotechnol 27(8):753–759

Chapter 14

Nanoarmored Multi-Enzyme Cascade Catalysis

Mansi Malhotra, Ankarao Kalluri, and Challa Vijaya Kumar

Abstract

This chapter reports a single-step preparation of nanoarmored bi-enzyme systems assembled on 1-D and 2-D nanomaterials, with glucose oxidase and peroxidase enzymes as model systems for cascade bio-catalysis. This is a simple and facile method to both exfoliate the bulk 1D (carbon nanotubes, CNT) and 2D nanomaterials (α-Zirconium phosphate, α-ZrP) and bind the enzymes in a single step. Exfoliation of the bulk material enhances the accessible surface area of the materials for the enzyme binding, and it also boosts the diffusion of reagents from the bulk phase to the active sites of the bio-catalysts. For example, a mixture of horseradish peroxidase, glucose oxidase, and bovine serum albumin (BSA) were adsorbed on the surfaces of the α-ZrP nanoplates or carbon nanotubes (CNT) as the bulk materials are exfoliated simultaneously, in a one-step process. The resulting bio-catalysts were thoroughly characterized by powder X-ray diffraction, electron microscopy, biochemical and biophysical methods, while enzyme activity studies proved successful binding of enzymes with retention of activities or even enhancements in their specific activities. For example, GOx/HRP/BSA/CNT displayed 6 times the activity of a mixture of GOx/HRP/BSA, under otherwise identical conditions. Similarly, GOx/HRP/BSA/ZrP had 3.5 times the activity of the corresponding mixture of GOx/HRP/BSA, in the absence of the nanoplates. These robust nano-dispersions worked extraordinarily well as active bio-catalysts. These two kinds of fabricated biocatalyst dispersions are also highly stable.

Key words Enzymology, Zirconium phosphate, Carbon nanotubes, Horseradish peroxidase, Glucose oxidase, Enzyme assays, Enzyme kinetics

1 Introduction

Elucidating the details of nanomaterial and protein/enzyme [1] interactions is of fundamental importance for the rational design of stable, functional, and robust enzyme-based catalysts for industrial bio-catalysis [2], bio-fuel cells [3], bio-sensors [4], and bioengineering [5] applications. Here, we report a simple and effective method for the nanoarmoring of cascade enzymes [6] by bio-nanomaterials without any supplementary treatment for enzyme attachment and enhanced activities.

We report facile exfoliation of the functional nanomaterials, the two-dimensional layered α-Zirconium phosphate [7] (α-ZrP, 2D)

Haralambos Stamatis (ed.), *Multienzymatic Assemblies: Methods and Protocols*,
Methods in Molecular Biology, vol. 2487, https://doi.org/10.1007/978-1-0716-2269-8_14,
© The Author(s), under exclusive license to Springer Science+Business Media, LLC, part of Springer Nature 2022

and the one-dimensional multiwalled carbon nanotubes [8] (MWCNT, 1D). Exfoliation was facilitated by a mixture of bovine serum albumin (BSA) and selected enzymes. Furthermore, exfoliation increases the surface area, surface-to-volume ratio, and allows binding of the protein/enzyme to the exposed surfaces of the nanomaterials. These biocatalyst dispersions were used to study the activity for cascade enzymology with glucose oxidase (GOx) and a peroxidase enzymes. The oxidation of glucose by GOx generates hydrogen peroxide which is utilized by the peroxidase to oxidize an organic substrate such as guaiacol and form a cascade couple. BSA works as a stabilizing agent and helps in both the exfoliation of nanomaterial and enzyme binding as well as stabilization of the enzymes. Attachment to nanomaterial surfaces is responsible for decreasing the overall entropy of enzymes rendering them more stable which is appealing for industrial catalysis [9, 10]. Enzyme binding depends on the charge, structure, and nature of the nanomaterials but pre-adsorption of BSA on these surfaces allows for both binding and stabilization of enzymes. By binding GOx and a peroxidase to the same nanomaterial, we compared the cascade enzymology of GOx/peroxidase bound to 1D vs 2D bio-nanomaterials. While CNTs are conductive, α-ZrP (2D) is an insulator [11], and in these experiments, the role of the nanomaterial in electronically coupling the redox centers of the cascade enzymes is investigated.

We hypothesize that the enzymes bound to 1D and 2D materials coated with BSA would improve their stability due to passification of high energy sites on the materials and stratification of the enzyme structure as a direct result of a reduction in enzyme conformational entropy [12]. Enzyme binding to exfoliated CNTs is expected to involve hydrophobic interactions and π-π stacking while H-bonding with BSA could enhance enzyme binding and stability [13]. Similarly, binding of the redox enzymes to the BSA-coated 2D material is expected to stabilize the biocatalyst while placing the redox couple in close proximity. In both cases, the proximity of the enzyme couple would enhance diffusion of the intermediates involved in the cascade catalysis.

Multi-enzymatic cascade reactions have recently been recognized as one of the most significant technologies to succeed in laboratory and industrial applications, such as the synthesis of pharmaceutical, cosmetic, and nutritional compounds [14]. The enzymes such as GOx and peroxidase enzymes like horseradish peroxidase (HRP), myoglobin (Mb), and hemoglobin (Hb) can facilitate electron transfer between the substrate and electrode [15, 16]. Thus, using a simple benchtop stirring method, the cascade enzymes were bound to the nanomaterials without the use of any other toxic reagent or material [17]. Enzymatic activities of not just GOx but also, the peroxidase enzymes bound to 1D and 2D materials were monitored by utilizing standard activity assay protocols. Interestingly, two-dimensional nanosheets of α-ZrP showed a much higher protein adsorption capacity than one-dimensional CNTs. We also found that GOx/HRP/BSA/ZrP had 3.5 times the activity of the corresponding mixture of GOx/HRP/BSA, in the absence of the nanoplates. Similarly, GOx/HRP/BSA/CNT had 6 times the activity of a mixture of GOx/HRP/BSA, under otherwise identical conditions. The details of our investigations are described below. Interestingly, specific activities of enzymes bound to the one-dimensional CNTs were higher than its two-dimensional counterpart, the nanosheets of α-ZrP and this might indicate the electronic coupling of the redox sites by the highly conducting CNTs.

2 Materials

Prepared all solutions using deionized water unless stated otherwise. Prepared the solutions at room temperature and stored in the refrigerator at 2–8 °C (unless indicated otherwise). Proteins and enzymes are sensitive to temperature (should be stored in the freezer at −10 °C), and we recommend that one should not leave them at room temperature for long times (*see* **Note 1**). Diligently followed all waste disposal regulations when disposing of the waste materials.

2.1 Preparation of ZrP-BSA-Enzyme Suspension

1. α-Zirconium phosphate (α-ZrP), synthesized following the well-known procedure [18].
2. Bovine Serum Albumin, BSA.
3. Glucose oxidase, GOx (Aspergillus niger).
4. Horseradish peroxidase, HRP.
5. Hemoglobin, Hb (human).
6. Myoglobin, Mb (bovine).
7. 5 ml glass vials as reactors
8. Magnetic stir bars (octagon- 1 cm × 0.2 cm) and benchtop magnetic stirrers.
9. Eppendorf tubes (1.5 mL) for centrifugation.
10. **Buffer**: 10 mM Sodium phosphate buffer, pH 7.2. Add 800 mL of distilled water in a 1 L beaker. Add 1.785 g of disodium phosphate and 0.461 g monosodium phosphate. Adjust the pH of the solution using HCl or NaOH to get the desired pH (7.2). Make up the volume to 1 L by adding distilled water and rechecking the pH.
11. KCl stock solution: 100 mM in distilled water. This is required for adjusting the conductivity and to measure the zeta potentials of samples.

12. TEM copper grid (Ted Pella) were used for TEM studies.
13. 20 μL, 200 μL, and 1000 μL micropipettes.

2.2 Preparation of MWCNT/BSA Dispersion

1. Safety equipment such as lab coat, safety goggles, and gloves.
2. Multiwalled carbon nanotubes, MWCNT (length 10–30 μm, outer diameter <8 nm, purity >95% purchased from Cheap Tubes, USA).
3. Bovine Serum Albumin, BSA.
4. Distilled water (18 MΩ.cm resistivity).
5. Round bottom flask (250 mL), magnetic stir bars of octagon shaped crossections (12.7 × 8 mm), and benchtop magnetic stirrers.
6. Eppendorf tubes (1.5 mL) and round bottom flask (250 mL).

2.3 Synthesis of BioCNT-GOx–HRP Nanomaterial Complex

1. Safety equipment such as lab coats, safety goggles, and gloves.
2. CNT stock dispersion: Prepare 5 mg/mL of bioCNT stock dispersion by mixing the concentrated, exfoliated, and debundled bioCNT (0.5 mL, 10 mg/mL) dispersion with 0.5 mL of phosphate buffer solution.
3. HRP stock solution: 114 μM of HRP (5 mg/mL) stock solution prepared by mixing HRP with 1 mL of buffer solution.
4. GOx stock solution: 5 mg/mL of GOx (31 μM) stock solution prepared by mixing GOx with 1 mL buffer solution.
5. Eppendorf tubes (1.5 mL).

2.4 Removing Unbound (Free) Enzymes and Proteins from the CNT Complex

1. Safety equipment such as lab coats, safety goggles, and gloves.
2. 15 mL glass vials.
3. UV-Visible spectrophotometer.
4. 100 μL and 1000 μL pipettes.
5. Beakers, conical flasks, measuring cylinders.

2.5 Enzyme Activity Studies of Bio-nanomaterials Bound to Enzymes

1. Safety equipment such as lab coats, safety goggles, and gloves.
2. Control GOx/HRP/BSA: Prepare stock solution of GOx (0.2 mg/mL), HRP (0.2 mg/mL), and BSA (2 mg/mL) from the main stock in 1 mL of the buffer solution (*see* **Note 2**).
3. Stock hydrogen peroxide (30%): 100 mM solution in buffer.
4. Stock guaiacol/o-methoxyphenol: 50 mM solution in buffer stored in dark (*see* **Note 3**).
5. Stock glucose: 30 mM solution in buffer solution.
6. Centrifuge (17,000 × *g*).
7. Volumetric micropipettes (20, 100, 200, and 1000 μL).

8. Glass vials (10 mL), centrifuge tubes (1.5 mL, 2 mL, and 15 mL).
9. UV-Visible spectrophotometer (Agilent) or plate reader instrument (Molecular devices).
10. Nylon or PES membrane filter (typical 180 μm thickness, 0.22 μm pore size, 4.7 cm diameter).

3 Methods

Enzymes are a naturally occurring class of proteins and nucleic acids, which have special properties like specificity, selectivity, and high catalytic activity. Enzymes are biofriendly, biocompatible, accelerate reaction rates by several orders of magnitude and produce fewer side-products than normal chemical reactions. Due to these fascinating properties, enzymes are used in the food industry, biofuel cells, wastewater treatment, and industrial bio-catalysis. Nanomaterials and MWCNTs are widely used as special supports for enzyme binding and stabilization. Due to the high aspect ratio and large surface area of these materials, they provide high enzyme loading capacity per unit area. Therefore, the enzymes immobilized on these robust nano scaffolds show increased enzyme stability and high volumetric enzyme activity. The surface functionalization process can further improve the stability of nanomaterials. GOx and peroxidase enzymes were immobilized onto the bioZrP and bioCNT's surfaces by physical binding via non-covalent interactions [3].

Performed all experiments at room temperature, unless mentioned otherwise.

3.1 Preparation of BioZrP-HRP-GOx Complex in a Single Step and Enzyme Nanoarmoring

3.1.1 Preparation of BioZrP-HRP-GOx Suspension

1. Weigh 5 mg bovine serum albumin (BSA), 1 mg of horseradish peroxidase/hemoglobin/myoglobin, and 1 mg of glucose oxidase and dissolve in 1 mL of phosphate buffer (10 mM, pH 7.2) (*see* **Note 4**).
2. Add 20 mg of crystalline α-Zirconium phosphate to the above solution.
3. Stir (100 mg) the α-ZrP, enzymes, and protein solution for 24 h on a benchtop magnetic stirrer in a 5 mL glass vial (*see* **Note 5**). The solutions will look as shown in Fig. 1.
4. Separate out the unexfoliated α-ZrP from the reaction mixture by centrifugation. Transfer the samples to a centrifuge tube and centrifuge them at 1500 × *g* for 10 min. Carefully isolate the supernatant from the bulk α-ZrP pellet [9] (*see* **Note 6**).
5. Calculate the final concentration of α-ZrP by drop-casting 200 μL of the exfoliated α-ZrP on a pre-weighed aluminum foil. Dry the samples overnight and weigh them (*see* **Note 7**).

	GOx/HRP/BSA/ZrP
GOx Bound	53. ± 2
HRP bound	82. ± 3

	GOx/Mb/BSA/ZrP
GOx Bound	43 ± 2
Mb bound	94 ± 2

	GOx/HRP/BSA/ZrP
GOx Bound	50 ± 1
HRP bound	86 ± 1

Fig. 1 Before and after pictures of the stirred GOx/peroxidase enzyme/BSA/ZrP. (**a**) GOx/HRP/BSA/ZrP (**b**) GOx/Mb/BSA/ZrP (**c**) GOx/Hb/BSA/ZrP and their respective percent binding affinities

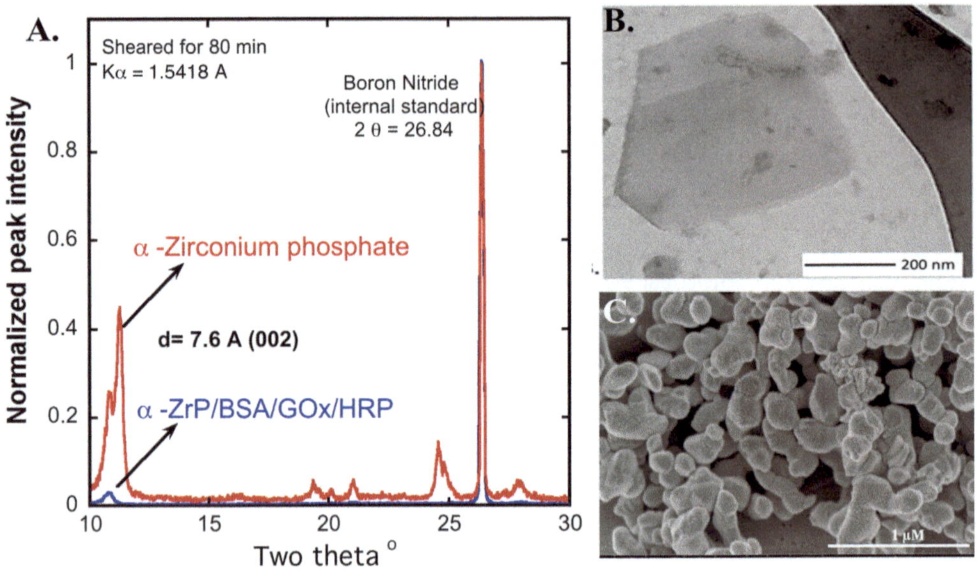

Fig. 2 Characterization of α-ZrP. (**a**) XRD (**b**). TEM and (**c**) SEM of GOx/HRP/BSA/ZrP

Also, characterize the exfoliated α-ZrP by X-ray diffraction [18] (XRD) and microscopy techniques as shown in Fig. 2.

3.1.2 Binding of Enzymes to α-ZrP

1. Centrifuge GOx/peroxidase enzyme/BSA/ZrP at 17,000 × *g* for 55 min to remove the excess protein and respective enzymes.
2. Save the supernatant to determine the concentration of bound enzymes carefully, without disturbing the pellet at the bottom, by monitoring the absorbance of heme (Soret absorbance at 403 nm) (*see* **Note 8**).
3. Determine the absorbance of supernatant at 403 nm (HRP) or 405 nm (Hb/Mb) and compare it with the control solution (1 mg GOx, 1 mg HRP/Hb/Mb, 5 mg BSA in 10 mM

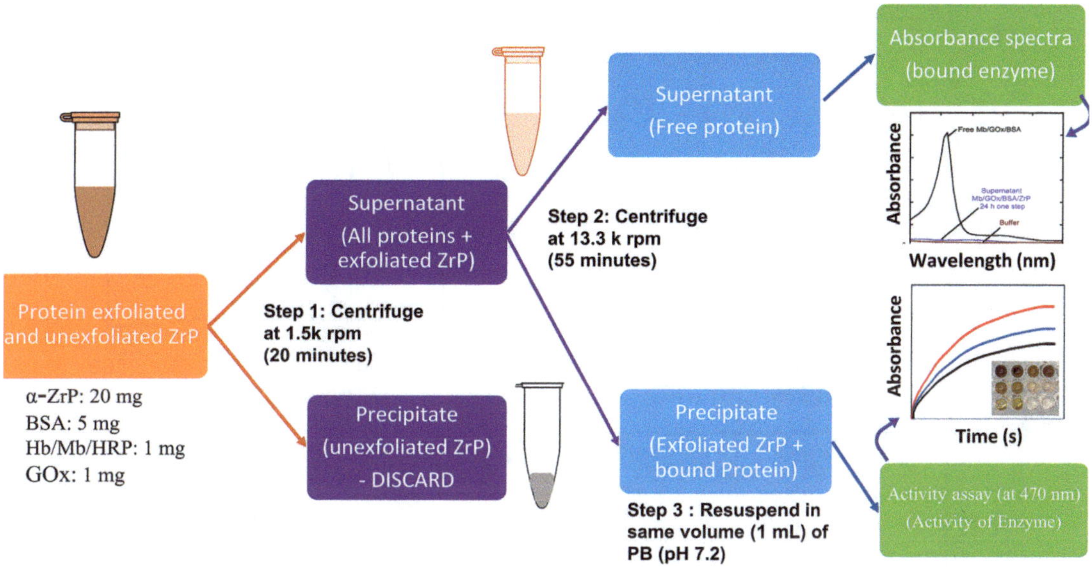

Scheme 1 Workflow for the binding of enzymes to α-ZrP

phosphate buffer, pH 7.2) to determine the amount of free enzyme.

4. Subtract the amount of calculated free (unbound) enzyme from the total amount of enzyme to estimate the amount of bound enzyme (*see* **Note 9**). For instance, in this case, the amount of free HRP in GOx/HRP/BSA/ZrP is 1 mg. Therefore, we subtracted the calculated 0.2 mg of HRP from 1 mg to get 8 mg of bound HRP (80% binding).
5. Resuspend the pellet in 1 ml of buffer solution to get exfoliated GOx/peroxidase enzyme/BSA/ZrP and determine the activities of enzymes present (*see* **Note 10**). Scheme 1 explains all the steps more clearly. The expected binding results are expressed in Fig. 3 wherein Mb shows the highest binding affinity followed by Hb and HRP. The highest GOx binding was observed with HRP but was around 40–50% for all samples [19].

3.1.3 TEM Characterization of Materials

1. Operate the Tecnai T12 instrument at an accelerating voltage of 120 kV (*see* **Note 11**).
2. Make aqueous samples of bioCNT (Subheading 3.2) and BioZrP dispersions at a concentration of ~ 1 nM in 1 mL distilled water. Use 10 μL of KCl to increase the conductivity of the samples.
3. Use a micropipette to drop cast 5 μL of the sample on the copper grid covered with a Formvar film (*see* **Note 12**).
4. Blot the excess solution using filter paper and air dry.

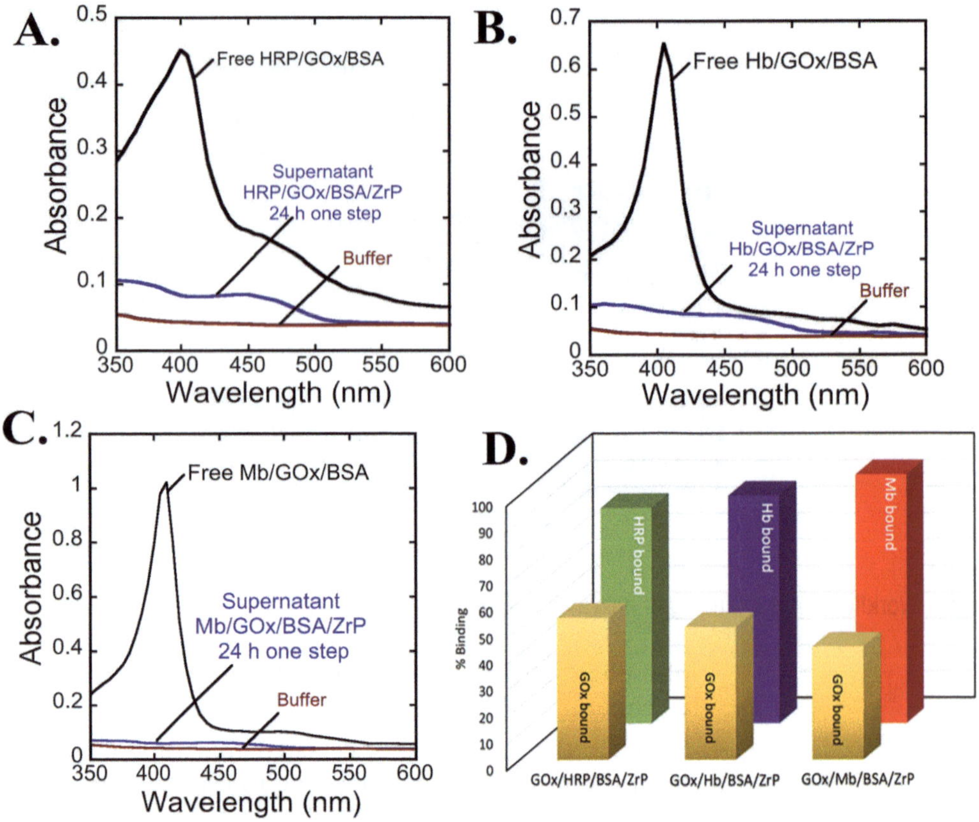

Fig. 3 Binding of enzymes in GOx/peroxidase enzyme/BSA/ZrP (10 mM sodium phosphate buffer at pH 7). Absorbance spectra for supernatant of (**a**) GOx/HRP/BSA/ZrP samples (depletion in peaks of HRP and GOx) (**b**) GOx/Hb/BSA/ZrP samples (depletion in peaks of Hb and GOx) (**c**) GOx/Mb/BSA/ZrP samples (depletion in peaks of HRP and GOx). (**d**) %binding of cascade enzymes in GOx/peroxidase enzyme/BSA/ZrP system

5. Dry the sample and image using Tecnai T12 TEM.
6. Alternatively, at the same time prepare the control ZrP samples in methanol solution by repeating the same steps.

3.1.4 Activity of GOx in GOx/Peroxidase Enzyme/BSA/ZrP Suspension

1. Dilute the exfoliated GOx/peroxidase enzyme/BSA/ZrP with buffer to make sure that the GOx and HRP/Hb/Mb concentrations are 1 μM each in the sample, respectively. Determine the absorbance of this solution at 470 nm (*see* **Note 13**).
2. Add 10 μL of the o-methoxyphenol stock solution to the above samples and mix them well.
3. Add 20 μL of the glucose stock solution to the above solution and monitor its activity at 470 nm for 1 min (*see* **Note 14**).
4. Determine the initial rate of the reaction from the slope of absorbance of the sample at 470 nm versus time (in minutes) for the first 20–30 s of the plot (linear portion of the plot) (*see*

Note 15). Correct the sample's absorbance by subtracting the initial absorbance at 470 nm.

5. Use the known extinction coefficient of 26,600/M/cm at 470 nm and the grams of enzyme present in the sample to calculate the activity of GOx [3].
6. Dilute the control solution with buffer to make sure that the GOx and HRP/Hb/Mb concentrations are 1 μM each in the sample. Calculate its activity by the following **steps 2–4**. Reference the activity of the control solution as 100% and use it to calculate the specific activities of other samples [20].

3.1.5 Activity of HRP in GOx/Peroxidase Enzyme/BSA/ZrP Suspension

1. Dilute the exfoliated BioZrP/peroxidase enzyme/GOx with buffer to make sure that the HRP concentration is 1 μM in the sample.
2. Add 10 μL of the o-methoxy phenol stock solution to the above samples and mix them well [21].
3. Add 5 μL of the hydrogen peroxide stock solution to the above solution and monitor its activity at 470 nm for 1 min.
4. Determine the initial rate of the reaction from the slope of absorbance of the sample at 470 nm versus time (in minutes) for the first 20–30 s of the plot (linear portion of the plot). Correct the sample's absorbance by subtracting the initial absorbance at 470 nm.
5. Use the known extinction coefficient of 26,600/M/cm at 470 nm and the grams of enzyme present in the sample to calculate the activity of HRP/Hb/Mb.
6. Dilute the control solution with buffer to make sure that the HRP/Hb/Mb concentration is 1 μM in the sample. Calculate its activity by the following **steps 2–4**. Reference the activity control solution as 100% and use it to calculate the specific activities of other samples. Our results from these experiment are displayed in Fig. 4. The activity of HRP was 3.1× times the activity of free HRP in GOx/HRP/BSA/ZrP.

3.2 Preparation of BioCNT/HRP/GOx Complex in One Step and Enzyme Nanoarmoring

3.2.1 Preparation of CNT-BSA Dispersion

1. Weigh 200 mg of MWCNT and transfer to 100 mL round bottom flask.
2. Dissolve BSA (200 mg/mL) in distilled water (18 mL) and adjust the pH of the solution to 7.2 by using 0.1 M HCl or 0.1 M NaOH solution and make up the final volume to 20 mL (*see* **Note 16**).
3. Transfer the above BSA solution into 100 mL of the round bottom flask.
4. Stir and heat the CNT and BSA mixture at 65 °C for 5 days (*see* **Note 17**).

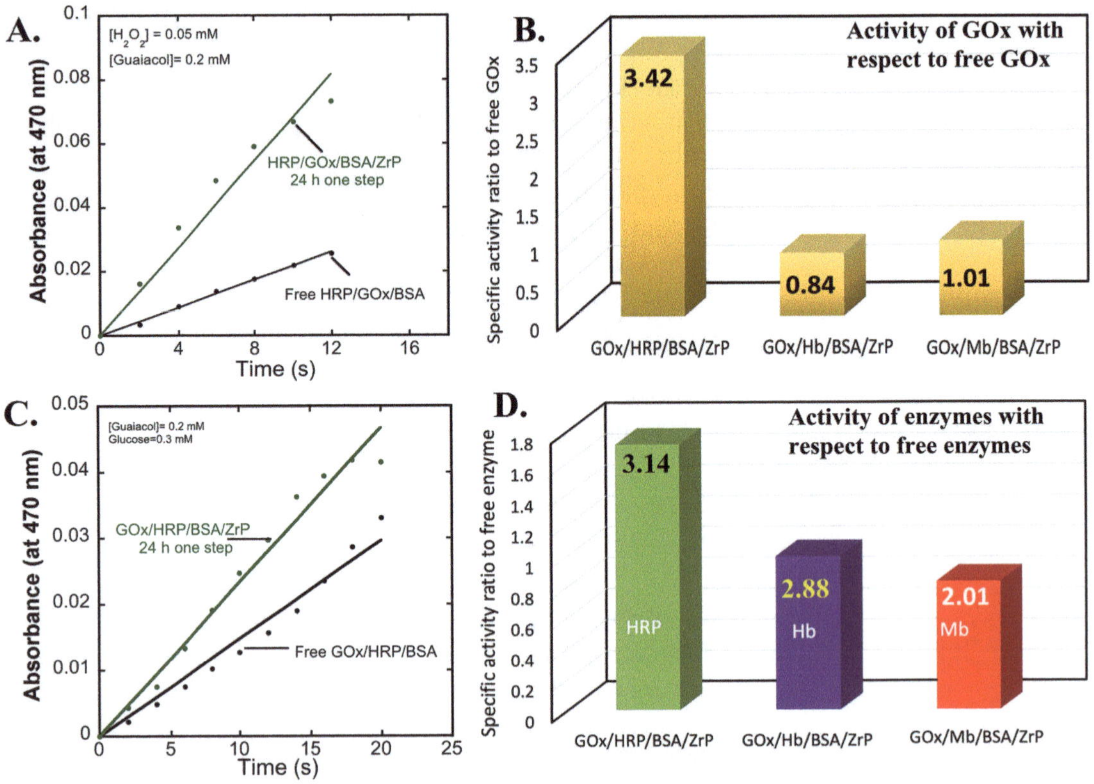

Fig. 4 Enzyme kinetics of the bound enzymes. (**a**) Activity of GOx in GOx/HRP/BSA/ZrP (**b**) Specific activity of GOx in GOx/peroxidase enzyme/BSA/ZrP (**c**) Peroxidase activity of HRP in GOx/HRP/BSA/ZrP (**d**) Specific activity of peroxidase enzymes in GOx/peroxidase enzyme/BSA/ZrP

5. Use a reflux condenser to prevent the evaporation of water in the round bottom flask.
6. By centrifugation method, remove the bundled MWCNT from the reaction mixture (*see* **Note 18**).
7. Transfer the MWCNT-BSA reaction mixture into a centrifuge tube (1.5 mL or 15 mL), and centrifuge at 900 × *g* for 5 min. Remove the supernatant carefully from the unexfoliated solid MWCNT pellet. Scheme 2 elaborates on the workflow of preparing the CNT suspensions.
8. Estimate the final concentration of MWCNT in the dispersion (supernatant) by using a UV-visible spectrophotometer [8].

3.2.2 Removal of Free BSA from CNT Dispersion

1. Take 1 mL of CNT-BSA dispersion (in triplicates) from the above step (after centrifugation at 900 × *g* for 5 min) and use for the following steps.
2. Centrifuge CNT-BSA dispersion at 17,000 × *g* for 3 h or more until a clear solution is obtained.

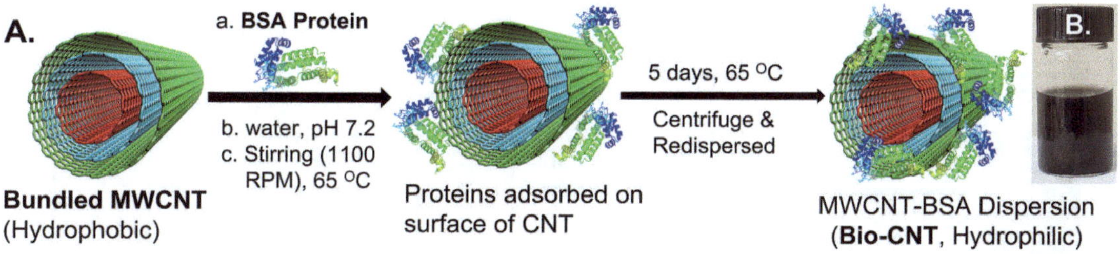

Scheme 2 Schematic representation for the preparation of exfoliated and debundled bioCNT dispersion by using BSA. (**a**) MWCNT and BSA solution mixed at pH 7.2 and stirred for 5 days at 65 °C followed by repeating the steps centrifuge, re-disperse, and disperse to make bioCNT dispersion. (**b**) Exfoliated and debundled bioCNT dispersion in a vial (image). BSA image taken from the protein data bank (PDB ID: 3V03)

3. Collect the supernatant solution, leaving only the BSA bound CNT pellet at the bottom.
4. Calculate the concentration of free BSA from the isolated supernatant (free/unbound BSA) solution.
5. Re-disperse the CNT-BSA pellet in distilled water and vortex well to disperse the solid pellet (*see* **Note 19**).
6. Again, centrifuge the new dispersion at 17,000 × *g* for 60 min to get a clear supernatant.
7. Remove clear supernatant from the centrifuge tube, leaving only the CNT pellet at the bottom. Save the removed supernatant protein solution for later to confirm the free BSA concentration (*see* **Notes 20** and **21**).
8. Resuspend the CNT pellet in distilled water to get the bioCNT dispersion. The bioCNT dispersion from this step can be used for further characterization and experiments.
9. By continuous washing of CNT and BSA dispersions, more than ~67% of free BSA is removed from the dispersions and ~33% of BSA is bound to the CNT in bioCNT dispersion. Due to this residual BSA present in the bioCNT dispersion, CNTs are readily resuspended in water due to the hydrophilic nature of bound BSA. Eventually, a fully exfoliated bioCNT (10 mg/mL) solution contains 6.7 mg of CNT with 3.3 mg of bound BSA (*see* **Note 22**). Scheme 3 represents the preparation of bound BSA CNT without the presence of unbound protein. Figures 5 and 6 display the bioCNT as characterized by UV-Visible spectroscopy and microscopy techniques [22].

3.2.3 Preparation of BioCNT/HRP/GOx Complex

1. Add Bio-CNT (2.5 mL, 25 mg), HRP (200 μL, 1 mg), and GOx (200 μL, 1 mg) and adjust the final volume to 5 mL with buffer (referred to as GOx/HRP/BSA/CNT). The workflow of this process is shown in Scheme 4.

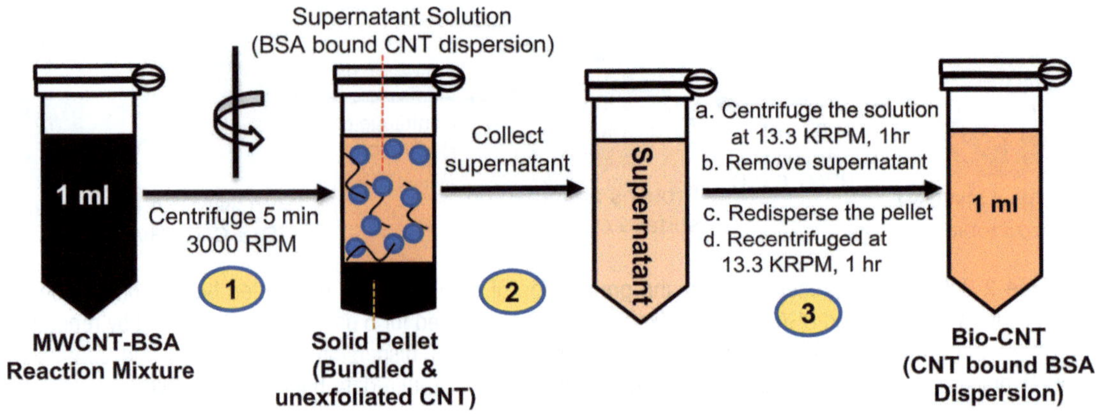

Scheme 3 Schematic representation of preparing bioCNT dispersion. Step 1: Centrifuge MWCNT with BSA reaction mixture for 5 min at 900 × *g*. Step 2: Remove the supernatant and recycle the solid pellet for the next reaction. Step 3: Collect the dispersion and redo the centrifugation step at 17,000 × *g* for 1 h to get the clear solution (free BSA). Repeat the steps of centrifugation and redispersion 2 times to remove the free BSA completely. Re-disperse the final solid pellet in deionized water to prepare the exfoliated bioCNT dispersion

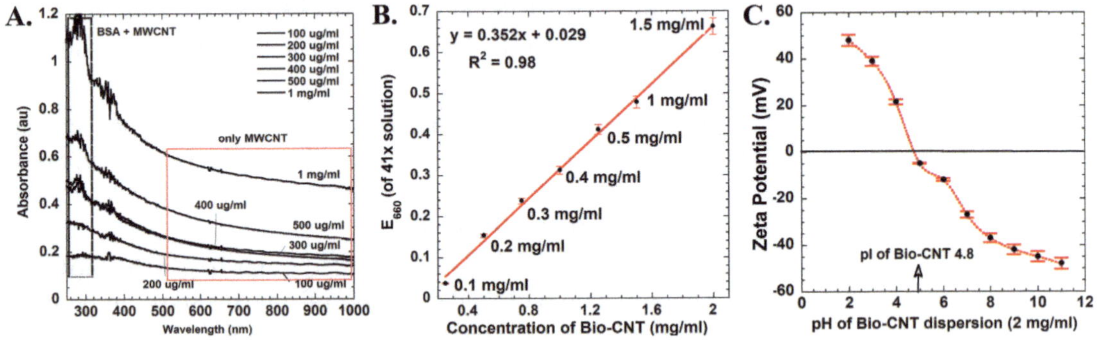

Fig. 5 (**a**) The absorption spectra of the bioCNT dispersion with the BSA and MWCNT combined absorbance at 250–300 nm. (**b**) The calibration curve used to calculate the extinction coefficient of the bioCNT dispersion at pH 7.2 in deionized water, all measurements were performed in triplicates. (**c**) Zeta potential of biofunctionalized CNT samples after incubation (1 day) at different pHs at room temperature

2. To stabilize and equilibrate the mixed samples, store them in the refrigerator for 2–8 °C for 24 h. By equilibration, the enzymes are adsorbed and immobilized on the surface of the bioCNT [23] (*see* **Note 23**).

3.2.4 Removal of the Excess Enzymes

1. Centrifuge bioCNT-HRP-GOx at 17,000 × *g* for 55 min to remove the excess enzymes/protein, repeat twice.
2. Save the supernatant to determine the concentration of bound enzymes carefully, without disturbing the pellet at the bottom, by monitoring the activity of supernatant (next section) (*see* **Note 24**).

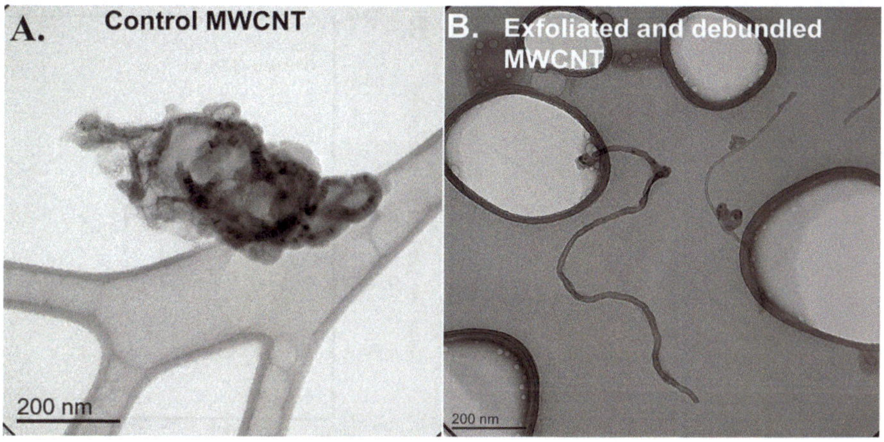

Fig. 6 TEM micrographs of (**a**) control MWCNT dispersed in methanol solution, (**b**) CNT-BSA (BioCNT) dispersion

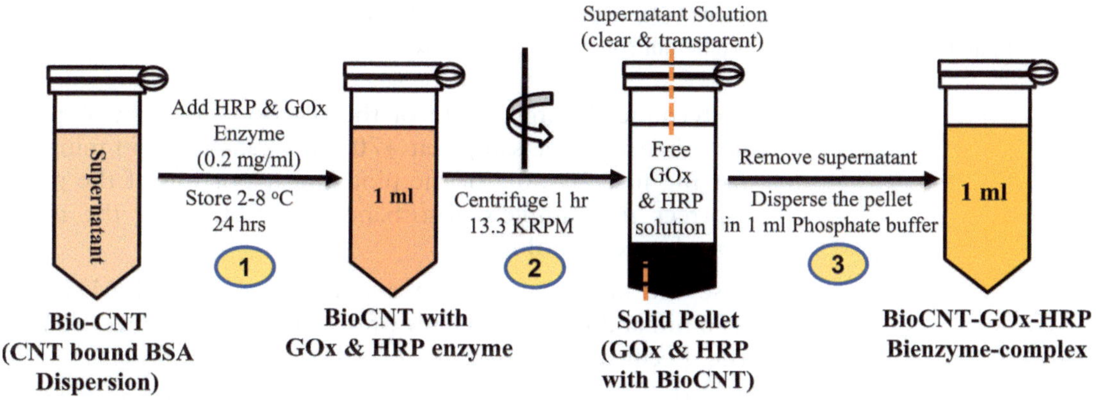

Scheme 4 Schematic illustration of BioCNT-GOx-HRP bi-enzymatic complex preparation and immobilization. Step 1: Incubation of the GOx and HRP enzymes with BioCNT dispersion at 2–8 °C for 24 h. Step 2: Centrifugation of the bi-enzymatic complex samples and collection the solid pellet (bound enzymes). Step 3: Redispersion of the solid pellet in 10 mM phosphate buffer

3. Resuspend the pellet in the buffer to get exfoliated BioCNT/HRP/GOx and determine the activities of dispersed enzymes (*see* **Note 25**).

3.2.5 Binding and Activity of GOx to BioCNT-HRP-GOx

1. Dilute the control solution in the buffer to obtain increasing concentrations of GOx ranging from 0.002 mg/mL to 0.02 mg/mL.
2. To determine the activities, add 20 μL of the stock O-methoxyphenol to the above samples (200 μL in the wells of a 96-well plate) and mix them well.
3. Add 20 μL of the stock solution of glucose to the above solution and monitor activity by following absorbance at 470 nm, each 2 seconds for 1 min.

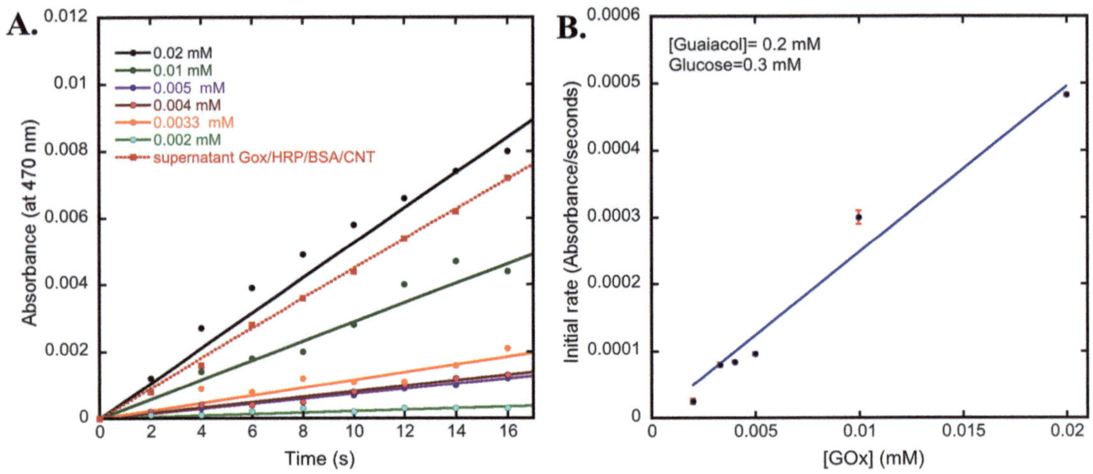

Fig. 7 Standard curve construction and its use for evaluating the binding of GOx to bioCNT/HRP/GOx. (**a**) Kinetic traces obtained for different concentrations of GOx ranging from 0.002 mg/mL to 0.02 mg/mL in phosphate buffer, pH 7.2, 10 mM and kinetic trace of GOx in the supernatant of bioCNT-HRP-GOx (red curve). (**b**) Standard curve created from the activities of different concentrations of GOx

4. Determine the initial rate of the reaction from the slope of absorbance of the sample at 470 nm versus time (in minutes) for the first 20–30 s of the plot (linear portion of the plot). Correct the sample's absorbance by subtracting the initial absorbance at 470 nm.
5. Use the known extinction coefficient of 26,600/M/cm at 470 nm and mass of enzyme present in the sample to calculate the specific activity of GOx.
6. Plot the activities versus the concentration of GOx to create the standard curve (Fig. 7a).
7. Dilute the supernatant from Subheading 3.2.4, **step 2** (×10) and determine its activity by following **steps 3–6**.
8. Use the activity of the supernatant and the slope of the standard curve to determine the concentration of free GOx in the supernatant (Fig. 7b).
9. Subtract the amount of free GOx from the initial concentration of GOx added to bioCNT dispersion to calculate the amount of bound GOx (Fig. 10a).
10. To determine the activity of GOx in bioCNT/HRP/GOx, dilute the suspensions by the buffer (20 times), add substrate solution and determine its absorbance at 470 nm (*see* **Note 26**).
11. Determine its activity by following **steps 3–6**. Reference the activity of the control solution as 100% and use it to calculate the relative specific activities. (As shown in Figs. 9a and 10b).

3.2.6 Binding and Activity of HRP in bioCNT-HRP-GOx

The peroxidase (HRP enzyme) activity is measured through the oxidation of the guaiacol (brown color). The activity of HRP in bioCNT dispersion is measured through conventional HRP assay. Next, the specific activity of the control enzyme (GOx/HRP/BSA) solution is compared with that of the GOx/HRP/BSA/CNT complex. The activity of GOx/HRP/BSA/CNT is also compared with that of GOx/HRP/BSA/ZrP.

1. Dilute the control solution in buffer to obtain different concentrations of HRP ranging from 0.002 mg/mL to 0.02 mg/mL.
2. To determine the activities, add 20 μL O-methoxyphenol to the above samples and mix them well.
3. Add 10 μL hydrogen peroxide reagent to the above solution and monitor the kinetic curve by recording absorbance at 470 nm, each 2 s for 1 min.
4. Determine the initial rate of the reaction from the initial slope of the absorbance of the sample at 470 nm versus time for the first 20–30 s of the linear portion of the plot. Correct the sample's absorbance by subtracting the initial absorbance at 470 nm (*see* **Note 27**).
5. Use the known extinction coefficient of 26,600/M/cm at 470 nm of the product and the mass of enzyme present in the sample to calculate the specific activity of HRP.
6. Plot the activities versus the concentration of HRP to create the standard curve (Fig. 8a).

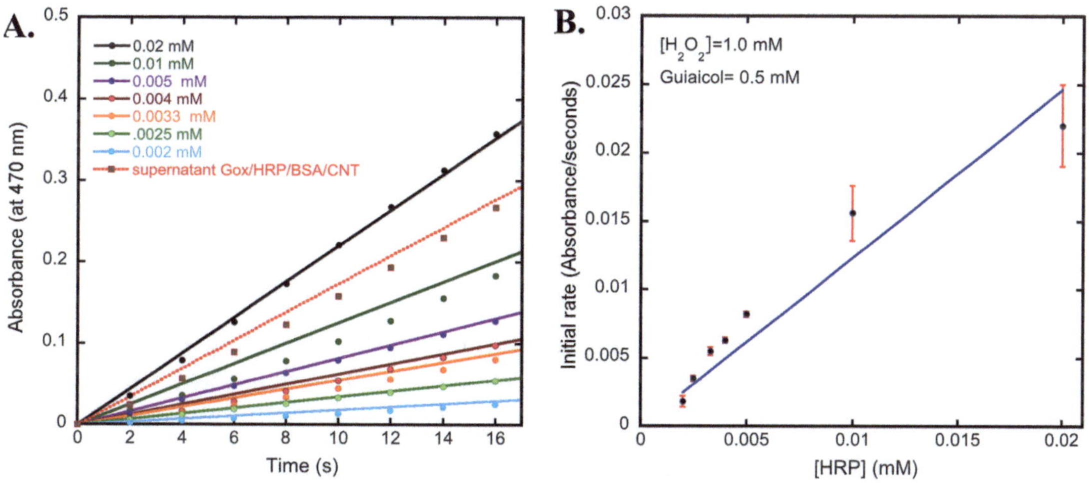

Fig. 8 Standard curve construction and its use for evaluating the binding of HRP to bioCNT/HRP/GOx. (**a**) Kinetic traces of different concentrations of HRP ranging from 0.002 mg/mL to 0.02 mg/mL in phosphate buffer, pH 7.2, 10 mM and kinetic trace of HRP in the supernatant of bioCNT/HRP/GOx (red curve). (**b**) standard curve created from the activities of different concentrations of HRP

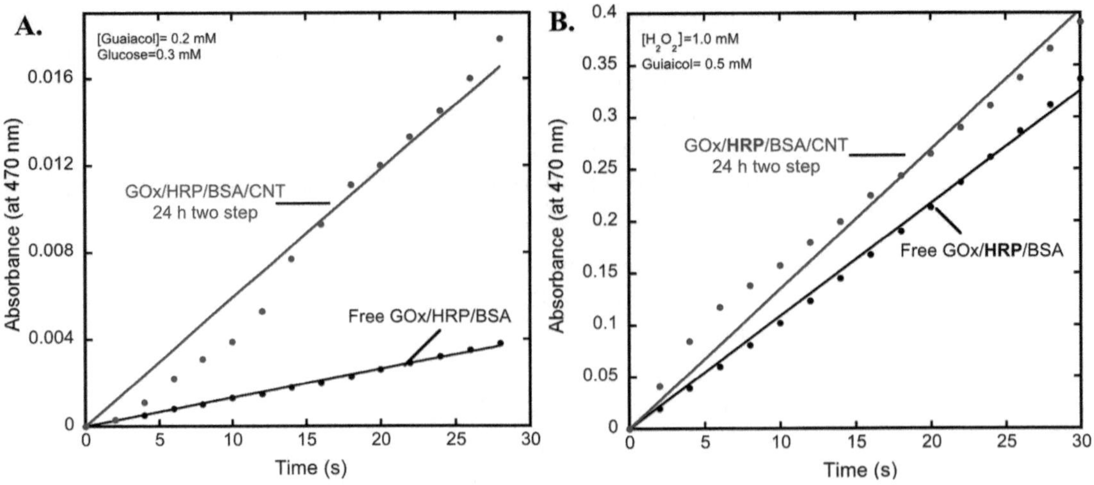

Fig. 9 Enzyme kinetic traces of GOx and HRP bound to bioCNT as compared to free GOx and HRP enzyme (**a**). Kinetic traces of activity of GOx in bioCNT-HRP-GOx (**b**). Kinetic traces of activity of HRP in bioCNT/HRP/GOx

7. Dilute the supernatant from Subheading 3.2.4, **step 2** (×10) and determine its activity by following **steps 3–6**.
8. Use the initial activity of the supernatant and the slope of the standard curve to determine the concentration of free HRP in the supernatant (Fig. 8b).
9. Subtract the amount of free HRP from the initial concentration of GOx added to bioCNT dispersion to calculate the amount of bound GOx (Fig. 9a).
10. To determine the activity of GOx in GOx/HRP/BSA/CNT, dilute the suspensions by using pH 7.2, 10 mM phosphate buffer (20 times), determine its absorbance at 470 nm (*see* **Note 28**).
11. Determine its activity by following **steps 3–6**. Reference the activity of the control solution as 100% and use it to calculate the relative specific activities of other samples (Figs. 9b and 10b).

In this chapter, we successfully demonstrated exfoliation of materials (1D and 2D) with proteins and their application in novel, enhanced cascade enzymology. The CNT, being a conductive material is expected to bridge the active sites of the two redox enzymes of the cascade, and binding of both partners of the cascade to enhance the activity by the proximity effect. In the case of α-ZrP, being an insulating material, it may only increase the activity of cascade enzymes due to the proximity effect only. Our experimental results indicated that GOx activity in bioCNT-HRP-GOx was increased significantly in comparison to that of GOx/HRP/BSA/ZrP. The three mechanisms that we suggest may be operating here are: (1) the transfer of electrons from GOx to the CNT,

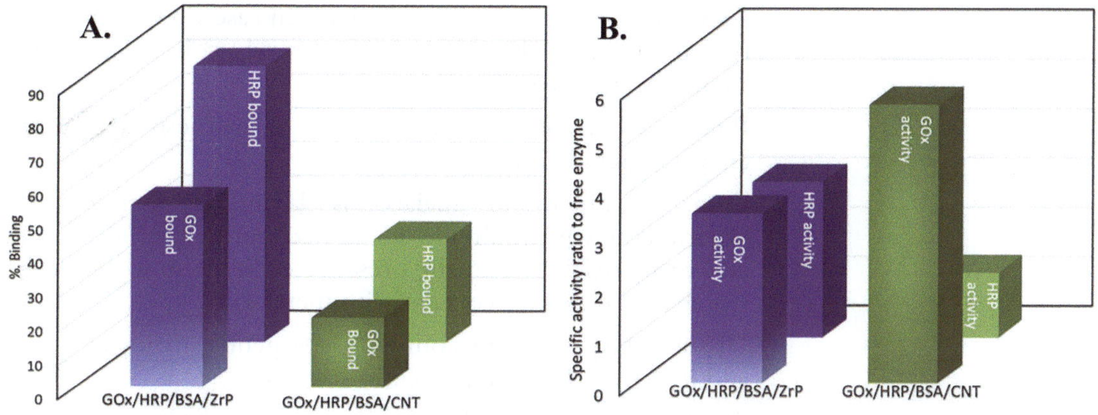

Fig. 10 Comparison of binding and activity of enzymes to insulating material, α-ZrP and conducting material, bioCNT. (**a**) % Binding of enzymes (GOX and HRP) in α-ZrP (first column) and bioCNT (second column). (**b**) Specific activities of enzymes (GOX and HRP) in α-ZrP (first column) and bioCNT (second column) with respect to free enzyme activity

from the CNT bridge to oxygen to form hydrogen peroxide which then reacts with HRP that is closest to it with enhanced activity and/or the oxidized HRP reacts with the substrate O-methoxyphenol via electron transfer facilitated by CNT bridge, (2) the proximity of redox enzymes where the hydrogen peroxide produced by GOx is quickly shuttled to a neighboring HRP due to the proximity effect, and (3) the change in enzyme structures after binding to expose the active sites with improved access by the substrates [24]. All three possibilities might work with CNTs while only steps (2) and (3) congtribute for enhanced activities with α-ZrP. The high specific activities and high percent binding values in both α-ZrP and CNT are a welcome change (Fig. 10) as well, but careful consideration of all the factors, parameters, and operating conditions performed during the experiment are to be examinedd in a detailed way in future work.

4 Notes

1. The stability of bioCNT and α-ZrP/BSA dispersions are important parameters to be analyzed for biological applications. BioCNT and α-ZrP/BSA dispersions can be stored at 2–8 °C for more than 3 months and 25 °C for up to 2 weeks with little or no loss of activities. Poor stability and aggregation of the CNT and α-ZrP dispersions are indicated by settling of the precipitate at the bottom of the container. If the sample needs to be stored for more than 3 months without any sterilization, there is a possibility of the growth of bacteria in bioCNT and α-ZrP/BSA dispersions due to degradation of

BSA and other proteins by microbes. In case we need to store the samples for a long time, we add benzyl alcohol (1%, 20 μL) to prevent the growth of bacteria.

2. Prepare fresh solutions at the time of synthesis to prevent enzyme degradation.
3. O-methoxyphenol or guaiacol is flammable, therefore, all experiments with it should be carried out in the hood. It is also easily oxidizable in the presence of light, so it should be stored in the dark or covered with metal foil.
4. Dissolve the proteins in a buffer first by gentle stirring and then add α-ZrP to the protein solution.
5. Since the solutions will be stirred overnight, the best practice is to either stick them on the magnetic stirrer with tape or use a retort stand.
6. Carefully transfer the supernatant without disturbing the precipitate. The precipitate can be dried out and subtracted from the starting amount of α-ZrP to calculate the amount of exfoliated α-ZrP.
7. Make sure that the mass of aluminum foil is comparable to the expected mass of exfoliated α-ZrP for maximizing the accuracy and precision and to minimize errors.
8. All the absorbance values should be under 1 to follow Beer's law. We dilute the supernatant and the control samples 4–5 times to achieve this.
9. We always save some of the control samples from the beginning of the experiment, so that we do not have to remake them later and save us from possible human error.
10. Resuspension is difficult to achieve, however, pipetting the solution up and down with a 1000 mL pipet helps a lot. Once all of the GOx/peroxidase enzyme/BSA/ZrP is dispersed, we move on to the next step.
11. TEMs are tricky to use. Please make sure you are fully trained before you plan to take images of your samples.
12. Make sure you drop cast the sample in a way that a thin layer is formed on the grid. We add 200 μL of the sample to achieve this.
13. We dilute the solutions at least 10 times to avoid scattering from α-ZrP. The absorbance spectra for these samples are collected in order to confirm the scattering and subtract the scattering effect from the activity assay. We sometimes use the instrument to blank the absorbance of GOx/peroxidase enzyme/BSA/ZrP before starting the activity assay to achieve this.

14. Glucose should be added last (since it is the first substrate that is required for the assay) and the sample should be stirred continuously while performing the assay to get better results.
15. Calculate the initial rate from the linear part of the absorbance versus time (which is generally the first the 20–30 s of the reaction).
16. We synthesize bioCNT at pH 7.2, so we generally add dilute NaOH to adjust the pH. Handle the pH meter carefully and wear gloves and other protective equipment before handling HCl and/or NaOH solutions.
17. To monitor the progress of exfoliation, aliquots of 50 μL samples can be taken out at specific intervals of time (every 24 h) and diluted with 2 mL of deionized water.
18. The reaction mixture MWCNT-BSA samples can be stored at 2–8 °C after the completion of final reaction step.
19. In case the MWCNT pellet does not dissolve well in water after the centrifugation step, sonicate the dispersion for 5 min to break up the pellets, and re-disperse the pellet in DI water.
20. CNTs are highly dispersible in water because of the BSA protein, so it is difficult to remove all the CNTs from the dispersions, even after prolonged periods (5–6 h) of centrifugation.
21. Alternatively, to overcome this centrifugation step (*see* **Note 20**), use nylon or PES membrane filter (0.22 μm) to remove the dispersed CNT followed by vacuum filtration set up to save time.
22. To obtain the dried mass of bioCNT (~10 mg/mL), drop cast 1 mL bioCNT dispersion on aluminum foil and dry in an oven at 50 °C for 2 h.
23. Shaking or gentle mixing of the bio-enzymatic complex will help in decreasing the enzyme loading time and might increase the binding of enzymes to the nanomaterial.
24. The bioCNT-HRP-GOx might not settle down completely after the centrifugation step of bioCNT-HRP-GOx since the CNTs get dispersed very well in the protein solution. Therefore, the percent binding of enzymes should be calculated from the activity of the free enzyme in the supernatant solution.
25. The mass of dry CNT pellet indicated that less than 5% of the total CNT was left in the supernatant solution confirming that the enzyme activity of the supernatant can be used for calculating the binding of enzymes to CNT.
26. Dilute the CNT suspension so that the absorbance at 470 nm is less than 1. We dilute them 20 times to make sure that the absorbance from CNT is as less as possible under the condition that enzyme activity is still measurable.

27. Both CNTs and proteins showed absorbance at a wavelength of 280 nm, so the HRP enzyme concentration calculated by using the Soret absorbance at 403 nm in the bioCNT-GOx-HRP enzymatic complex, which is more reliable and accurate compared to absorbance at 280 nm.
28. High speed of shaking or mixing, prolonged centrifugation, and storing them for long times at room temperature (>20 °C) might decrease the enzyme activities by denaturation.

References

1. Nel AE et al (2009) Understanding biophysicochemical interactions at the nano-bio interface. Nat Mater 8:543–557
2. Zore OV, Pattammattel A, Gnanaguru S, Kumar CV, Kasi RM (2015) Bienzyme–polymer–graphene oxide quaternary hybrid biocatalysts: efficient substrate channeling under chemically and thermally denaturing conditions. ACS Catal 5:4979–4988
3. Chung Y, Tannia DC, Kwon Y (2018) Glucose biofuel cells using bi-enzyme catalysts including glucose oxidase, horseradish peroxidase and terephthalaldehyde crosslinker. Chem Eng J 334:1085–1092
4. Wang C, Wang Q, Tan R (2018) Preparation of enzyme-functionalized carbon nanotubes and their application in glucose and Fe(2+) detection through "turn on" and "turn off" approaches. Analyst 143:4118–4127
5. Jin L et al (2012) Functionalized graphene oxide in enzyme engineering: a selective modulator for enzyme activity and thermostability. ACS Nano 6:4864–4875
6. Hwang ET, Lee S (2019) Multienzymatic cascade reactions via enzyme complex by immobilization. ACS Catal 9:4402–4425
7. Baveghems CL, Anuganti M, Pattammattel A, Lin Y, Kumar CV (2018) Tuning enzyme/alpha-Zr(IV) phosphate nanoplate interactions via chemical modification of glucose oxidase. Langmuir 34:480–491
8. Kalluri A, Puglia MK, Malhotra M, Kumar CV (2020) Exfoliated and water dispersible biocarbon nanotubes for enzymology applications. Methods Enzymol 630:407–430
9. Pattammattel A, Deshapriya IK, Chowdhury R, Kumar CV (2013) Metal-enzyme frameworks: role of metal ions in promoting enzyme self-assembly on alpha-zirconium(IV) phosphate nanoplates. Langmuir 29:2971–2981
10. Kumar CV, Chaudhari A (2000) Proteins immobilized at the galleries of layered α-zirconium phosphate: structure and activity studies. J Am Chem Soc 122:830–837
11. Puglia MK, Malhotra M, Kumar CV (2020) Engineering functional inorganic nanobiomaterials: controlling interactions between 2D-nanosheets and enzymes. Dalton Trans 49:3917–3933
12. Deshapriya IK, Kumar CV (2013) Nanobio interfaces: charge control of enzyme/inorganic interfaces for advanced biocatalysis. Langmuir 29:14001–14016
13. Chen M, Zeng G, Xu P, Lai C, Tang L (2017) How do enzymes 'meet' nanoparticles and nanomaterials? Trends Biochem Sci 42: 914–930
14. Novak MJ et al (2015) "Stable-on-the-table" enzymes: engineering the enzyme–graphene oxide interface for unprecedented kinetic stability of the biocatalyst. ACS Catal 6:339–347
15. Magyar M et al (2016) Real-time sensing of hydrogen peroxide by ITO/MWCNT/horseradish peroxidase enzyme electrode. J Nanomater 2016:1–11
16. Lin Y, Ren J, Qu X (2014) Catalytically active nanomaterials: a promising candidate for artificial enzymes. Acc Chem Res 47:1097–1105
17. Azevedo RM et al (2015) A strategy for improving peroxidase stability via immobilization on surface modified multi-walled carbon nanotubes. J Chem Technol Biotechnol 90: 1570–1578
18. Sun L, Boo WJ, Sue H-J, Clearfield A (2007) Preparation of α-zirconium phosphate nanoplatelets with wide variations in aspect ratios. New J Chem 31:39–43
19. Pattammattel A et al (2013) Tuning the activities and structures of enzymes bound to graphene oxide with a protein glue. Langmuir 29: 15643–15654
20. Neves-Petersen MT et al (2007) Photophysics and photochemistry of horseradish peroxidase A2 upon ultraviolet illumination. Biophys J 92: 2016–2027

21. Zhang Y et al (2012) Assembly of graphene oxide-enzyme conjugates through hydrophobic interaction. Small 8:154–159
22. Majumder A et al (2013) Bio-functionalization of multi-walled carbon nanotubes. Phys Chem Chem Phys 15:17158–17164
23. Prasad KP, Chen Y, Chen P (2014) Three-dimensional graphene-carbon nanotube hybrid for high-performance enzymatic biofuel cells. ACS Appl Mater Interfaces 6:3387–3393
24. Kim K-W, Kim BC, Lee HJ, Kim J, Oh M-K (2011) Enzyme logic gates based on enzyme-coated carbon nanotubes. Electroanalysis 23: 980–986

Chapter 15

Implementing Multi-Enzyme Biocatalytic Systems Using Nanoparticle Scaffolds

Joyce C. Breger, Gregory A. Ellis, Scott A. Walper, Kimihiro Susumu, and Igor L. Medintz

Abstract

Interest in multi-enzyme synthesis outside of cells (in vitro) is becoming far more prevalent as the field of cell-free synthetic biology grows exponentially. Such synthesis would allow for complex chemical transformations based on the exquisite specificity of enzymes in a "greener" manner as compared to organic chemical transformations. Here, we describe how nanoparticles, and in this specific case—semiconductor quantum dots, can be used to both stabilize enzymes and further allow them to self-assemble into nanocomplexes that facilitate high-efficiency channeling phenomena. Pertinent protocol information is provided on enzyme expression, choice of nanoparticulate material, confirmation of enzyme attachment to nanoparticles, assay format and tracking, data analysis, and optimization of assay formats to draw the best analytical information from the underlying processes.

Key words Assay, Catalysis, Enzyme, Multi-Enzyme, Kinetics, Michaelis–Menten, Quantum dot, Nanoparticle, Spectrophotometry, Cascade, Synthetic Biology

1 Introduction

Understanding how to employ multi-enzyme cascades for small molecule synthesis is becoming a robust and exciting research area in synthetic biology [1, 2]. The ability to mix and match enzymes from different organisms such as from plants, yeast [3], bacteria, etc. [4] to yield pathways achieving synthesis of complex molecules [5–8] is tantalizingly close with far reaching consequences in both academia [9] and industry [10–12]. Soon biochemists will be freed from the strictures associated with traditional organic synthesis methods such as low yields, 100-step synthetic processes, and producing excess hazardous waste to create compounds that are commonly made with exquisite precision in nature using water, salts, other natural elements, and sugars [1, 6, 7, 13, 14]. The possibilities are literally endless as this technology matures. To

Haralambos Stamatis (ed.), *Multienzymatic Assemblies: Methods and Protocols*,
Methods in Molecular Biology, vol. 2487, https://doi.org/10.1007/978-1-0716-2269-8_15,
© The Author(s), under exclusive license to Springer Science+Business Media, LLC, part of Springer Nature 2022

that end, much research and effort still needs to be performed to understand and overcome the pitfalls of multi-enzyme cascades [15–17]. This includes identifying the fewest number of enzymes needed to achieve the synthesis of a desired compound efficiently, identifying their origin in nature, and how to isolate active forms [7, 18, 19]. Most of this can be accomplished through the development of sophisticated genomic sequencing software like those provided by NIH through the iSeqTools Portal [20] and machine learning algorithms to sort through the endless enzymatic combinations for candidate enzyme systems such as Retropath by the Faulon group [21–23]. The reaction conditions, the number of each enzyme needed to achieve maximum flux through the system, and how to increase the stability of enzymes must be done through more classical empirical research avenues [2, 24].

Real world application of enzymatic cascades to replace organic synthesis methods of high-value compounds hinges on creating stable enzyme systems. Numerous efforts have been expended on increasing the stability of enzymes through a variety of techniques such as immobilization on solid supports [25, 26], fused together via SpyTag/SpyCatcher then assembled to another support [26, 27], DNA [28], etc. [29, 30]. However, these techniques have their drawbacks such as limiting the movement of the enzyme, incorrect enzyme positioning, cost, and complexity. Although gaining long-term stability, enzymes commonly lose activity when attached to macro surfaces. Nanoparticles (NPs), defined as having spatial dimensions in the nanometer range, offer a way to increase enzyme stability without sacrificing functionality if they are used as the enzyme display scaffold [9, 31–34]. Methods have matured to create a wide range of NPs and functionalities that can be customized to the needs of a particular system such as NPs coated in stabilizing ligands that have a predominate pKa or are colloidally stable in nonpolar or polar environments [35–38]. In previous papers, we have shown how the nanoparticle (NP) colloidal stabilizing ligands can be engineered to have a specific pKa and other properties which directly influences enzyme kinetics [39–41]. This current methodology paper focuses on the use of inorganic, colloidal NPs known as semiconductor quantum dots (QDs) but there are a wide range of materials that can be employed including iron oxide and noble metal (Au, Ag) particulates [42–44].

NPs have intrinsic properties that are inherently different from their bulk equivalents due to their nanometer size. These properties include increased surface to volume ratio, modified quantum confined electronic, optical, and magnetic properties as well as restructuring of their local environment in comparison to the bulk solution which has direct consequences for enzyme activity [9, 45, 46]. In this methodology paper, we will specifically focus on the use of QDs as prototypical nanoscaffolds due to their ease of functionalization [47–50]. Semiconductor QDs with a ZnS shell

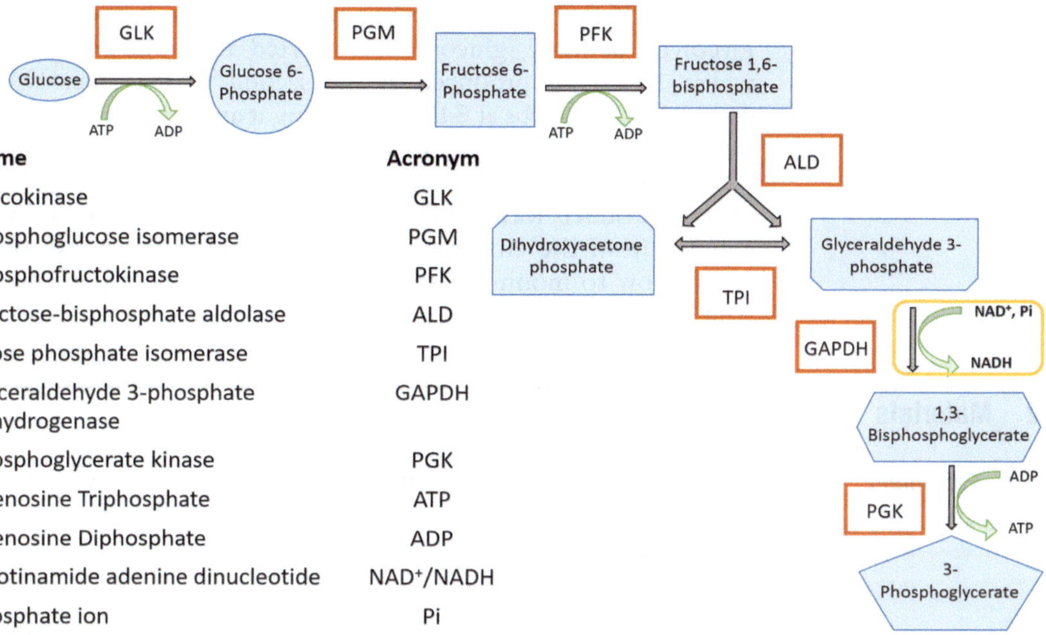

Name	**Acronym**
Glucokinase	GLK
Phosphoglucose isomerase	PGM
Phosphofructokinase	PFK
Fructose-bisphosphate aldolase	ALD
Triose phosphate isomerase	TPI
Glyceraldehyde 3-phosphate dehydrogenase	GAPDH
Phosphoglycerate kinase	PGK
Adenosine Triphosphate	ATP
Adenosine Diphosphate	ADP
Nicotinamide adenine dinucleotide	NAD^+/NADH
Phosphate ion	Pi

Fig. 1 An example of a multi-enzymatic pathway that can be studied on NPs. Shown are seven of the enzymes that are part of the oxidative glycolytic pathway. Highlighted in the yellow square is the conversion of NAD^+ to NADH which can be used to directly monitor flux and product formation over time by simply measuring the change in absorbance at 340 nm with a commercial plate reader. The last enzyme is included to drive the reaction in the forward reaction. Two moles of 1,3-bisphosphoglycerate are produced per one mole glucose consumed which corresponds to 1 mol of ATP consumed and then regenerated in equal amount

enables easy self-assembly of His-tagged proteins or enzymes mediated through metal affinity coordination between neighboring imidazole side chains of the polyhistidine tag and the Zn^{2+} rich surface of the QDs [47, 48, 51]. The dissociation coefficient is on the order of magnitude of 10^{-9} M, which for all intents and purposes, suggest instant and strong binding of the enzyme to the QD surface [41, 49]. As some of the enzymes are multimeric, containing multiple polyhistidine tags, they can bind with multiple NPs creating nanoclusters which have a unique and complex microenvironment compared to the bulk solution [52–55]. The close proximity of multiple enzymes together with altered microenvironment can yield increases in product formation due to "channeling" [31, 52, 56].

In this methods paper, we employ seven of the enzymes associated with the oxidative glycolytic pathway as a prototypical example of a multi-enzyme cascade attached to NPs (Fig. 1). However, the interested scientist could use as few as two enzymes or up to any number of enzymes that are part of a pathway to study. There are very few limitations to this avenue of research. The most important aspect to keep in mind is how the scientist will detect product formation over time. Indirect or direct methods can be employed and we will briefly talk about each in later sections. In the case of the glycolytic pathway, glucose is converted to 3-phosphoglycerate where NADH is produced at the second to last step. NADH conveniently absorbs at 340 nm, making it an ideal signal transduction modality to monitor this assay and study how experimental factors influence overall enzymatic flux through the system and overall product formation over time [57]. The methodology extends from the enzyme production through the assays themselves, how to monitor assays, and interpretation of the resulting data.

2 Materials

2.1 Equipment

1. Personal safety equipment: safety glasses, laboratory coat, latex or nitrile gloves.
2. Chemical waste receptacles.
3. Plastic ware: 1.5 mL, 2.0 mL, and 5.0 mL microcentrifuge tubes along with 15 mL and 50 mL conical centrifuge tubes.
4. 1–10 μL pipette tips, 20–200 μL pipette tips, 10–1000 μL pipette tips.
5. Microcentrifuge tube racks, racks to hold 15 mL and 50 mL conical centrifuge tubes.
6. 1–10 μL, 20–100 μL, 100–1000 μL pipettes.
7. 12-well pipette such as 5–50 μL Finnpipette™ Novus Electronic Multichannel Pipette.
8. Automatic pipettes 10 μL, 100 μL, and 1000 μL volumes such as Eppendorf Xplorer® electronic pipettor.
9. Glassware: Certified autosampler vials preferably with low dead volume (e.g., total recovery or max recovery vials) with silicone/PTFE septum caps; alternative is a LC/MS sample plate.
10. Glassware: Graduated cylinders.
11. Ultrapure water with a resistivity of 18.2 MΩ cm, for example, from a Milli-Q purification system (Millipore, Billerica, MA).
12. UV-visible spectrophotometer such as a Nanodrop 2000 Spectrophotometer (Thermo Scientific).
13. Gel electrophoresis apparatus.
14. Gel imaging system with ultraviolet transillumination such as BioRad ChemiDoc™ XRS+ with Image Lab™ software.
15. Benchtop minicentrifuge.
16. Benchtop vortexer.
17. Spectrophotometric multiwall plate reader such as Tecan Sapphire (Tecan).

18. pH meter.
19. Liquid chromatography instrumentation such as a Waters Acquity H-class UltraPerformance Liquid Chromatography® (UPLC) system and included computer and software.
20. Depending on detection from LC: Mass spectrometer (MS) such as Waters SQ Detector 2 Mass Spectrometer (including computer, software, and roughing pump such as Sogevac SV40 BI), UV-Vis detector such as eλPDA detector, and/or fluorescence detector.
21. Source of clean nitrogen gas such as Parker Balston® Nitro-FlowLab N_2 generator.
22. Transfer pipettes.
23. Serological pipettes.
24. Stir plate.
25. Refrigerator such as a deli style refrigerator or 4 °C cold room.
26. Centrifuge.
27. Sterile culture tubes.
28. Sterile baffled flasks.
29. Polypropylene centrifuge bottles (500 mL).
30. Oak Ridge 3119 screw tip centrifuge tubes.
31. Shaking incubator.
32. Cell lysis equipment (sonifier, press, or, tissue).
33. Tube rotator.
34. Ring stand.
35. Three finger column clamp.
36. Disposable or reusable gravity flow columns.

2.2 Reagents

1. Water soluble QDs coated with compact ligand CL4 or dihydrolipoic acid (DHLA) [38], commercially available QDs such as ITK-COOH or other nanoparticulate material that allows for self-assembly of polyhistidine-appended enzymes to the surface (*see* Subheading 3.1).
2. Enzymes (*see* Subheading 3.2 and **Note 1**) or purchase with polyhistidine his-tag, hexahistidine His_6 terminally appended to each protein.
3. 1× Phosphate Buffered Saline (PBS): 137 mM NaCl, 2.7 mM KCl, 10 mM Na_2HPO_4, 1.8 mM KH_2PO_4.

2.2.1 Reagents for Bacterial Production of Enzymes

1. Terrific broth: 24 g/L yeast extract, 20 g/L tryptone, 4 mL/L glycerol, 100 mL phosphate buffer (0.17 M KH_2PO_4, 0.072 M K_2HPO_4).

2. Luria Bertani broth (LB): 10 g/L tryptone, 10 g/L NaCl, 5 g/L yeast extract.
3. Immobilized metal affinity chromatography (IMAC) resin.
4. Cell lysis buffer: ½× PBS, 1 mM EDTA, 2 mg/mL hen egg white lysozyme, 0.005% Triton X-100.
5. 5× Equilibration buffer: 50 mM phosphate (pH 6.5), 5 M NaCl, 100 mM imidazole.
6. Elution buffer: 50 mM phosphate buffer (pH 6.5), 500 mM NaCl, 100 mM imidazole.

2.2.2 Reagents to Confirm Enzyme-QD Assembly via Gel Electrophoresis

1. Electrophoresis running buffer (1× TBE buffer): 90 mM tris-borate ethylenediaminetetraacetic acid (EDTA; 2 mM), pH 8.0–8.5 Add 100 mL of 10× TBE buffer to a graduate cylinder and pour into a 1 L clean, glass bottle. Pour 900 mL of deionized water (d.i. H_2O) to a graduated cylinder and add to the 1 L glass bottle. Add a stir bar and place the bottle on a magnetic stir plate. Choose a moderate magnetic stir setting to ensure complete mixing.
2. Dissolve low EEO agarose powder into 1× TBE, pH = 8 running buffer at a concentration ranging from 0.8 to 2% w/v in an Erlenmeyer flask or something similar with a Kimwipe® in the neck to prevent the solution from boiling over (*see* **Note 2**).
3. Gel loading buffer: 0.05% (w/v) bromophenol blue, 40% (w/v) sucrose, 0.1 M EDTA; pH 8.0, 0.5% (w/v) sodium lauryl sulfate (SDS).

2.2.3 Reagents for Enzyme Cascade Reactions

1. Prepare a 1 M HEPES (4-(2-hydroxyethyl)-1-piperazineethanesulfonic acid) buffer, pH = 8–8.5. For a 250 mL solution, weigh out 59.575 g of HEPES, and add to a 250 mL or larger glass bottle. Measure 250 mL of d.i.H_2O into a graduated cylinder. Add the water to the glass bottle with a stir bar. Stir until HEPES is dissolved. Adjust pH utilizing a pH meter and adding NaOH pellets one at time until completely dissolved and the desired pH is achieved. Store at 4 °C.
2. Prepare a 250 mM HEPES, pH = 8–8.5 solution from 1 M HEPES, pH = 8–8.5 stock solution in a similar manner as described above.
3. Enzyme substrate stock solution: 30 mM $MgCl_2 \cdot 6H_2O$, 8 mM glucose, 8 mM KH_2PO_4, 8 mM K_2HPO_4, 2.25 mM NAD^+, 15 mM ATP, 15 mM ADP, and 250 mM HEPES, pH = 8–8.5.

2.2.4 Reagents for LCMS Analysis

For buffers used with LCMS make sure the reagents are LCMS grade and use glassware that has never seen soap. An example of buffers to use for LC with a Waters Bridged Ethylene Hybrid (BEH) Amide column would be [58]:

1. Buffer A: 95:5 H_2O:acetonitrile, 30 mM ammonium formate, pH = 3.
2. Buffer B: 15:85 H_2O:acetonitrile, 30 mM ammonium formate, pH = 3.
3. Seal wash solution: 80:20 H_2O:methanol.
4. Wash/Purge solution: 50:50 H_2O:acetonitrile.

3 Methods

3.1 Nanoparticles

The breadth of available colloidal NPs and their use in biological applications is vast and beyond the current scope [35, 59]. As such we do not address their synthesis or biofunctionalization here beyond describing the properties that allow their utility in the current context. The materials utilized here are colloidal semiconductor QDs that consist of a CdSe/ZnS core/shell structure. As the QDs are not intrinsically dispersible in aqueous media as synthesized or on their own, they were surface exchanged with compact zwitterionic ligands that provide colloidal stability [37, 38, 60]. The small size of these ligands also allow the enzyme's terminal hexahistidine sequences to penetrate through and coordinate to the QD's ZnS surface via metal affinity coordination. *See* Fig. 2 for a representative image schematically depicting one such QD-protein bioconjugate. Such self-assembly just requires mixing of the QDs with the enzymes and occurs rapidly. Moreover, this is a stoichiometric assembly that follows a process that mimics a Poisson distribution on a per QD basis [49]. Other biologicals that are appropriately modified with a hexahistidine terminus, including DNA oligonucleotides and peptides, will assemble to QDs in a similar manner. Indeed this approach can act as an intermediary attachment to allow display of highly insoluble molecules around a QD [61, 62]. It has also been demonstrated that commercial QDs functionalized with an amphiphilic surface polymer terminating in a carboxyl group can also allow protein coordination in a similar manner even within intracellular environments [49, 63]. Here, the carboxyl groups are believed to mimic the function of a nitrilotriacetic acid (NTA) group and this notion is supported by the fact that adding a small amount of Ni^{2+} to such reactions enhances binding. Note that the polymer coating will make the hydrodynamic coating of such QDs far larger. To allow similar coordination to other nanoparticulates beyond QDs, they need to be surface

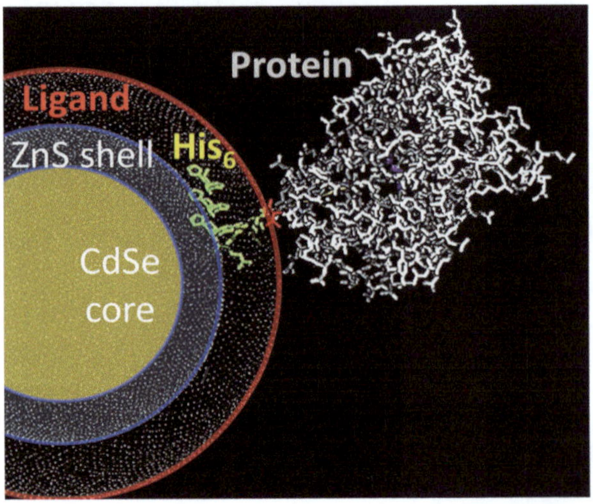

Fig. 2 Schematic of a model protein self-assembled to a QD. In this schematic representation, yellow fluorescent protein (YFP) is used as a model and shown as displayed on a CdSe/ZnS core-shell QD by His_6 metal affinity coordination driven self-assembly [104]. The QD core/shell structure has a radius of ~30 Å with core/shell components arbitrarily designated. The YFP is shown attached to the QD ZnS by its terminal sequence His_6. YFP is a ~30 kDa 7-helix barrel structured fluorescent protein derived from green fluorescent protein. The putative QD ligand is shown extending from the QD surface by ~10 Å

functionalized with Ni-NTA groups and this has already been shown to work well for gold NPs [42].

3.2 Protein Expression and Purification

3.2.1 Protein Expression

While historically the work horses for recombinant molecule production, bacterial microbes such as *Escherichia coli* and yeast strains such as *Pichia pastoris* and *Saccharomyces cerevisiae* have been supplemented by a growing list of microbial species, plants, and even cell-free systems. The growing diversity of production platforms allows researchers to tailor production systems to improve biomolecule yield, ensure post-translational modification, and modulate other biophysical properties based on experimental needs and downstream applications. Each of these systems will have its own protocols and requirements for biomolecule production as well as suitable genetic systems that will enable recombinant protein production.

Biomolecule production requires the design and assembly of a synthetic genetic construct that controls the on-demand production of the biomolecule of interest. Often, the gene of interest is incorporated into an extra-chromosomal piece of DNA that it transformed into the chassis organism. Typically, these constructs are referred to as plasmids or expression vectors, each of which conveys properties that impact efficiency of expression. Plasmids are often chosen for their abundance or copy number within the host, antibiotic selection marker, and other properties [64, 65]. In addition to these, bacterial expression vectors, particularly those of Gram-negative bacteria, allow for protein export from the cell or localization within the cell to either the cytoplasm or periplasm. Localization of recombinant proteins within the confines of the cell allows proteins to fold in either an oxidizing (periplasm) or reducing (cytoplasm) environment which can have significant impacts on protein folding, stability, and activity in some cases. Additionally, cellular localization also allows researchers to utilize specific methods of protein isolation including complete cell lysis to free cytoplasmic contents or any of several methods to either permeabilize or remove the outermost membrane to free periplasmic components as will be discussed below. Once a suitable vector has been decided, plasmid vectors can be purchased from common providers of molecular biology reagents or designed and assembled based on experimental needs and requirements.

Regulation of expression of recombinant proteins is typically accomplished using a chemical inducer which activates the promoter controlling transcription and subsequent translation of the gene of interest encoding the target protein. Common inducers in *E. coli* include lactose, arabinose, and rhamnose; however, there are numerous other induction/expression systems in *E. coli* many of which are summarized in these reviews [65, 66]. Inducers have their own distinct properties that contribute to protein expression. Typically inducers are selected based on strength of induction and inducer longevity in the culture media. Some systems such as the commonly used *lac* operon allow researchers to use natural lactose as the inducer which is consumed by the bacterial host or alternatively use a chemical compound such as β-1,D-thiogalactopyranoside (IPTG) which also activates the *lac* promoter but is not degraded and allows for continuous production of recombinant proteins [67].

For brevity, this section will describe methods for recombinant protein expression in laboratory strains of *E. coli*, specifically, *E. coli* strain BL21 (De3), a commercially available bacterial species which maintains a lactose inducible gene encoding the DNA-dependent RNA polymerase from bacteriophage T7 [68, 69]. In this system of recombinant protein production, the gene encoding the target protein is cloned into a plasmid vector which possesses a T7-specific promoter and terminator flanking the gene of interest (Fig. 3). The methods below detail a typical protein expression process using the vector pET28 which lacks a leader sequence (cytoplasmic targeting) and carries a gene for kanamycin resistance.

1. *E. coli* strain BL21(De3) (New England Biolabs: C25271) transformed with the plasmid containing the gene of interest are plated to LB media containing 50 μg/mL kanamycin then incubated at 37 °C.

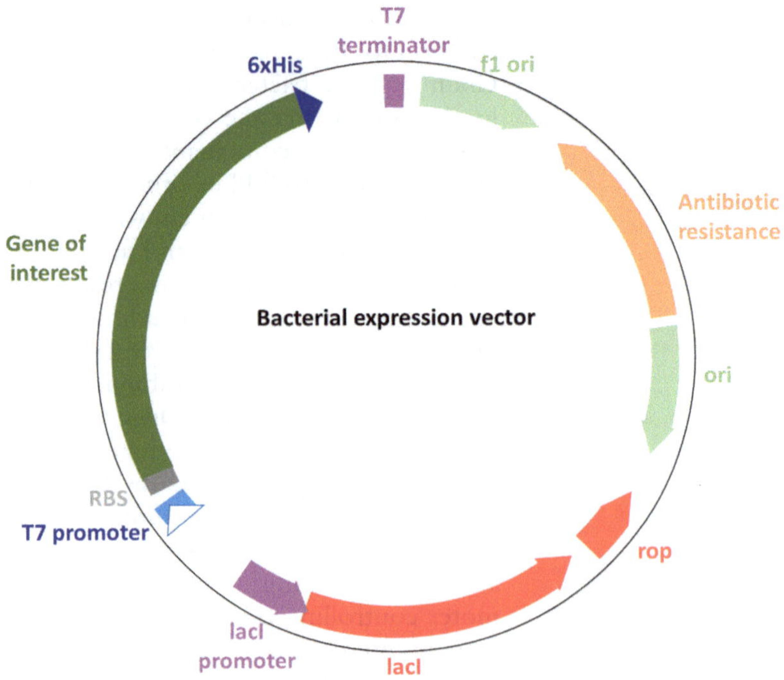

Fig. 3 Schematic map of a bacterial expression vector or plasmid. The origin of replication (ori) allows for cellular replication of the plasmid DNA. Antibiotic resistance or selection markers can vary based on experimental needs. Here, bacterial expression of the gene of interest (dark green) is controlled by components of the *lac* operon including the transcriptional repressor (lacI), regulatory protein (rop), and lac operator. Transcription is driven by the T7 RNA polymerase as directed by the similarly labeled T7 promoter and T7 terminator sequences. RBS – ribosome binding sight, ori – origin of replication

2. Single colonies from selection plates are transferred to 50 mL of Terrific Broth in a 250 mL flask containing 50 μg/mL kanamycin.
3. Incubate cultures overnight at 37 °C shaking at 200–250 rpm.
4. Inoculate 500 mL TB containing 50 μg/mL of kanamycin in a baffled 2 L flask with a 1:100 dilution of the overnight culture (*see* **Note 3**).
5. Incubate cultures at 37 °C shaking at 200–250 rpm for 2–4 h until an optical density at 600 nm (OD_{600}) of 0.6–0.8 is attained.
6. Add IPTG to a final concentration of 0.5 mM to induce protein expression.
7. Incubate culture 12–16 h at 30 °C shaking at 200–250 rpm.
8. Transfer culture media to 500 mL polypropylene centrifuge bottles.

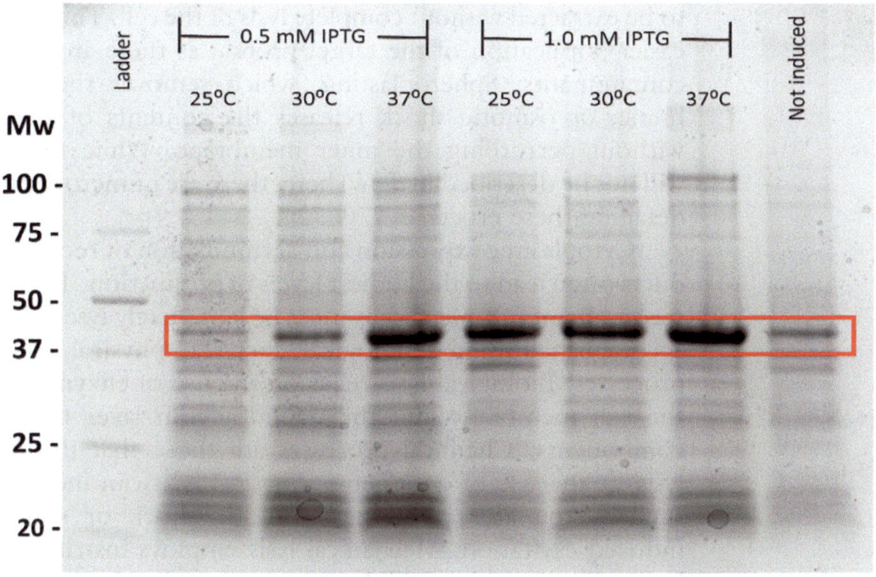

Fig. 4 Optimization of protein expression conditions. A range of protein expression conditions are often examined to determine optimal expression conditions. Here, inducer concentration and temperature are tested after overnight growth. Soluble proteins from cell lysates are resolved via 4–15% gradient SDS-PAGE and visualized *via* Coomassie staining. The representative protein of interest glycine oxidase (~42 kDa) is defined by the red block. Relative intensity of this protein band is reflective of the level of induction and protein accumulation

9. Centrifuge cell suspension at ≥4000 × *g* for 20 minutes to pellet cells.
10. Decant supernatant to vessel for sterilization and disposal.
11. Freeze cell pellet overnight or until ready for cell lysis and protein purification (*see* **Note 4**).

3.2.2 Cellular Lysis

Growth of the cells and the exact time and protocol used for induction of the protein in the bacteria typically require empirical optimization, *see* Fig. 4 for an example gel showing different growth conditions and time for an induced protein. Unless the recombinant proteins are exported from the cell, membranes must be permeabilized or the entire cell ruptured to free the recombinant protein and allow for subsequence purification. The most effective methods are often chassis or platform specific due to variabilities in cell morphologies including the presence or absence of a cell wall, membrane composition, or properties of the peptidoglycan layers.

Expression in *E. coli* and other Gram-negative bacteria allows researchers to direct protein accumulation to the periplasmic space or the cytoplasm during expression. In addition to being able to form disulfide bonds within the oxidizing environment of the periplasm, localization to this cellular compartment allows proteins to be extracted without complete lysis of the cell. This often leads to easier purification of the target protein as there are fewer cellular contaminants. Spheroplasting, which removes the outer membrane, or osmotic shock releases the contents of the periplasm without perturbing the inner membrane. While these methods will not be described in detail here, there are numerous publications detailing these processes [70–72].

Cytoplasmic expression and accumulation of recombinant products often lead to the highest levels of production. To extract these recombinant products, cells must be completely lysed which can be accomplished using a biological, chemical, physical, or mechanical processes. Biological lysis relies on the use of enzymes to degrade the cell membrane and the peptidoglycan layer to free cellular components. Chemical processes are those that include alkaline lysis or the use of detergents. Physical lysis can include methods such as the aforementioned osmotic shock or sound energy-induced cavitation. Mechanical lysis employs instruments such as bead mills/beaters or high pressure homogenizers to disrupt the cell, Fig. 5. Methods are often chosen based on experimental needs, availability of equipment, cost, limitation due to downstream applications, and other factors. Additionally, in many instances, these methods are combined as researchers develop protocols to satisfy their experimental needs. The protocol below is one that can combine or be substituted by several other methods of cell lysis as

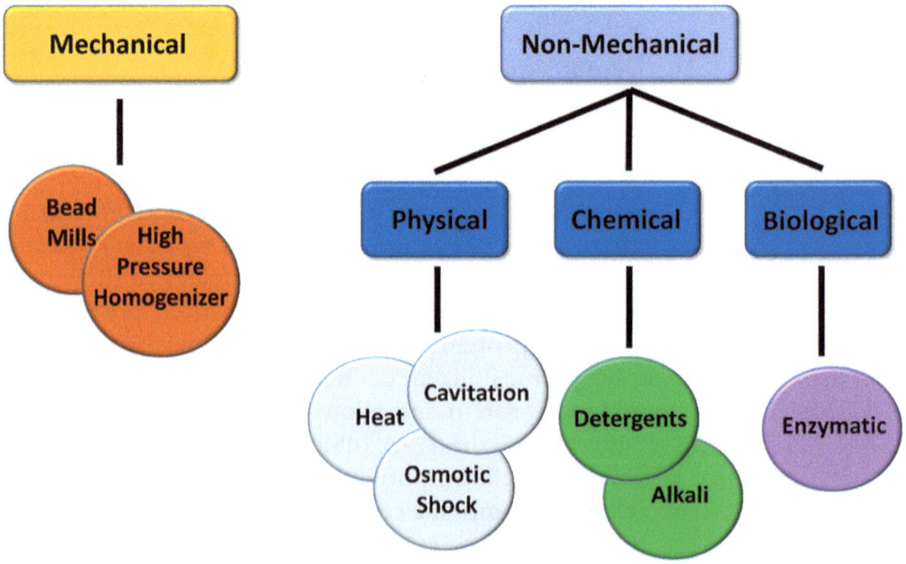

Fig. 5 Methods of cell lysis. Following expression of recombinant proteins in a microbial chassis organism, proteins are extracted either from the periplasmic space or the cytoplasm of the microbe. Cell lysis can be completed with rupture of the cell or partial where the membranes are permeabilized selectively to free periplasmic components. Methods are often chosen based on availability of equipment (mechanical methods) or experimental needs

needed. Additional information on alternate methods of cell lysis can be found in an excellent review by Islam et al. [73].

1. Remove cell pellet from −20 °C storage and thaw on ice.
2. Add 30 mL of cold cell lysis buffer per 10 g of cell pellet.
3. Resuspend cell pellet in lyse buffer using a metal spatula to dislodge cells and mix to homogeneity.
4. Incubate on ice with agitation using a platform rocker or equivalent for 30 min.
5. Transfer cell suspension to a polypropylene beaker (based on total volume) which is suspended in ice or ice water.
6. Using a Branson sonifier or equivalent, lyse cells using an 80% duty cycle and adjusting the output to a frequency of 20 kHz.
7. Rest for 1 minute on ice to ensure temperature of the cell lysate does not rise significantly.
8. Repeat sonication cycle 2 more times (*see* **Note 5**).
9. Transfer cell lysate to Oak ridge 3119 centrifugation tubes (or equivalent).
10. Centrifuge lysate 15,000 × *g*, 30 min, and 4 °C to separate soluble and insoluble materials.
11. Decant supernatant to conical tubes for subsequent protein purification (*see* **Note 6**).

3.2.3 Protein Purification

Separation of recombinant proteins from the plethora of cellular proteins and biomolecules released upon cell lysis can prove a daunting task. Fortunately, proteins are highly diverse biomolecules with unique biophysical properties. Amino acid composition, folding pattern, multimerization, subunit associations, post-translational modifications, and others all contribute to the size, charge, and hydrophilicity/hydrophobicity of the protein. These unique biophysical properties of proteins, once identified, can then be targeted and used to design an appropriate enrichment strategy.

The most common method of separating proteins is column chromatography. Chromatography is not a single process, rather a sweeping term that covers methods that include size exclusion chromatography, ionic exchange, reverse phase, and affinity chromatography to name a few. These chromatographic methods and others have been described in the literature and are the subject of many books and commercial reference guides [74, 75]. While chromatographic methods can often be used as a single step purification process, some proteins can prove particularly challenging to isolate and may require the combination of multiple methods in order to obtain a product with a high degree of purity. Upstream and downstream protocols such as salting out, membrane filtration, immunoprecipitation, and others are routinely employed in protein purification to concentrate proteins of interest both before and after chromatography [76–78].

Immobilized metal affinity chromatography (IMAC) has become a common method of purification due to the simplicity and efficiency of the method [79]. Here, proteins of interest are modified through genetic engineering to contain a C-terminal hexahistidine sequence, His_6. This 6 amino acid sequence will coordinate a divalent metal ion, typically Ni^{2+}, which is itself immobilized to a solid support/resin functionalized with a nitrilotriacetic acid group (NTA). Using a high salt buffer, non-target proteins are washed from the column resin while the His_6-tagged protein remains bound to the Ni-NTA resin. After sufficient washing, His_6-tagged proteins are eluted using a competing small molecule such as imidazole or chelating agent such as ethylenediaminetetraacetic acid (EDTA). Oftentimes, the column-bound protein is first washed with a low concentration imidazole solution to remove any non-specific proteins that are high in histidine residues that may also be attached to the resin. As with most chromatographic methods, the protocol detailed below can be modified through buffer modification including raising and lowering the pH, increasing the salt concentration, or raising or lowering the imidazole concentration in buffers to limit non-specific protein binding to the Ni-NTA resin.

1. Resuspend Ni-NTA to ensure homogenous slurry.
2. Transfer desired volume of slurry to 2 mL microfuge tube.
3. Gently centrifuge 300 × *g* for 3 min to pellet resin.
4. Decant supernatant.
5. Resuspend resin in cold 1× equilibration buffer filling microfuge tube.
6. Repeat centrifugation and resuspension in column wash buffer a minimum of 3 additional times to remove residual storage buffer and equilibrate the resin to optimal buffer conditions for the target protein's purification.
7. Transfer resin to soluble fraction isolated from the cell lysis protocol (Subheading 3.2.2).
8. Add volume of 5× equilibration buffer to resin/lysate mixture ensuring a final 1× concentration equivalent.
9. Incubate at 4 °C on a rotary wheel for a minimum of 3 h to allow His_6-tagged proteins to bind Ni-NTA resin.
10. Gently centrifuge to pellet resin, 800 × *g*, 3 min, at 4 °C.
11. To wash the resin, decant supernatant and resuspend resin in 30 bed volumes of ice cold equilibration buffer (*see* **Note 7**).
12. Repeat resuspension, centrifugation, and supernatant decantation a minimum of two additional times.

13. Resuspend resin in 5 mL of ice cold equilibration buffer.
14. Assemble ring stand, three finger clamp and gravity flow column (*see* **Note 8**).
15. Transfer resin suspension to a gravity flow column.
16. Allow resin bed to form as excess equilibration buffer flows through the column.
17. Trying to not disturb the resin bed, add 5–10 bed volumes of elution buffer to the column.
18. Collect 1 mL fractions to microfuge tubes. Store fractions at 4 °C.
19. Assess the purity of chromatography fractions via SDS-PAGE or similar method (*see* **Notes 9** and **10**).
20. Enzyme-containing fractions are then pooled and concentration determined with measurement of absorbance at 280 nm or other commercial quantitation assay such as a Bradford assay or a bicinchoninic acid assay (BCA).
21. Pooled samples are then dialyzed to remove the excess imidazole and then are diluted to the desired concentration using a relevant buffer.
22. Add sterile glycerol or other cryoprotectant to a final concentration of 20% v/v and aliquot to solution to microfuge tubes (50–100 μL volume). Note this new enzyme concentration for future reference.
23. Snap freeze samples using an ethanol/dry ice bath.
24. Transfer tubes to freezer storage box labeled with the enzyme name, date synthesized, final concentration, and aliquot volume.
25. Store at -80 °C or colder until ready to use.

3.3 Confirming Assembly of Enzymes to NPs

3.3.1 Electrophoretic Mobility Shift Assay

Individual enzyme assembly or multi-enzyme cascade assembly to NPs such as QDs via metal affinity coordination can be confirmed utilizing electrophoretic mobility shift assays. The reader is directed to the selected references for other examples of enzymes attached to NPs such as QDs and AuNPs [38, 42, 52, 54, 80]. Typically a 0.8–2% w/v low EEO agarose gel is made with 1× TBE, pH = 8 buffer. Some trouble shooting will have to be attempted to determine the best conditions for sufficient separation of unbound NPs and NPs bound with increasing number of enzyme. This will depend on the type and size of NP used, the size of the enzyme, and the overall charge of the enzyme-QD bioconjugate.

1. Typically, a 100 mL to 150 mL of agarose solution is more than sufficient for casting one or two gels. Cast a gel by pouring hot 0.8–2% w/v agarose into a gel tray. The ends of the gel tray should be blocked by placing the gel tray in the gel chamber

perpendicularly. It is helpful to rub a bit of the running buffer over the rubber gaskets to act as a lubricant to help slide the gel tray in place. Use a pipette tip to remove any bubbles that might be floating around. Place a 10 well to 12 well comb into the gel and let rest for at least 30 min or until gel has solidified. The scientist can cover the gel tray with a piece of plastic wrap to prevent any dust or debris from falling into the gel as it cools.

2. Prepare stock solutions of enzyme and QDs in 1× PBS, pH = 8. For example, use GLK which is a dimer with a MW of 36.9 kDa. The total volume of each enzyme-QD bioconjugate should be 20 μL. Therefore, adjust stock concentrations based on the molar ratio of enzyme to QD. Choose a wide range of enzyme to QD molar ratios to clearly visualize the impedance of bioconjugate migration.
3. Prepare enzyme-QD bioconjugates. Typically, 5 pmol of QD is sufficient to see under UV light with a bioimager and will conserve material. If using another type of NP such as AuNPs, the amount of NP will need to be determined that allows for sufficient visualization. Refer to Table 1 for an example recipe for preparing enzyme-QD bioconjugates.
4. After adding enzyme to NP, gently vortex or flick with your finger to ensure adequate mixing and spin down with a table top microcentrifuge. Allow enzyme-QD bioconjugates to assemble on ice or at 4 °C for at least 30 min and with multiple enzyme additions at least an hour.
5. After assembly is complete, add 5 μL of gel loading buffer which is composed of 1× gel loading buffer, 1× PBS, and 5%

Table 1
Electrophoretic mobility shift assay. Example recipe for creating enzyme-QD bioconjugates for imaging NP assembly

Lane number	1	2	3	4	5	6	7	8	9	10
Ratio of enzyme to QD	0	0.25	0.5	1	2	4	6	8	12	0
Picomoles of QD	5	5	5	5	5	5	5	5	5	5
Picomoles of enzyme	0	1.25	2.5	5	10	20	30	40	60	0
QD (1 μM) (μL)	5	5	5	5	5	5	5	5	5	5
Enzyme (1 or 10 μM) (μL)	0	1.25	2.5	5	10	2*	3*	4*	6*	0
Buffer (1× PBS, pH = 8) (μL)	15	13.75	12.5	10	5	13	12	11	9	15
Loading buffer (μL)	5	5	5	5	5	5	5	5	5	5

Note: to achieve a wide range of ratios of enzyme to NP, two concentrations of stock enzyme should be used – 1 μM for lanes 2–5 and 10 μM for lanes 6–9 w/v glycerol. The faint contrast in color helps the user visualize sample loading and the glycerol ensures the sample sinks to the bottom of the well.

6. Gently pull comb from gel and set aside. Rotate the gel tray in the buffer chamber and submerge the gel tray with 1× TBE, pH = 8 buffer.
7. Add 20 μL of each enzyme-QD bioconjugate to each well in the gel according to Table 1. Pipette slowly and deliberately to avoid introducing any bubbles into the well.
8. Connect the leads to the buffer chamber such that QDs flow to the red positive anode (*see* **Note 11**).
9. Set the voltage to 90 mV and wait 5 min for the enzyme-QD bioconjugates to fully enter the gel. Take a fluorescent picture with a gel imaging device such as the BioRad ChemiDoc™ XRS+ with Image Lab™ software. Return the gel to the running buffer, maintaining the same orientation of the gel tray as before, restart the voltage, and repeat every 5 min for up to 20 min. Over time, the fluorescent bands associated with the QDs will shift with higher molar ratio enzyme-QD bioconjugate being retarded in their movement compared to QDs solely.
10. Figure 6 shows an example of a time sequence of images with typical results. The changes in QD migration and the rate of change as a function of enzyme ratio present confirm protein assembly to the QD in a stoichiometric manner.

3.3.2 Other Methods

Other methods are available to verify enzyme attachment to the QDs. These include dynamic light scattering to look at the increases in hydrodynamic diameter as the proteins assemble to the QDs [81, 82]. Other possibilities are size-separation/size exclusion chromatography, differential dialysis, and binding/release columns if the proteins or another protein on the QD assembly binds to some type of functionalized media such as biotin to soft-avidin media for example [83].

3.4 Kinetic Assay for Observing Modified Flux with Enzyme–QD Conjugates

3.4.1 Multi-Enzyme Assembly onto QDs

Multi-enzyme assembly onto the surface of QDs can be achieved by simply mixing each enzyme with QDs at the desired stoichiometry [40, 52–54]. Care should be taken to ensure that the final enzyme concentration is sufficiently high to yield progress curves, i.e., kinetic traces, that have sharp linear regions with a sufficient number of data points at the substrate concentration tested [84, 85]. The progress curves should have sufficiently linear beginning to the curve, if there are too few linear data points, calculating the initial rate can be inaccurate. This is done to meet the classic Michaelis–Menten or Briggs–Haldane conditions of diffusion limited maximum initial rates. The number of stat points can also be

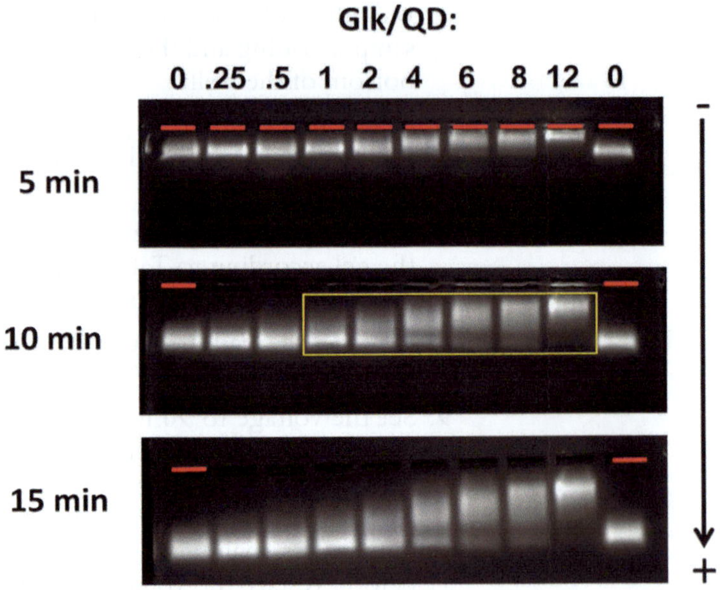

Fig. 6 Representative image of agarose gel electrophoretic separation of increasing assembly of glucokinase (GLK) to QDs capped with a small zwitterionic ligand in 1× PBS and then separated on a 1× TBE gel in TBE buffer. Pictures were taken every 5 min to show the migration and separation of increasing enzyme-NP bioconjugates. QDs with no enzyme travel the furthest while increasing the ratio of enzyme attached to NP decreases the distance traveled through the gel (slower migration rate). Red lines indicate the well/starting positions while the yellow box indicates where the moderate ratios of enzyme attached to QDs separate from QDs with no enzyme attached

influenced by the number of time points you take as well as how active the enzyme is. Enzymes can be purchased or made in-house but need to contain a terminal, polyhistidine tag to ensure assembly to the Zn-rich QD surface through metal affinity coordination. However, the reader should be aware that there are other conjugation strategies available that can be adopted in a similar manner [32, 43, 86, 87]. The placement of the multi-His tag should not interfere with the enzyme's active site, impeding the overall enzymatic rate [88].

1. Determine the order of enzyme addition to NP. The order of attachment is important as it will affect the overall rate of the reaction. Examples include assembling in the forward direction of the reaction scheme, the backward direction of the reaction scheme, or random assembly.
2. In general, the NPs will be added first followed by sequential addition of each enzyme. In the case of random assembly, the enzymes are added first, lightly mixed, followed by QDs.

Table 2
Example of ratios of enzymes to nanoparticles

Enzyme name	Ratio to QD
Glucokinase (GLK)	5.5
Phosphoglucose isomerase (PGM)	1
Phosphofructokinase I (PFK)	9
Fructose-bisphosphate aldolase (ALD)	12
Triose phosphate isomerase (TPI)	1
Glyceraldehyde-3-phosphate dehydrogenase (GAPDH)	27
Phosphoglycerate kinase (PGK)	7.5
Nanoparticle	1

3. Determine the molar ratio of each enzyme to NP to be tested. Preliminary ratios can be estimated utilizing the initial rates of individual enzymes. (*See* Table 2 as an example).
4. Some of the enzymes are multimeric with multiple His_6 tags [56]. Therefore, the enzymes can crosslink multiple NPs together forming a NP-Enzyme cluster. Dynamic light scattering (DLS) can be utilized to determine the average number of NPs in a cluster which can be incorporated into the optimization protocol for determining individual enzyme ratios [42, 52].
5. The initial rates, the maximum number of enzymes that can fit on to a NP based on NP size, and the average cluster size can all be incorporated into an optimization protocol to determine the ideal molar ratio of each enzyme to QD.
6. Gel electrophoresis can be utilized to confirm enzyme assembly to the QD (*see* Subeading 3.3.1).
7. Utilizing seven of the enzymes associated with glycolysis, GLK to PGK, as an example of a multi-enzymatic reaction scheme, determine the molar ratios of each enzyme to the QD and assembly in the forward direction of the reaction (Table 3). First calculate the final concentration of QD in the well and then work backwards in the calculation. If doing a 1 to 2 dilution where half the volume in the reaction well will be the multi-enzyme-QD construct and the other half of the substrate solution multiply the final concentration by 2 to achieve the stock concentration of the QDs.
8. Multiply the final concentration of QDs by the desired molar ratio of each enzyme to achieve the final concentration of each enzyme. For example, if the molar ratio of GLK to QD is 5.5, the molar ratio of PGM is 1, and the QD final concentration is

Table 3
Sample assembly list. Suggested stock enzyme concentrations for easy pipetting

Order of addition	Name	Ratio	Stock concentration (nM)	Volume (μL)
1	QD	1	100	50
2	Glucokinase (GLK)	5.5	550	50
3	Phosphoglucose isomerase (PGM)	1	100	50
4	Phosphofructokinase I (PFK)	9	900	50
5	Fructose-bisphosphate aldolase (ALD)	12	1200	50
6	Triose phosphate isomerase (TPI)	1	100	50
7	Glyceraldehyde 3-phosphate dehydrogenase (GAPDH)	27	2700	50
8	Phosphoglycerate kinase (PGK)	7.5	750	50
9	250 mM HEPES buffer, pH = 8 to 8.5			100
	Total Volume:			500

5 nM, then the final concentration of GLK is 27.5 nM and PGM is 5 nM. Multiply the final concentration by 2 if employing a 1 to 2 dilution with the substrate solution to achieve a stock concentration of each enzyme, in this case 55 nM GLK, 10 nM PGM, and 10 nM QD.

9. Determine the total volume needed of the multi-enzyme-NP construct. To do this, determine the number of replicates needed for each test condition, the number of test conditions, and the volume needed for each replicate. For example, if there are 3 replicates for each condition tested, 6 conditions being tested, and 25 μL of sample needed, then the minimum volume of sample needed would be $3 \times 6 \times 25 = 450$ μL. Determine the dead volume in your multichannel pipette. As an example, assume it is about 25 μL plus another 25 μL for extra measure. Therefore the total volume needed for the multi-enzyme-NP construct is at least 500 μL (*see* Table 3).

10. Prepare stock concentrations of each enzyme from frozen enzyme stocks using 250 mM HEPES buffer, pH = 8–8.5 (*see* **Notes 12** and **13**, Table 3). Prepare at least 10 μL extra of the enzyme stock solution from frozen enzyme stocks than needed to account for pipetting error.

11. Prepare a stock concentration of QDs using 250 mM HEPES buffer, pH = 8–8.5. First sonicate QDs for a couple of minutes then vortex to ensure the QDs are dispersed. Add QDs to HEPES, vortex, and spin down with a tabletop microcentrifuge for a few seconds.

12. Prepare the multi-enzyme-QD construct in the forward direction of the reaction scheme. Be sure to lightly vortex then spin down each stock enzyme solution to ensure complete mixing before adding the enzyme to the QD. To a 1.5 mL microcentrifuge tube, add the appropriate volume of QD from the stock solution. Then follow in order of the reaction scheme each enzyme. Make sure the volume added goes to the bottom of the tube, vortexing gently or flicking the tube with your fingers followed by briefly centrifuging the tube can accomplish this task. Wait a minute or so before adding the next enzyme to aid assembling to the NP surface.
13. After adding the final enzyme, close the cap, vortex lightly or flick with finger a few times and spin down with table top microcentrifuge.
14. Prepare a sample with no QDs as a control in the same manner.
15. Let the samples assemble on ice or at 4 °C for at least an hour, preferably two to three to ensure complete assembly to the QDs.
16. Add 250 mM HEPES, pH = 8–8.5 to bring the sample up to its final volume.
17. This method assumes using a 384-well plate with a final volume of 50 μL, take this in consideration for scaling up to a 96-well plate or scaling down to 1536-well plate. Use an automatic pipette to efficiently dispense 25 μL of enzyme-NP construct to each well. Place a piece of sticky film on top and spin down for a few seconds using a centrifuge, spinning briefly up to 300 rpm. If a centrifuge is not available, gently knock or tap plate to ensure droplets are at the bottom of the well (*see* **Note 14**).
18. Prepare the substrate solution. In this experiment there is only one substrate solution and the concentrations of all the components are maintained throughout. Calculate the total number of wells, multiply by 25 μL, add the volume needed to prime the injector (usually 1.5–2 mL) if utilizing one plus approximately 5–10% to ensure enough volume. The substrate solution should have a final concentration of the following: 15 mM $MgCl_2$, 4 mM glucose, 4 mM monobasic phosphate, 4 mM dibasic phosphate, 1.125 mM NAD^+, 7.5 mM ATP, and 7.5 mM ADP. Use 1 M HEPES, pH=8 to 8.5 to achieve a final concentration of 250 mM HEPES in the substrate solution (*see* **Note 15**).
19. If using a plate reader with an injector, prime the injector according to the instrument instructions and dispense 25 μL of substrate solution to each sample well. If not, use a solution trough and a multichannel pipette to efficiently add 25 μL of substrate solution to the bottom of each well. A 12-well

multichannel pipette works well as it will naturally correspond to every other well of a 384-well plate horizontally. After the substrate has been added, place a piece of sticky film atop the plate and immediately place in a plate reader. After the substrate has been added, the reaction has started so care should be taken to work confidently and quickly (*see* **Note 16**).

20. Immediately start the kinetic program of the plate reader. The kinetic program should consist of mixing for a short period of time, mixing prior the measurement and measuring the absorbance or fluorescence at a specific wavelength every specified time point. The plate reader should also be pre-equilibrated to run at the desired temperature (*see* **Note 17**).
21. As an example, if using a Tecan Sapphire plate reader a program was constructed that included the following steps:
22. Set the plate reader incubation temperature to 30 °C.
23. Add 25 μL to the highlighted wells using the injector that has already been primed according to instrument instructions.
24. Plate is pushed out by instrument, asks user to add cover.
25. Plate is pushed in by instrument, shakes plate at a medium speed for 5 s to ensure adequate mixing.
26. Shake the plate for 5 s prior to each absorbance measurement to ensure adequate mixing.
27. Measure the absorbance of each well at 340 nm every 3 min for up to 18 h. This monitors the enzymatic conversion of NAD^+ to NADH.
28. Once the kinetic program is finished, subtract the first time point from subsequent time points for each run, convert the absorbance measurements to concentration values utilizing Beer's Law with a molar absorptivity of 6220 $M^{-1}cm^{-1}$ for NADH for example. Determine the path length utilizing the manufacturer's measurements of one well of a 384-well plate.
29. Replicates can be averaged together and plotted utilizing any standard graphing software. Data analysis can be automated by writing scripts in MATLAB®, R, or another computer programming software. Figure 7 is an example of the type of data typically generated. In this example the amount of NP was varied while the enzyme concentration remained constant.
30. Given that this is a multi-enzyme reaction, typical kinetic parameters usually calculated for single enzyme assays such as catalytic rate (k_{cat}) and affinity or Michaelis constant (K_M) are not directly applicable. However, they are sometimes used to compare flux under different conditions [57, 84]. It is important to note that these are all apparent values and usually reflect the rate of the slowest, rate-limiting enzyme in the reaction.

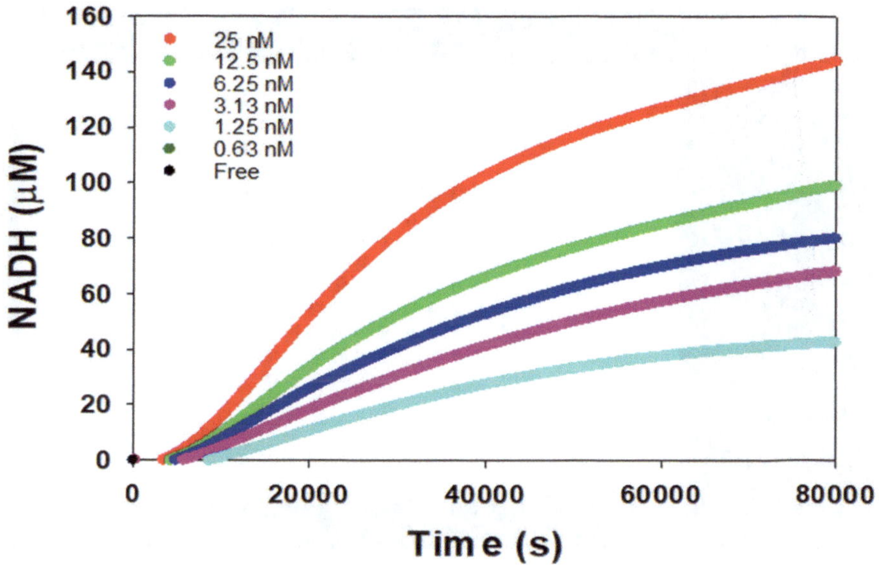

Fig. 7 Typical data generated from multi-enzyme-NP bioconjugate experiments. As the amount of NP increases (enzyme constant) so does the overall initial rate of the reaction and cumulative amount of NADH formed. Product formed from free enzyme and 0.63 nM QD are essentially 0

3.4.2 Serial Dilution of Enzyme-QD Bioconjugates Experiment

As stated previously, a lot of different factors are involved with the optimization of multi-enzyme cascades assembled to NPs. Overall, acceleration or flux through the system can change significantly based on the concentration of the enzyme or the NP concentration. Therefore, it can be insightful to make a stock solution of enzyme-QD bioconjugate and then perform a series of different serial dilutions, testing them side-by-side. This is especially helpful for determining conditions when reactions are diffusion limited or where the activity of the corresponding free enzymes in solution becomes essentially null.

1. As an example, make a double batch of the multi-enzyme-QD bioconjugates in Table 3.
2. Let assemble as described previously. After adding the final amount of HEPES to bring up the construct to its final volume perform three successive dilutions by taking 500 μL of the pervious dilution and diluting it with 500 μL of 250 mM HEPES, pH = 8–8.5.
3. Aliquot out the samples into a 384-well plate. Figure 8 shows an example of how the samples might be delivered. In this figure, there are 8 different conditions being tested. This might be due to different molar ratios of enzymes to QDs and/or different types of NPs. Each condition is tested in triplicate and diluted 3 times to test four different enzyme/NP concentrations while the molar ratio remains constant.

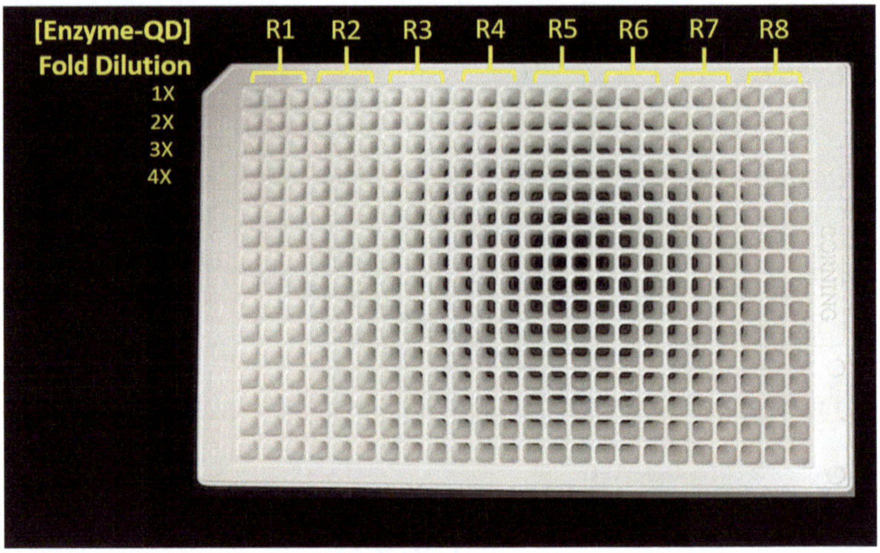

Fig. 8 A 384-well plate where 8 different conditions (R1–R8) are being being tested in the horizontal direction and 4 different dilutions (1x–4x) are being being tested in the vertical direction. It is helpful to think about the experimental design and setup as a matrix. This also highlights how many experiments in triplicate can be accomplished in this format

4. Add substrate, perform the kinetic assay, and analyze the results as described previously. A 384-well plate allows the user to test many different variations and combinations, generating a massive amount of data. Figure 8 shows a sample 384-well plate where each condition (R_i) was replicated 3 times and an enzyme dilution was performed four times.

3.5 Mass Spectral Analysis

When enzymatic activity is not easily detected through plate reader analysis, it may be monitored through liquid chromatography (LC) separation of the product followed by its detection. This detection can be accomplished through a UV-VIS detector or a fluorescence detector if the product (or a derivative of the product) contains a chromophore or fluorophore. Alternatively, the product may be detected by mass spectral analysis, which is discussed here. An exhaustive protocol for mass spectrometry (MS) analysis is beyond the scope of this paper; for more information, including extensive validation of methods (including intra- and inter-day precision, accuracy, lower limits of detection and quantification, stability, carryover, matrix effects, recovery measurement) and other mass spectral techniques besides LC/MS (e.g., GC-MS, and direct infusion) the interested reader is directed to reference [89] and references therein including FDA guidance [90] as well as references [91–94]. For this discussion, we assume the product is a small, polar compound that will undergo a type of LC-termed hydrophilic interaction chromatography (HILIC) and will be

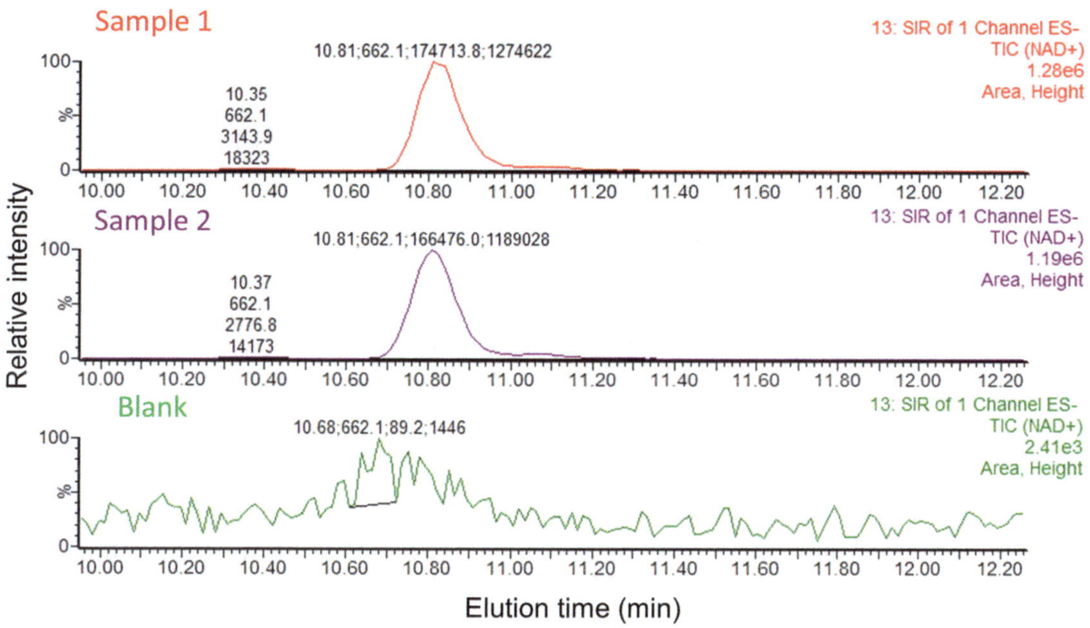

Fig. 9 Representative MS results from 7 enzyme reactions where the amount of NAD^+ was being monitored to cross correlate with UV-vis monitoring of the same substrate in the plate reader. Samples were measured by programming the instrument to isolate and monitor the total ion chromatogram (TIC) corresponding to just the NAD^+ component (mass ~ 664.1). SIR – selective reagent ion

detected on a single-quadrupole MS, though we note that other analytes and methods may rely on other chromatography (e.g., C18 reverse phase) and other instrumentation (e.g., a triple-quadrupole MS). The real beauty and wide applicability of MS-based analysis above other types of analytical techniques is that it can help analyze multiple different substrate/product components in a given reaction mixture. For example, *see* Fig. 9 where just the NAD^+ component is isolated from the reaction. This can provide insight into the rate of reactions and where particular bottlenecks are present or if enzymes are minimally functional. The following is a somewhat generic protocol that can be used as a basis for developing the scientist's own MS approach.

1. Determine a suitable internal standard (IS) compound. This is often an isotopically labeled version of the enzymatic product (analyte) or similar compound (e.g., containing ^{2}H, ^{13}C, or ^{15}N).
2. Make buffers for LC (*see* Subheading 2.2.4) (if no onboard degasser on the instrument, degas before use) and prime the system. Make sure to use glassware that has not seen soap and to use LC/MS-grade solvents. If non-LC/MS-grade salts must be used, syringe filter (0.2 μm) (*see* **Note 18**) a concentrated solution first before adding.

3. Make sure the mass spectrometer has been on the recommended time by manufacturer (i.e., "warmed-up") and that clean N_2 is supplied before use. Using direct infusion and/or combined infusion and LC, preferably infuse a "blank" pure solution to ensure the mass spectrometer is clean. Then, tune the mass spectrometer for your analyte and the IS (using direct infusion) to determine the best parameters for detection (e.g., polarity and cone voltage). If possible, tune while combining direct infusion with the expected LC conditions (e.g., with direct infusion at 10% flow rate of LC). Use these values when developing MS methods for selected ion monitoring (e.g., SIR) or other MS modes depending on instrument and application (e.g., selected or multiple reaction monitoring (SRM, MRM) with triple quadruples mass spectrometers or mass scanning).
4. Setup a LC gradient to elute the analyte and IS such that they do not elute too early (e.g., void volume), do not elute very late (e.g., > ~10 column volumes [95]), and are preferentially separated. For HILIC, the gradient should go from a high to low percentage organic buffer (e.g., Buffer B, *see* Subheading 2.2.4). For reverse phase (RP), the gradient should go from a low to high percentage organic buffer. Especially for HILIC, the sample solvent should match closely to starting conditions [96]. Ensure backpressure does not go above that recommended by instrumentation or column (*see* **Note 19**).
5. Select conditions to quench the enzymatic reaction and to extract the analyte and IS from the buffer, enzymes, and NP as much as possible. One example is to add an organic solvent (e.g., cold methanol or acetonitrile) to precipitate the enzymes. Another example is to use 40:40:20 acetonitrile:methanol:water with 0.1 M formic acid at −20 °C, vortex 10 s, place on ice for 2 min, followed by neutralizing with 15% (w/v) ammonium bicarbonate in H_2O and placing at −20 °C for 20 min [93]. If multiple samples are collected over time, samples can be left frozen until use (e.g., −20 °C or −80 °C, with care to use a freezer amenable to storing organic solutions). For either example, centrifuge (e.g., 10 min at 16,000 rcf, preferably at 4 °C) or filter (0.2 μm) to remove precipitants (*see* **Note 20**). Watch for and avoid emulsions as part of the quench/extract process. If non-volatile salts remain after quenching/extracting, they can be diverted to waste during the LC gradient instead of into the MS, as long as they do not co-elute with the analyte or IS (*see* **Notes 21** and **22**).
6. Test LC/MS analysis of analyte and IS extracted from reaction conditions (*vs.* from pure solution) to ensure detection still adequate. Note that the LC gradient may need to be adjusted

to minimize interference from any remaining matrix components (*see* **Note 22**).

7. Perform LC/MS analysis of a range of analyte and IS concentrations extracted from reaction buffer to determine linear range of detection, lower limit of detection (LLOD), and lower limit of quantification (LLOQ). There are different definitions regarding these limits, such as a minimum LLOQ defined as a signal to noise ratio of 3:1 to 5:1 (or more) [89].
8. If the reaction conditions during linear enzymatic response allow for the analyte concentration after quenching and extraction to fall in the linear detection range, LC/MS acquisition of reaction samples can proceed. If the reaction conditions produce analyte which is too concentrated, dilute the sample. If reaction conditions produce analyte which is too dilute, the quenched/extracted reaction can be dried under vacuum (e.g., with a speedvac) or undergo concentration through solid-phase extraction, as long as the analyte (and IS) is stable throughout any process (e.g., non-volatile for vacuum). Ensure that the final sample has enough volume for injection and remaining dead volume for the LC instrumentation.
9. Before acquiring LC/MS data on reaction samples, prepare a quality control (QC) sample (for more robust validation procedures, multiple QCs may be needed, *see* Refs. 89–93). A QC sample can be prepared by removing and combining small aliquots from each reaction sample (either before or after quenching/extracting). A QC sample can also be prepared from a standard sample of analyte at a concentration not used for the standard curve (*see* **step 10**).
10. Prepare a standard curve of analyte under the same conditions as the reaction spanning above and below the maximum/minimum concentration anticipated in the reaction samples. Preferably also prepare a sample identical to the standard curve but with no analyte as a "0" control. A standard curve can also be prepared of the IS to ensure that IS responses in reaction samples are in the linear range, as well as a "00" control sample of no analyte and no IS. Quench/extract standards using same method as determined for reaction samples (*see* **step 5**). For preparation of a blank sample, one can either use the "00" control or use pure solvents (e.g., water) such that the aqueous/organic constitution of the blank matches the reaction samples.
11. For batch acquisition of LC/MS samples, make sure the column is pre-equilibrated (*see* **Note 19**). Ideally, perform a system suitability test before any large batch of acquisitions.
12. Preferably acquire data on samples in the following order:
13. Blank, to ensure system is clean.

14. QC, as start of batch.
15. Blank.
16. The analyte standard curve, including “0” control.
17. Blank.
18. Randomized samples, preferably interspersed with QC samples. Randomization is important to minimize batch effects / effects of drift.
19. Blank.
20. A second analyte curve if desired, especially if drift is anticipated.
21. Blank.
22. QC, as end of batch.
23. Blank, to ensure system is clean for next use.
24. After LC/MS data acquisition, analyze the data. Normalize the analyte response to the IS response. Ensure there is minimal drift over the batch acquisition by comparing QC samples. Ensure that the determined analyte concentrations fall within the standard curve. If applicable, one can also compare UV-VIS or fluorescence data with MS data. MS data is typically interpreted from observing the mass of the target compound or some multiple of its mass-to-charge ratio depending upon net charge in the MS. This typically requires a thorough working knowledge of MS techniques.

4 Notes

1. The scientist is encouraged to utilize any multi-enzyme system they are interested in studying. As an example, this methods paper utilizes seven of the enzymes in glycolysis: GLK, PGM, PFK, ALD, TPI, GAPDH, and PGK. Each of these enzymes was designed to have a terminal His-tag to facilitate assembly on to QDs. However, enzymes with terminal His-tags can also be purchased from a variety of reputable suppliers.
2. Proceed to microwave 30 s for up to 2 min until agarose is completely dissolved. The flask can be swirled in between 30 s intervals to ensure adequate mixing. Be very careful, microwaving the agarose too long can cause it to boil over or flash boil when it is swirled. Please use an oven mitt or extra paper towels to protect hands.
3. Successful protein expression in *E. coli* requires adequate oxygenation of the culture. Use baffled flasks for expression and ensure that the total volume does not exceed 25% of the total volume of the flask or vessel.

4. Protein expression can require a great deal of optimization for each protein of interest. Typically an expression panel is first performed to identify "best" expression conditions. Variables such as inducer concentration, incubation temperature during expression, and duration of expression are often initially examined as shown in Fig. 4. Efficiency of expression can also be affected by media composition, induction point (OD_{600}), and many others. The interested reader is referred to other Methods protocols printed in this same series [97–100].
5. Additional sonication cycles may be necessary to ensure complete cell lysis.
6. Recombinant proteins will often accumulate as insoluble inclusion bodies. Analysis of soluble and insoluble fractions via sodium dodecyl sulfate polyacrylamide gel electrophoresis (SDS-PAGE) will allow for a quick confirmation/analysis of protein solubility and fractionation. Insoluble proteins can often be extracted using denaturing purification methods or through detergent treatment to further separate proteins from insoluble cell debris [101, 102].
7. The scientist is encouraged to collect wash fractions for later SDS-PAGE evaluation of purification protocol.
8. Protein purification should proceed at 4 °C in a cold room or deli fridge if possible.
9. Additional purification methods (size exclusion chromatography, ionic exchange purification) may be necessary to remove non-target proteins.
10. If sufficiently pure after IMAC purification, fractions containing the protein of interest can be pooled and dialyzed to remove imidazole.
11. A helpful mnemonic is "Run Towards Red."
12. When deciding what concentration to make the stock enzyme solutions consider a multiple of the molar ratio of the enzyme of interest to QD for easy pipetting. Using GLK as an example, if the molar ratio of GLK to QD is 5.5, prepare the stock concentration as 0.55 μM or a multiple of that, the volume of the GLK stock to add to the QD would then be (5 nM QD × 5.5 ratio × 2 dilution factor) = 55 nM, (55 nM/1000) × 500 μL/0.55 μM = 50 μL. Keeping this in mind to prepare the other enzyme stock solutions such that the volumes to add to the QD can be maintained at 50 μL saving pipetting time.
13. Keep all reagents and enzymes on ice. This is very important and helps ensure active enzyme.

14. No droplets should cling to the sides of the well, if they drop during the time course of the experiment, the results will be skewed for those wells.

15. The stock solution will be double each component's concentration because of a 1 to 2 dilution with the multi-enzyme-QD bioconjugate. The ratio of substrate volume to bioconjugate volume and total volume can be adjusted dependent on your experimental design.

16. The scientist should have plenty of room to work with no distractions, tell lab mates not to interrupt, all materials needed are close at hand, and the kinetic program should already be set up ready to go. The scientist will need a steady hand. It is helpful to line up the first tip of the 12-well pipette to the first well and then rock your hand to the right side, the rest of the tips will naturally correspond with the every other well. To do the second half of the row, the scientist can line up the last pipette tip with the last well of the row and rock the pipette to the left side.

17. Experiments are typically set up to run overnight but the duration of the time course will rely on the kinetics of the individual system. An overnight run maximizes time in the lab as the next day can be devoted to analyzing data and planning the next experiment.

18. If filtering, ensure filter will not result in contaminants for MS and will not capture analyte/IS. For example, nylon is easily detected by MS. PTFE is typically a good choice [103].

19. Column equilibration for HILIC can take a significant time; check instructions with column. HILIC may also require several blanks or QCs at the beginning of a batch to ensure column is ready for entire batch (*i.e.*, to minimize batch effects/effects of drift). Between or within each acquisition should include a wash-out phase (preferably) and a re-equilibration phase, unless using an isocratic method.

20. The IS can be either added to the reaction (if non-interfering), post-reaction prior to quench/extraction, or added to the same component used to quench/extract, though earlier in the process is preferable to monitor as much pipetting error as possible.

21. There are other extraction methods available, such as solid-phase extraction or liquid-liquid extraction; the interested reader is directed to references [91, 93]. These can be particularly beneficial if sample concentration is needed.

22. Ideally, perform several tests to ensure analysis robustness, such as quenching and extraction efficiency. *See* Refs. 89–93 as indicated above. For example, quenched solutions can be tested repeatedly over time to ensure no change in product

concentration (i.e., adequate quenching). Extraction efficiency can be estimated by comparing the concentration of analyte and IS detected from extracted complex sample matrix vs. extracted from pure solvent.

Acknowledgments

The authors acknowledge ONR, NRL Base Funding, a NRL Nanoscience Institutes Program. I.L.M. acknowledges USDA NIFA Award #2020-67021-31254 and SERDP Award # WP21-1073 New Start Project (W74RDV03497375).

References

1. Reetz MT (2016) What are the limitations of enzymes in synthetic organic chemistry? Chem Rec 16(6):2449–2459. https://doi.org/10.1002/tcr.201600040
2. Martinez-Rodriguez S, Torres JM, Sanchez P, Ortega E (2020) Overview on multienzymatic cascades for the production of non-canonical alpha-amino acids. Front Bioeng Biotechnol 8:36. https://doi.org/10.3389/fbioe.2020.00887
3. Galanie S, Thodey K, Trenchard IJ, Interrante MF, Smolke CD (2015) Synthetic biology complete biosynthesis of opioids in yeast. Science 349(6252):1095–1100. https://doi.org/10.1126/science.aac9373
4. Shahidi F, Janak Kamil YVA (2001) Enzymes from fish and aquatic invertebrates and their application in the food industry. Trends Food Sci Technol. 12(12):435–464. https://doi.org/10.1016/S0924-2244(02)00021-3
5. Friedrich S, Hahn F (2015) Opportunities for enzyme catalysis in natural product chemistry. Tetrahedron 71(10):1473–1508. https://doi.org/10.1016/j.tet.2014.12.026
6. Erdmann V, Lichman BR, Zhao JX, Simon RC, Kroutil W, Ward JM, Hailes HC, Rother D (2017) Enzymatic and chemoenzymatic three-step cascades for the synthesis of stereochemically complementary trisubstituted tetrahydroisoquinolines. Angew Chem Int Ed 56(41):12503–12507. https://doi.org/10.1002/anie.201705855
7. Wang J, Jain R, Shen XL, Sun XX, Cheng MY, Liao JC, Yuan QP, Yan YJ (2017) Rational engineering of diol dehydratase enables 1,4-butanediol biosynthesis from xylose. Metab Eng 40:148–156. https://doi.org/10.1016/j.ymben.2017.02.003
8. Robescu MS, Serra I, Terreni M, Ubiali D, Bavaro T (2020) A multi-enzymatic cascade reaction for the synthesis of vidarabine 5'-monophosphate. Catalysts 10(1):12. https://doi.org/10.3390/catal10010060
9. Vranish JN, Ancona MG, Walper SA, Medintz IL (2018) Pursuing the promise of enzymatic enhancement with nanoparticle assemblies. Langmuir 34(9):2901–2925. https://doi.org/10.1021/acs.langmuir.7b02588
10. Zaks A (2001) Industrial biocatalysis. Curr Opin Chem Biol. 5(2):130–136. https://doi.org/10.1016/S1367-5931(00)00181-2
11. Wu SL, Liu QS, Tan HD, Zhang FY, Yin H (2020) A novel 2,5-furandicarboxylic acid biosynthesis route from biomass-derived 5-hydroxymethylfurfural based on the consecutive enzyme reactions. Appl Biochem Biotechnol 191(4):1470–1482. https://doi.org/10.1007/s12010-020-03290-1
12. Galanie S, Entwistle D, Lalonde J (2020) Engineering biosynthetic enzymes for industrial natural product synthesis. Nat Prod Rep 37(8):1122–1143. https://doi.org/10.1039/c9np00071b
13. Valliere MA, Korman TP, Arbing MA, Bowie JU A bio-inspired cell-free system for cannabinoid production from inexpensive inputs. Nat Chem Biol 16. https://doi.org/10.1038/s41589-020-0631-9
14. Wiltschi B, Cernava T, Dennig A, Casas MG, Geier M, Gruber S, Haberbauer M, Heidinger P, Acero EH, Kratzer R, Luley-Goedl C, Muller CA, Pitzer J, Ribitsch D, Sauer M, Schmolzer K, Schnitzhofer W, Sensen CW, Soh J, Steiner K, Winkler CK, Winkler M, Wriessnegger T (2020) Enzymes revolutionize the bioproduction of value-added compounds: from enzyme discovery to special applications. Biotechnol Adv 40:51. https://doi.org/10.1016/j.biotechadv.2020.107520

15. France SP, Hepworth LJ, Turner NJ, Flitsch SL (2017) Constructing biocatalytic cascades: in vitro and in vivo approaches to de novo multi-enzyme pathways. ACS Catal 7(1): 710–724. https://doi.org/10.1021/acscatal.6b02979
16. Jaroensuk J, Intasian P, Wattanasuepsin W, Akeratchatapan N, Kesornpun C, Kittipanukul N, Chaiyen P (2020) Enzymatic reactions and pathway engineering for the production of renewable hydrocarbons. J Biotechnol 309:1–19. https://doi.org/10.1016/j.jbiotec.2019.12.010
17. Sheldon RA, Brady D (2019) Broadening the scope of biocatalysis in sustainable organic synthesis. ChemSusChem 12(13): 2859–2881. https://doi.org/10.1002/cssc.201900351
18. Morgado G, Gerngross D, Roberts TM, Panke S (2018) Synthetic biology for cell-free biosynthesis: fundamentals of designing novel in vitro multi- enzyme reaction networks. In: Zhao H, Zeng AP (eds) Synthetic biology - metabolic engineering, Advances in biochemical engineering-biotechnology, vol 162. Springer, Cham, pp 117–146. https://doi.org/10.1007/10_2016_13
19. Marchand JA, Neugebauer ME, Ing MC, Lin CI, Pelton JG, Chang MCY (2019) Discovery of a pathway for terminal-alkyne amino acid biosynthesis. Nature 567(7748):420–424. https://doi.org/10.1038/s41586-019-1020-y
20. National Human Genome Research Institute Genome sequencing informatics tools. https://iseqtools.org/
21. Carbonell P, Parutto P, Baudier C, Junot C, Faulon JL (2014) Retropath: Automated pipeline for embedded metabolic circuits. ACS Synth Biol 3(8):565–577. https://doi.org/10.1021/sb4001273
22. Delépine B, Duigou T, Carbonell P, Faulon J-L (2018) RetroPath2.0: A retrosynthesis workflow for metabolic engineers. Metab Eng 45:158–170. https://doi.org/10.1016/j.ymben.2017.12.002
23. Koch M, Duigou T, Faulon J-L (2020) Reinforcement learning for bioretrosynthesis. ACS Synth Biol 9(1):157–168. https://doi.org/10.1021/acssynbio.9b00447
24. Muschiol J, Peters C, Oberleitner N, Mihovilovic MD, Bornscheuer UT, Rudroff F (2015) Cascade catalysis - strategies and challenges en route to preparative synthetic biology. Chem Commun 51(27):5798–5811. https://doi.org/10.1039/c4cc08752f
25. Zhong C, Duic B, Bolivar JM, Nidetzky B (2020) Three-enzyme phosphorylase cascade immobilized on solid support for biocatalytic synthesis of cello-oligosaccharides. ChemCatChem 12(5):1350–1358. https://doi.org/10.1002/cctc.201901964
26. Peng F, Ou XY, Guo ZW, Zeng YJ, Zong MH, Lou WY (2020) Co-immobilization of multiple enzymes by self-assembly and chemical crosslinking for cofactor regeneration and robust biocatalysis. Int J Biol Macromol 162: 445–453. https://doi.org/10.1016/j.ijbiomac.2020.06.141
27. Liu JL, Zabetakis D, Breger JC, Anderson GP, Goldman ER (2020) Multi-enzyme assembly on T4 phage scaffold. Front Bioeng Biotechnol 8:6. https://doi.org/10.3389/fbioe.2020.00571
28. Klein WP, Thomsen RP, Turner KB, Walper SA, Vranish J, Kjems J, Ancona MG, Medintz IL (2019) Enhanced catalysis from multienzyme cascades assembled on a DNA origami triangle. ACS Nano 13(12):13677–13689. https://doi.org/10.1021/acsnano.9b05746
29. Liu XL, Formanek P, Voit B, Appelhans D (2017) Functional cellular mimics for the spatiotemporal control of multiple enzymatic cascade reactions. Angew Chem Int Ed 56(51):16233–16238. https://doi.org/10.1002/anie.201708826
30. Azuma Y, Zschoche R, Tinzl M, Hilvert D (2016) Quantitative packaging of active enzymes into a protein cage. Angew Chem Int Ed 55(4):1531–1534. https://doi.org/10.1002/anie.201508414
31. Ellis GA, Klein WP, Lasarte-Aragones G, Thakur M, Walper SA, Medintz IL (2019) Artificial multienzyme scaffolds: pursuing in vitro substrate channeling with an overview of current progress. ACS Catal 9(12): 10812–10869. https://doi.org/10.1021/acscatal.9b02413
32. Johnson BJ, Russ Algar W, Malanoski AP, Ancona MG, Medintz IL (2014) Understanding enzymatic acceleration at nanoparticle interfaces: approaches and challenges. Nano Today 9(1):102–131. https://doi.org/10.1016/j.nantod.2014.02.005
33. Kim J, Grate JW, Wang P (2006) Nanostructures for enzyme stabilization. Chem Eng Sci 61(3):1017–1026. https://doi.org/10.1016/j.ces.2005.05.067
34. Wu M, Algar WR (2015) Acceleration of proteolytic activity associated with selection of thiol ligand coatings on quantum dots. ACS Appl Mater Interfaces 7(4):2535–2545. https://doi.org/10.1021/am507466b

35. Heuer-Jungemann A, Feliu N, Bakaimi I, Hamaly M, Alkilany A, Chakraborty I, Masood A, Casula MF, Kostopoulou A, Oh E, Susumu K, Stewart MH, Medintz IL, Stratakis E, Parak WJ, Kanaras AG (2019) The role of ligands in the chemical synthesis and applications of inorganic nanoparticles. Chem Rev 119(8):4819–4880. https://doi.org/10.1021/acs.chemrev.8b00733
36. Rameshwar T, Samal S, Lee S, Kim S, Cho J, Kim IS (2006) Determination of the size of water-soluble nanoparticles and quantum dots by field-flow fractionation. J Nanosci Nanotechnol 6(8):2461–2467. https://doi.org/10.1166/jnn.2006.544
37. Susumu K, Oh E, Delehanty JB, Blanco-Canosa JB, Johnson BJ, Jain V, Hervey WJ, Algar WR, Boeneman K, Dawson PE, Medintz IL (2011) Multifunctional compact zwitterionic ligands for preparing robust biocompatible semiconductor quantum dots and gold nanoparticles. J Am Chem Soc 133(24):9480–9496. https://doi.org/10.1021/ja201919s
38. Susumu K, Oh E, Delehanty JB, Pinaud F, Gemmill KB, Walper S, Breger J, Schroeder MJ, Stewart MH, Jain V, Whitaker CM, Huston AL, Medintz IL (2014) A new family of pyridine-appended multidentate polymers as hydrophilic surface ligands for preparing stable biocompatible quantum dots. Chem Mater 26(18):5327–5344. https://doi.org/10.1021/cm502386f
39. Diaz SA, Sen S, Gemmill KB, Brown CW, Oh E, Susumu K, Stewart MH, Breger JC, Aragones GL, Field LD, Deschamps JR, Kral P, Medintz IL (2017) Elucidating surface ligand-dependent kinetic enhancement of proteolytic activity at surface-modified quantum dots. ACS Nano 11(6):5884–5896. https://doi.org/10.1021/acsnano.7b01624
40. Díaz SA, Malonoski AP, Susumu K, Hofele RV, Oh E, Medintz IL (2015) Probing the kinetics of quantum dot-based proteolytic sensors. Anal Bioanal Chem 407(24): 7307–7318. https://doi.org/10.1007/s00216-015-8892-y
41. Algar WR, Malonoski A, Deschamps JR, Blanco-Canosa JB, Susumu K, Stewart MH, Johnson BJ, Dawson PE, Medintz IL (2012) Proteolytic activity at quantum dot-conjugates: kinetic analysis reveals enhanced enzyme activity and localized interfacial "hopping". Nano Lett 12(7): 3793–3802. https://doi.org/10.1021/nl301727k
42. Breger JC, Oh E, Susumu K, Klein WP, Walper SA, Ancona MG, Medintz IL (2019) Nanoparticle size influences localized enzymatic enhancement-A case study with phosphotriesterase. Bioconjug Chem 30(7): 2060–2074. https://doi.org/10.1021/acs.bioconjchem.9b00362
43. Sapsford KE, Algar WR, Berti L, Gemmill KB, Casey BJ, Oh E, Stewart MH, Medintz IL (2013) Functionalizing nanoparticles with biological molecules: developing chemistries that facilitate nanotechnology. Chem Rev 113(3):1904–2074. https://doi.org/10.1021/cr300143v
44. Algar WR, Prasuhn DE, Stewart MH, Jennings TL, Blanco-Canosa JB, Dawson PE, Medintz IL (2011) The controlled display of biomolecules on nanoparticles: a challenge suited to bioorthogonal chemistry. Bioconjug Chem 22(5):825–858. https://doi.org/10.1021/bc200065z
45. Bilal M, Cui JD, Iqbal HMN (2019) Tailoring enzyme microenvironment: state-of-the-art strategy to fulfill the quest for efficient bio-catalysis. Int J Biol Macromol 130: 186–196. https://doi.org/10.1016/j.ijbiomac.2019.02.141
46. Pfeiffer C, Rehbock C, Hühn D, Carrillo-Carrion C, De Aberasturi DJ, Merk V, Barcikowski S, Parak WJ (2014) Interaction of colloidal nanoparticles with their local environment: the (ionic) nanoenvironment around nanoparticles is different from bulk and determines the physico-chemical properties of the nanoparticles. J R Soc Interface 11(96):13. https://doi.org/10.1098/rsif.2013.0931
47. Prasuhn DE, Blanco-Canosa JB, Vora GJ, Delehanty JB, Susumu K, Mei BC, Dawson PE, Medintz IL (2010) Combining chemoselective ligation with polyhistidine-driven self-assembly for the modular display of biomolecules on quantum dots. ACS Nano 4(1): 267–278. https://doi.org/10.1021/nn901393v
48. Prasuhn DE, Deschamps JR, Susumu K, Stewart MH, Boeneman K, Blanco-Canosa JB, Dawson PE, Medintz IL (2010) Polyvalent display and packing of peptides and proteins on semiconductor quantum dots: predicted versus experimental results. Small 6(4):555–564. https://doi.org/10.1002/smll.200901845
49. Blanco-Canosa JB, Wu M, Susumu K, Petryayeva E, Jennings TL, Dawson PE, Algar WR, Medintz IL (2014) Recent progress in the bioconjugation of quantum dots. Coord Chem Rev 263-264:101–137. https://doi.org/10.1016/j.ccr.2013.08.030

50. Boeneman K, Delehanty JB, Blanco-Canosa JB, Susumu K, Stewart MH, Oh E, Huston AL, Dawson G, Ingale S, Walters R, Domowicz M, Deschamps JR, Algar WR, DiMaggio S, Manono J, Spillmann CM, Thompson D, Jennings TL, Dawson PE, Medintz IL (2013) Selecting improved peptidyl motifs for cytosolic delivery of disparate protein and nanoparticle materials. ACS Nano 7(5):3778–3796. https://doi.org/10.1021/nn400702r
51. Algar WR, Blanco-Canosa JB, Manthe RL, Susumu K, Stewart MH, Dawson PE, Medintz IL (2013) Synthesizing and modifying peptides for chemoselective ligation and assembly into quantum dot-peptide bioconjugates. Methods Mol Biol 1025. https://doi.org/10.1007/978-1-62703-462-3_5
52. Breger JC, Ancona MG, Walper SA, Oh E, Susumu K, Stewart MH, Deschamps JR, Medintz IL (2015) Understanding how nanoparticle attachment enhances phosphotriesterase kinetic efficiency. ACS Nano 9(8): 8491–8503. https://doi.org/10.1021/acsnano.5b03459
53. Brown CW, Oh E, Hastman DA, Walper SA, Susumu K, Stewart MH, Deschamps JR, Medintz IL (2015) Kinetic enhancement of the diffusion-limited enzyme beta-galactosidase when displayed with quantum dots. RSC Adv 5:93089–93094
54. Claussen JC, Malanoski A, Breger JC, Oh E, Walper SA, Susumu K, Goswami R, Deschamps JR, Medintz IL (2015) Probing the enzymatic activity of alkaline phosphatase within quantum dot bioconjugates. J Phys Chem C 119(4):2208–2221. https://doi.org/10.1021/jp5110467
55. Lata JP, Gao L, Mukai C, Cohen R, Nelson JL, Anguish L, Coonrod S, Travis AJ (2015) Effects of nanoparticle size on multilayer formation and kinetics of tethered enzymes. Bioconjug Chem 26(9):1931–1938. https://doi.org/10.1021/acs.bioconjchem.5b00354
56. Vranish JN, Ancona MG, Oh E, Susumu K, Lasarte Aragonés G, Breger JC, Walper SA, Medintz IL (2018) Enhancing coupled enzymatic activity by colocalization on nanoparticle surfaces: kinetic evidence for directed channeling of intermediates. ACS Nano 12(8):7911–7926. https://doi.org/10.1021/acsnano.8b02334
57. Díaz SA, Choo P, Oh E, Susumu K, Klein WP, Walper SA, Hastman DA, Odom TW, Medintz IL (2021) Gold nanoparticle templating increases the catalytic rate of an amylase, maltase, and glucokinase multienzyme cascade through substrate channeling independent of surface curvature. ACS Catal 11:627–638. https://doi.org/10.1021/acscatal.0c03602
58. Danaceau JP, Chambers EE, Fountain KJ (2012) Hydrophilic interaction chromatography (HILIC) for LC-MS/MS analysis of monoamine neurotransmitters. Bioanalysis 4(7):783–794. https://doi.org/10.4155/bio.12.46
59. Lu HD, Tang SY, Yun GL, Li HY, Zhang YX, Qiao RR, Li WH (2020) Modular and integrated systems for nanoparticle and microparticle synthesis-A review. Biosensors-Basel 10(11):34. https://doi.org/10.3390/bios10110165
60. Susumu K, Field LD, Oh E, Hunt M, Delehanty JB, Palomo V, Dawson PE, Huston AL, Medintz IL (2017) Purple-, blue-, and green-emitting multishell alloyed quantum dots: synthesis, characterization, and application for ratiometric extracellular pH sensing. Chem Mater 29(17):7330–7344. https://doi.org/10.1021/acs.chemmater.7b02174
61. Nag OK, Stewart MH, Deschamps JR, Susumu K, Oh E, Tsytsarev V, Tang QG, Efros AL, Vaxenburg R, Black BJ, Chen YC, O'Shaughnessy TJ, North SH, Field LD, Dawson PE, Pancrazio JJ, Medintz IL, Chen Y, Erzurumlu RS, Huston AL, Delehanty JB (2017) Quantum dot-peptide-fullerene bioconjugates for visualization of in vitro and in vivo cellular membrane potential. ACS Nano 11(6):5598–5613. https://doi.org/10.1021/acsnano.7b00954
62. Stewart MH, Huston AL, Scott AM, Oh E, Algar WR, Deschamps JR, Susumu K, Jain V, Prasuhn DE, Blanco-Canosa J, Dawson PE, Medintz IL (2013) Competition between forster resonance energy transfer and electron transfer in stoichiometrically assembled semiconductor quantum dot-fullerene conjugates. ACS Nano 7(10):9489–9505. https://doi.org/10.1021/nn403872x
63. Boeneman K, Delehanty JB, Susumu K, Stewart MH, Medintz IL (2010) Intracellular bioconjugation of targeted proteins with semiconductor quantum dots. J Am Chem Soc 132(17):5975–5977. https://doi.org/10.1021/ja100201w
64. Del Solar G, Giraldo R, Ruiz-Echevarria MJ, Espinosa M, Diaz-Orejas R (1998) Replication and control of circular bacterial plasmids. Microbiol Mol Biol Rev 62(2):434–464
65. Rosano GL, Ceccarelli EA (2014) Recombinant protein expression in Escherichia coli: advances and challenges. Front Microbiol 5: 17. https://doi.org/10.3389/fmicb.2014.00172

66. Marschall L, Sagmeister P, Herwig C (2017) Tunable recombinant protein expression in E. coli: promoter systems and genetic constraints. Appl Microbiol Biotechnol 101(2): 501–512. https://doi.org/10.1007/s00253-016-8045-z
67. Emrich CA, Medintz IL, Chu WK, Mathies RA (2007) Microfabricated two-dimensional electrophoresis device for differential protein expression profiling. Anal Chem 79(19): 7360–7366. https://doi.org/10.1021/ac0711485
68. Debendorff JW, Studier FW (1991) Controlling basal expression in an inducible T7 expression system by blocking the target T7 promoter with lac repressor. J Mol Biol 19(1):45–59
69. Studier FW, Rosenberg AH, Dunn JJ, Dubendorff JW (1990) Use of T7 RNA polymerase to direct expression of cloned genes. Methods Enzymol 185:60–89
70. Giannini E, Gonzalez LJ, Vila AJ (2019) A simple protocol to characterize bacterial cell-envelope lipoproteins in a native-like environment. Protein Sci 28(11):2004–2010. https://doi.org/10.1002/pro.3728
71. Malherbe G, Humphreys DP, Dave E (2019) A robust fractionation method for protein subcellular localization studies in *Escherichia coli*. BioTechniques 66(4):171–178
72. Shriver-Lake LC, Zabetakis D, Goldman ER, Anderson GP (2017) Evaluation of anti-botulinum neurotoxin single domain antibodies with additional optimization for improved production and stability. Toxicon 135:51–58. https://doi.org/10.1016/j.toxicon.2017.06.002
73. Shehadul Islam M, Aryasomayajula A, Selvaganapathy P (2017) A review on macroscale and microscale cell lysis methods. Micromachines 8(3):83. https://doi.org/10.3390/mi8030083
74. Coskun O (2016) Separation techniques: chromatography. North Clin Istanb 3(2): 156–160. https://doi.org/10.14744/nci.2016.32757
75. Walls D, Loughran ST (eds) (2011) Protein chromatography, Methods in molecular biology, vol 681. Humana Press, Totowa, NJ
76. Feroz H, Vandervelden C, Ikwuagwu B, Ferlez B, Baker C, Lugar DJ, Grzelakowski M, Golbeck JH, Zydney AL, Kumar M (2016) Concentrating membrane proteins using ultrafiltration without concentrating detergents. Biotechnol Bioeng 10(2): 122–130
77. Goldring JPD (2019) Concentrating proteins by salt, polyethylene glycol, solvent, SDS precipitation, three-phase partitioning, dialysis, centrifugation, ultrafiltration, lyophilization, affinity chromatography, immunoprecipitation or increased temperature for protein isolation, drug interaction, and proteomic and peptidomic evaluation. Methods Mol Biol 1855:41–59
78. Hyde AM, Zultanski SL, Waldman JH, Zhong Y-L, Shevlin M, Peng F (2017) General principles and strategies for salting-out informed by the Hofmeister series. Org Process Res Dev 21(9):1355–1370. https://doi.org/10.1021/acs.oprd.7b00197
79. Bornhorst JA, Falke JJ (2000) Purification of proteins using polyhistidine affinity tags. Methods Enzymol 326:245–254
80. Díaz SA, Breger JC, Malanoski A, Claussen JC, Walper SA, Ancona MG, Brown CW, Stewart MH, Oh E, Susumu K, Medintz IL (2015) Modified kinetics of enzymes interacting with nanoparticles. In: SPIE Nanoscience + Engineering, San Diego, USA, 2015. International Society for Optics and Photonics, pp 95500K-95500K-95509
81. Buckhout-White S, Ancona M, Oh E, Deschamps JR, Stewart MH, Blanco-Canosa JB, Dawson PE, Goldman ER, Medintz IL (2012) Multimodal characterization of a linear DNA-based nanostructure. ACS Nano 6(2):1026–1043. https://doi.org/10.1021/nn204680r
82. Jennings TL, Becker-Catania SG, Triulzi RC, Tao GL, Scott B, Sapsford KE, Spindel S, Oh E, Jain V, Delehanty JB, Prasuhn DE, Boeneman K, Algar WR, Medintz IL (2011) Reactive semiconductor nanocrystals for chemoselective biolabeling and multiplexed analysis. ACS Nano 5(7):5579–5593. https://doi.org/10.1021/nn201050g
83. Medintz IL, Anderson GP, Lassman ME, Goldman ER, Bettencourt LA, Mauro JM (2004) General strategy for biosensor design and construction employing multifunctional surface-tethered components. Anal Chem 76(19):5620–5629. https://doi.org/10.1021/ac049419o
84. Cornish-Bowden A (2012) Fundamentals of enzyme kinetics, 4th edn. Wiley-Blackwell, Weinheim
85. Duggleby RG (2001) Quantitative analysis of the time courses of enzyme-catalyzed reactions. Methods. 24(2):168–174. https://doi.org/10.1006/meth.2001.1177
86. Gawande MB, Goswami A, Asefa T, Guo HZ, Biradar AV, Peng DL, Zboril R, Varma RS (2015) Core-shell nanoparticles: synthesis and applications in catalysis and electrocatalysis. Chem Soc Rev 44(21):7540–7590. https://doi.org/10.1039/c5cs00343a

87. Kim Y, Ho SO, Gassman NR, Korlann Y, Landorf EV, Collart FR, Weiss S (2008) Efficient site-specific labeling of proteins via cysteines. Bioconjug Chem 19(3):786–791. https://doi.org/10.1021/bc7002499
88. Walper SA, Turner KB, Medintz IL (2015) Enzymatic bioconjugation of nanoparticles: developing specificity and control. Curr Opin Biotechnol 34:232–241. https://doi.org/10.1016/j.copbio.2015.04.003
89. Colangelo JL (2017) The validation of quantitative mass spectrometry assays for clinical chemistry assessments in animal models. Toxicol Pathol 45(7):977–982. https://doi.org/10.1177/0192623317737240
90. Food and Drug Administration (2018) Bioanalytical method validation guidance for industry
91. RSC Analytical Methods Committee (2014) Guide to achieving reliable quantitative LC-MS measurements. https://www.rsc.org/images/AMC%20LCMS%20Guide_tcm18-240030.pdf
92. Kapoore RV, Vaidyanathan S (2016) Towards quantitative mass spectrometry-based metabolomics in microbial and mammalian systems. Philos Trans A Math Phys Eng Sci 374(2079). https://doi.org/10.1098/rsta.2015.0363
93. Lu W, Su X, Klein MS, Lewis IA, Fiehn O, Rabinowitz JD (2017) Metabolite measurement: pitfalls to avoid and practices to follow. Annu Rev Biochem 86:277–304. https://doi.org/10.1146/annurev-biochem-061516-044952
94. Neue UD (2002) HPLC troubleshooting guide. https://www.waters.com/waters/library.htm?locale=en_US&lid=1528445
95. HPLC back to basics: a laboratory companion for liquid chromatographers #1. https://learning.sepscience.com/hubfs/Guides/HPLC%20Solutions/HPLC_BtB_v2.pdf
96. How to avoid common problems with HILIC methods. https://www.restek.com/Technical-Resources/Technical-Library/General-Interest/general_GNAR2716-UNV
97. Correa A, Oppezzo P (2015) Overcoming the solubility problem in E. coli: available approaches for recombinant protein production. In: GarciaFruitos E (ed) Insoluble proteins: methods and protocols, Methods in molecular biology, vol 1258. Springer, Cham, pp 27–44. https://doi.org/10.1007/978-1-4939-2205-5_2
98. Dodson CA (2017) Production of protein kinases in e-coli. In: BurgessBrown NA (ed) Heterologous gene expression in E. coli: methods and protocols, Methods in molecular biology, vol 1586. Springer, Cham, pp 251–264. https://doi.org/10.1007/978-1-4939-6887-9_16
99. Rues RB, Henrich E, Boland C, Caffrey M, Bernhard F (2016) Cell-free production of membrane proteins in Escherichia coli lysates for functional and structural studies. In: Mus-Veteau I (ed) Heterologous expression of membrane proteins: methods and protocols, Methods in molecular biology, vol 1432, 2nd edn. Springer, Cham, pp 1–21. https://doi.org/10.1007/978-1-4939-3637-3_1
100. Ukkonen K, Neubauer A, Pereira VJ, Vasala A (2017) High yield of recombinant protein in shaken E. coli cultures with enzymatic glucose release medium EnPresso B. In: Burgess-Brown NA (ed) Heterologous gene expression in E. coli: methods and protocols, Methods in molecular biology, vol 1586. Springer, Cham, pp 127–137. https://doi.org/10.1007/978-1-4939-6887-9_8
101. Wingfield PT (2015) Overview of the purification of recombinant proteins. Curr Protoc Protein Sci 80:1–50. https://doi.org/10.1002/0471140864.ps0601s80
102. Yang Z, Zhang L, Zhang Y, Zhang T, Feng Y, Lu X, Lan W, Wang J, Wu H, Cao C, Wang X (2011) Highly efficient production of soluble proteins from insoluble inclusion bodies by a two-step-denaturing and refolding method. PLoS One 6(7):1–8. https://doi.org/10.1371/journal.pone.0022981
103. Tran JC, Doucette AA (2006) Cyclic polyamide oligomers extracted from nylon 66 membrane filter disks as a source of contamination in liquid chromatography/mass spectrometry. J Am Soc Mass Spectrom 17(5):652–656. https://doi.org/10.1016/j.jasms.2006.01.008
104. Medintz IL, Pons T, Susumu K, Boeneman K, Dennis AM, Farrell D, Deschamps JR, Melinger JS, Bao G, Mattoussi H (2009) Resonance energy transfer between luminescent quantum dots and diverse fluorescent protein acceptors. J Phys Chem C 113(43):18552–18561. https://doi.org/10.1021/jp9060329

Chapter 16

A Four-enzyme Nanoassembly Consisting of Hydrolases and Oxidoreductases for Multi-step Cascade Reactions

Archontoula Giannakopoulou, Michaela Patila, Nikolaos Chalmpes, Angeliki C. Polydera, Dimitrios Gournis, and Haralambos Stamatis

Abstract

Cascade reactions catalyzed by multi-enzymatic systems have attracted enormous scientific interest over the last decade. They are an emerging technology that significantly expands the applicability of biocatalysts in several biotechnological processes, such as the synthesis of high value-added products. Immobilization of enzymes on a solid carrier is a commonly used strategy to improve the stability and reuse of multiple enzyme systems. Magnetic nanoparticles have been applied as promising nanocarriers for either the immobilization of one enzyme or the co-immobilization of multiple enzymes. In this chapter, we describe the preparation of magnetic iron oxide nanoparticles γ-Fe_2O_3 modified with 3-(aminopropyl)-triethoxysilane (APTES), for the simultaneous covalent co-immobilization of oxidoreductases and hydrolytic enzymes, such as cellulase, β-glucosidase (bgl), glucose oxidase (GOx), and horseradish peroxidase (HRP). Several spectroscopic techniques that are used to characterize the structure and the catalytic performance of such systems are also described.

Key words Multi-enzymatic assemblies, Nanobiocatalyst, Amino-functionalized magnetic nanoparticles, Cellulase, β-glucosidase, Glucose oxidase, Horseradish peroxidase, Four-step cascade reaction

1 Introduction

Over the last decade, nanobiocatalysis has emerged as an extremely promising field for the development of innovative biocatalytic systems [1]. Cascade reactions, the integration of multiple enzyme transformations, are an important emerging technology due to the need in exploring more sustainable alternative ways of producing chemicals and useful bioproducts [2], as well as to develop effective biosensors [3]. The highly effective catalytic mechanisms of cascade reactions combined with the benefits of enzyme immobilization have promoted the development of innovative strategies to combine a variety of enzyme activities through the immobilization of enzymes in nanocarriers and to create artificial nanosystems [4, 5].

Haralambos Stamatis (ed.), *Multienzymatic Assemblies: Methods and Protocols*,
Methods in Molecular Biology, vol. 2487, https://doi.org/10.1007/978-1-0716-2269-8_16,
© The Author(s), under exclusive license to Springer Science+Business Media, LLC, part of Springer Nature 2022

Co-immobilization of multiple enzymes can generally be accomplished by five chemical methods: (1) non-specific immobilization, (2) sequential immobilization, (3) site immobilization, (4) specific site immobilization, and (5) cross-linking [6]. However, in order to maintain the enzymatic activity of all individual enzymes, many parameters should be considered, such as the appropriate immobilization technique, the physical properties of the nanocarrier, as well as the structure and function of each enzyme [7, 8] Therefore, the selection of the appropriate carrier for enzyme immobilization is of high priority. The main characteristics of carriers widely used in co-immobilization approaches are chemical and thermal stability, biocompatibility, easy recovery, and reuse, as well as availability and price [9–11].

The rapid development of nanotechnology in recent decades has led to the development of nanoparticles used as highly promising carriers, both in the immobilization of one enzyme and in the co-immobilization of multiple enzymes. Especially magnetic nanoparticles, due to their magnetic core, can be easily separated from the reaction mixture only by applying an external magnetic field, so that they can be reused effectively. In addition, in order to reduce their toxicity and increase their biocompatibility, magnetic nanoparticles can be coated with different functional groups, such as amino groups [12–14].

In the present chapter, we describe the methodology required for the simultaneous covalent co-immobilization of different classes of enzymes, such as oxidoreductases and hydrolases. Four enzymes, namely cellulase from *Trichoderma reesei*, β-glucosidase from *Thermotoga maritima* (bgl), glucose oxidase (GOx), and horseradish peroxidase (HRP) are co-immobilized on amino-functionalized magnetic nanoparticles (γ-Fe-Aptes MNPs) in order to produce a nanobiocatalytic system capable of performing a four-step cascade reaction, such as the breakdown of cellulose. A schematic representation of the overall cascade process described above is briefly depicted in Fig. 1. Finally, in the present chapter, we describe various spectroscopic methods required for the characterization of the structural and catalytic behavior of such systems.

2 Materials

Prepare all solutions using ultra-pure water (prepared by purifying deionized water, to attain a sensitivity of 18 MΩ-cm at 25 °C) and analytical grade reagents. Prepare and store all reagents at room temperature (unless indicated otherwise). Diligently follow all waste disposal regulations when disposing of waste materials.

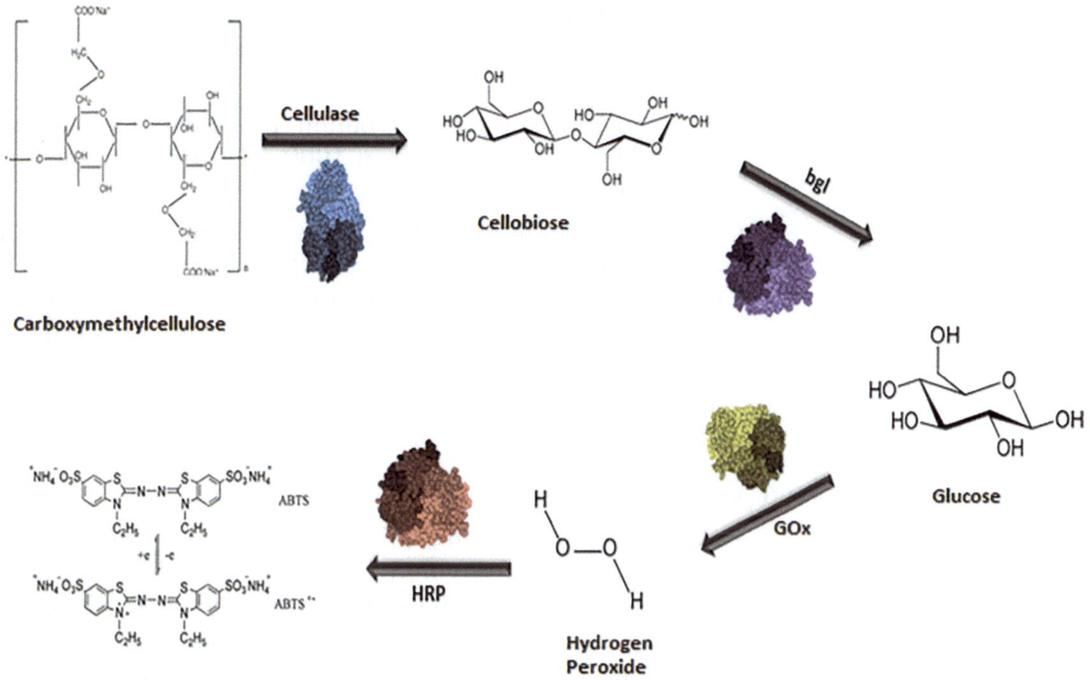

Fig. 1 A schematic representation of the overall enzymatic process involving the breakdown of cellulose through a four-step cascade reaction

2.1 Synthesis of Amino-functionalized Magnetic Nanoparticles (γ-Fe-Aptes-MNPs)

1. Beakers.
2. Spherical bottle.
3. Volumetric cylinders.
4. Magnetic stirrer.
5. Reflux accessories.
6. Centrifuge.
7. Vacuum desiccators.
8. Pipette Pasteur.
9. Ultrasonic bath.
10. Chloroform (≥99.0%).
11. Ethanol (≥99.8%).
12. Iron (II) chloride tetrahydrate (≥99.0%).
13. (3-Aminopropyl) triethoxysilane (99%).
14. Oleylamine (≥98.0%).

2.2 Co-immobilization of Cellulase, bgl, GOx, and HRP on γ-Fe-Aptes-MNPs

1. 0.2 M solution of monobasic sodium phosphate (A): Weigh 27.8 g of $NaH_2PO_4{\cdot}H_2O$ in 1000 mL ultra-pure water in a glass beaker.
2. 0.2 M solution of dibasic sodium phosphate (B): Weigh 53.65 g of $Na_2HPO_4 \cdot 7H_2O$ in 1000 mL ultra-pure water in a glass beaker.
3. 0.1 M phosphate buffer, pH 6.0: Add 87.7 mL of solution A and 12.3 mL of solution B to a 100-mL graduated cylinder. Then, add 100 mL of ultra-pure water to reach a final volume of 200 mL. Mix and adjust pH either with HCl or NaOH, if necessary. Store at 4 °C.
4. Amino-functionalized magnetic nanoparticles.
5. Tween-20 solution: 1% v/v.
6. Glutaraldehyde solution: 25% v/v (*see* **Note 1**).
7. Cellulase solution: 3 mg/mL. Add 20 μL of commercially available stock cellulase solution and transfer to 980 μL buffer phosphate 0.1 M, pH 6.0 in an Eppendorf tube (*see* **Note 2**).
8. GOx solution: 3 mg/mL. Weight 3 mg enzyme solid powder and transfer to 1 mL buffer phosphate 0.1 M, pH 6.0 in an Eppendorf tube.
9. HRP solution: 9 mg/mL. Weight 9 mg enzyme solid powder and transfer to 1 mL buffer phosphate 0.1 M, pH 6.0 in an Eppendorf tube.
10. bgl solution: 0.3 mg/mL. Add 30 μL of commercially available stock β-glucosidase enzyme solution to 970 μL of 0.1 M buffer phosphate, pH 6.0.
11. Ultrasonic bath.
12. SpeedVac vacuum concentrator.

2.3 Determination of Enzyme Activity Recovery

1. Spectrophotometer Shimadzu, UV-1601, UV-visible.
2. 0.2 M solution of acetic acid (A): Dilute 11.55 mL concentrated acetic acid (17.4 M) in 1000 mL of distilled water in a glass beaker (*see* **Note 3**).
3. 0.2 M solution of sodium acetate (B): Weight 16.4 g of sodium acetate powder in 1000 mL of distilled water in a glass beaker.
4. 0.1 M acetate buffer, pH 5.0: Add 14.8 mL of solution A and 35.2 mL of solution B to a 100-mL graduated cylinder. Then, add 50 mL of ultra-pure water to reach a final volume of 100 mL. Mix and adjust pH either with HCl or NaOH, if necessary. Store at 4 °C.
5. 0.1 M solution of citric acid (A): Weigh 19.21 g of citric acid in 1000 mL ultra-pure water in a glass beaker.

6. 0.1 M solution of dibasic sodium phosphate (B): Weigh 53.65 g of $Na_2HPO_4 \cdot 7H_2O$ in 1000 mL ultra-pure water in a glass beaker.
7. 0.1 M citrate phosphate buffer, pH 6.5: Add 13.6 mL of solution A and 36.4 mL of solution B to a 100-mL graduated cylinder. Then, add 50 mL of ultra-pure water to reach a final volume of 100 mL. Mix and adjust pH with HCl. Store at 4 °C.
8. 0.2 M solution of monobasic sodium phosphate (A): Weigh 27.8 g of $NaH_2PO_4 \cdot H_2O$ in 1000 mL ultra-pure water in a glass beaker.
9. 0.2 M solution of dibasic sodium phosphate (B): Weigh 53.65 g of $Na_2HPO_4 \cdot 7H_2O$ in 1000 mL ultra-pure water in a glass beaker.
10. 0.05 M phosphate buffer, pH 7.0: Add 19.5 mL of solution A and 30.5 mL of solution B to a 100-mL graduated cylinder. Then, add 50 mL of ultra-pure water to reach a final volume of 100 mL. Mix and adjust pH either with HCl or NaOH. Store at 4 °C.
11. Carboxymethylcellulose (CMC) solution 1% w/v: Weigh 1 g of CMC and add 100 mL of 0.1 M acetate buffer, pH 5.0 in a glass beaker (*see* **Note 4**).
12. Dinitrosalicylic Acid (DNSA) reagent: Weigh 1.0 g of 3,5-Dinitrosalicylic acid and add 20 mL of 2 M sodium hydroxide (NaOH) solution in a glass beaker. To prepare the NaOH solution, weigh 1.6 g NaOH powder and add 20 mL of distilled water in a glass beaker (*see* **Note 5**). Then add slowly 30 g of sodium potassium tartrate to the above solution (*see* **Note 6**) and finally dissolve in a final volume of 100 mL using distilled water (*see* **Note** 7). Store in a dark bottle for further use at room temperature.
13. *p*-nitrophenyl-β-D-glucopyranoside (*p*-NPG) solution, 5 mM: Weigh 1.5 mg of *p*-NPG in an Eppendorf tube and add 1 mL of citrate phosphate buffer 0.1 M, pH 6.5.
14. Glucose solution 1 M: Weigh 0.9 g of glucose and add 5 mL of distilled water in a glass beaker (*see* **Note 8**).
15. 2,2′-Azino-bis (3-ethylbenzothiazoline-6-sulfonic acid) diammonium salt (ABTS) solution, 20 mM: Weigh 5.5 mg of ABTS in an Eppendorf tube and add 0.5 mL of 0.1 M phosphate buffer, pH 6.0 (*see* **Note 9**).
16. Aqueous solution of hydrogen peroxide (H_2O_2), 1 M: Add 112 μL of 8.82 M of H_2O_2 solution in 882 μL of high purity water (HPLC).
17. Cellulase solution: As described in Subheading 2.2 but add 0.1 M acetate buffer, pH 5.0, instead.

18. bgl solution: 0.3 mg/mL. As described in Subheading 2.2.
19. GOx solution: 0.5 mg/mL. Weigh 0.5 mg enzyme solid powder and transfer to 1 mL buffer phosphate 0.05 M, pH 7.0 in an Eppendorf tube.
20. HRP solution: 0.5 mg/mL. Weigh 0.5 mg enzyme solid powder and transfer to 1 mL buffer phosphate 0.05 M, pH 7.0 in an Eppendorf tube.
21. Nanobiocatalyst solution 1 mg/mL: Weigh 1 mg of nanobiocatalyst and add 1 mL of 0.1 M citrate phosphate buffer, pH 6.5.

2.4 Characterization of γ-Fe-Aptes-MNPs and γ-Fe-Aptes-MNPs-4enzymes - Fourier-transform Infrared Spectroscopy (FTIR)

1. FTIR-8400 infrared spectrometer (Shimadzu, Tokyo, Japan) equipped with a deuterated triglycine sulfate (DTGS) detector.
2. Potassium Bromide (KBr, spectroscopic grade) powder.
3. Agate mortar.
4. Pike Hand-Press.
5. Die Set.
6. Spatula.
7. Solid Sample Holder.
8. Nanobiocatalyst powder (approximately 0.5 mg).

2.5 Activity and Kinetic Studies of Free and Co-immobilized Enzymes

1. Spectrophotometer Shimadzu, UV-1601, UV-visible.
2. 0.1 M acetate buffer, pH 5.0: As described in Subheading 2.3.
3. 0.1 M citrate phosphate buffer, pH 6.5: As described in Subheading 2.3.
4. 0.05 M phosphate buffer, pH 7.0: As described in Subheading 2.3.
5. CMC solution 1% w/v: As described in Subheading 2.3.
6. DNSA reagent: As described in Subheading 2.3.
7. *p*-NPG solution, 5 mM: As described in Subheading 2.3.
8. Glucose solution 1 M: As described in Subheading 2.3.
9. ABTS solution: 20 mM. As described in Subheading 2.3.
10. H_2O_2: 1 M. As described in Subheading 2.3.
11. Cellulase, bgl, GOx and HRP solutions: As described in Subheading 2.3.
12. Nanobiocatalyst solution: As described in Subheading 2.3.

2.6 Monitoring the Reaction Progress of the 4-step Cascade Biotransformation of Cellulose

1. CMC solution 1% w/v: As described in Subheading 2.3.
2. ABTS solution: 20 mM. As described in Subheading 2.3 but in 0.1 M phosphate buffer, pH 6.0.

3. Nanobiocatalyst solution: 1 mg/mL. Weigh 1 mg of nanobiocatalyst in an Eppendorf tube and then add 1 mL buffer citrate phosphate 0.1 M, pH 6.0.

3 Methods

3.1 Synthesis of γ-Fe-Aptes-MNPs

A thermolysis method was used based on the thermolytic decomposition of an iron chloride precursor using oleylamine as a solvent and a capping agent for the preparation of the organophilic iron nanoparticles (Fig. 2). The protocol below describes in detail the synthesis of γ-Fe_2O_3 and their subsequent functionalization with aminosilane [15].

1. Disperse 2 g of Iron (II) chloride tetrahydrate in 160 mL oleylamine at 80 °C and reflux for 30 min till the mixture becomes a clear solution.
2. Then raise the temperature at 250 °C for 1 h.
3. Wash and centrifuge three times (1800 × *g*, 10 min); the first wash should be with 50 mL ethanol and the subsequent ones with an ethanol: chloroform solution (10:1 v/v).
4. Collect the precipitate and leave the iron oxide nanoparticles to dry.
5. Add 140 mg of freshly prepared γ-Fe_2O_3 nanoparticles in a solution of 0.8 g (3-aminopropyl) triethoxysilane and 2 mL of dry chloroform.
6. Leave in sonication (160 W) for 20 min in order to avoid aggregations.
7. Reflux the mixture for 72 h at 100 °C.
8. Follow with washings and centrifugations by dry chloroform (1800 × *g*, 10 min).
9. Finally dry in vacuum.

3.2 Co-immobilization of Cellulase, bgl, GOx, and HRP on γ-Fe-Aptes-MNPs

Cellulase, bgl, GOx, and HRP were covalently attached onto the surface of γ-Fe-Aptes-MNPs by employing glutaraldehyde as a cross-linking agent. The optimization of several parameters affecting the co-immobilization yield such as the cross-linker concentration, the mass ratio of γ-Fe-Aptes-MNPs to total enzyme, the mass ratio between the enzymes, and the incubation time, were examined and optimized in our previous work [15]. Based on the optimized preparation conditions (glutaraldehyde concentration, 10% v/v; a mass ratio of γ-Fe-Aptes-MNPs to enzyme content, 1: 1; a mass ratio of cellulase:bgl:GOx:HRP, 1:1:1:3 and incubation time, 2 h), we describe in detail the co-immobilization protocol followed for the preparation of the four-enzymatic magnetic nanobiocatalyst.

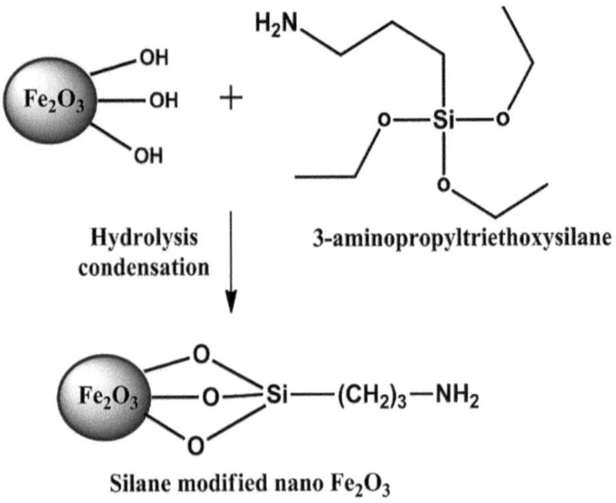

Fig. 2 A schematic illustration of the synthesis and functionalization of magnetic nanoparticles (γ-Fe-Aptes)

1. Weigh 15 mg of γ-Fe-Aptes-MNPs and disperse them in 20 mL phosphate buffer 0.1 M, pH 6.0 using an ultrasonic bath.
2. Then, add slowly 0.36 mL of Tween-20 (1% v/v) and leave in the ultrasonic bath for about 30 min.
3. Add glutaraldehyde solution (to a final concentration 10% v/v) and let for mix in an incubator at 30 °C for 30 min.
4. After incubation, centrifugate the solution at 1200 × *g* for 10 min thrice. After each centrifugation remove the supernatant and then add 5 mL of phosphate buffer 0.1 M, pH 6.0.
5. After the last centrifugation, re-disperse the nanomaterials in 10 mL buffer (the same as above) for 10 min.
6. Then, transfer the enzyme cocktail (total protein content of 15 mg) containing cellulase, bgl, GOx, and HRP (in a typical mass ratio of 1:1:1:3) to the nanomaterial solution.
7. Leave the final solution for incubation at 30 °C for 2 h.
8. Centrifugate the final mixture at 15,000 × *g* for 10 min and consequently remove the supernatant.
9. The final precipitate is left to air dry in a SpeedVac vacuum concentrator.

3.3 Determination of Enzymes Activity Recovery

1. CMC is used as a substrate to determine the enzymatic activity of cellulase before and after the co-immobilization procedure (*see* **Note 10**).
   (a) A reaction mixture consisting of 1% *w/v* CMC and 3 μg/mL of free enzyme (or 0.1 mg/mL of co-immobilized enzyme) is incubated at 50 °C for 15 min.

(b) Subsequently, an equal volume of DNSA is added to terminate the reaction and then the reaction mixture is boiled for 5 min.

(c) Then collect the samples, dilute them with distilled water and cool in an ice bath.

(d) The released reducing sugar (estimated as glucose) can be measured spectrophotometrically at 540 nm (Shimadzu, Tokyo, Japan).

(e) One unit of cellulase activity is typically defined as the amount of the enzyme required for releasing one μmole of reducing sugars (glucose) at optimum conditions.

2. *p*-NPG is used as a substrate to determine the enzymatic activity of bgl before and after the co-immobilization procedure (*see* **Note 11**).

   (a) Briefly, 0.12 μg/mL of free enzyme solution (or 0.1 mg/mL of co-immobilized nanobiocatalyst) are mixed with 2.0 mM *p*-NPG in 100 mM citrate phosphate buffer, pH 6.5 at 50 °C.

   (b) The reaction is followed by measuring the increase in absorbance of *p*-NP at 410 nm.

   (c) Absorption values are taken every 1 min for a total of 5 min and were converted to *p*-NP concentration using the molar extinction coefficient of *p*-NP at 410 nm.

   (d) Enzymatic activity is estimated as the amount (μmoles) of *p*-NP transformed per minute.

3. Glucose is used as a substrate to determine the enzymatic activity of GOx before and after the co-immobilization procedure (*see* **Note 12**).

   (a) Briefly, 0.5 μg/mL of free enzyme solution (or 0.1 mg/mL of co-immobilized GOx) are mixed with 250 mM glucose solution, 0.4 mM ABTS, and 0.3 μg/mL HRP. The enzymatic activity is determined as the amount (μmoles) of ABTS transformed per min.

   (b) The reaction is followed by measuring the increase in absorbance of $ABTS^+$ at 405 nm.

   (c) Absorption values are taken every 1 min for a total of 5 min.

4. Hydrogen peroxide is used as a substrate to determine the enzymatic activity of HRP before and after the co-immobilization procedure (*see* **Note 13**).

   (a) Briefly, 0.5 mM ABTS are mixed with 0.1 mM hydrogen peroxide and 0.4 μg/mL of free enzyme solution (or 0.1 mg/mL of co-immobilized HRP) to a final total volume of 1 mL supplemented with 100 mM sodium phosphate buffer, pH 6.0.

(b) Absorbance values are taken every 1 min for a total of 10 min and can be converted to ABTS concentration using the molar extinction coefficient of oxidized ABTS at 405 nm.

(c) Enzymatic activity can be determined as the amount (μmol) of ABTS transformed per minute.

3.4 Characterization of γ-Fe-Aptes-MNPs the Nanobiocatalyst-FTIR

Fourier-transform infrared spectroscopy (FTIR) is a commonly applied technique to confirm the successful enzyme immobilization on nanoparticles. The steps followed to attain FTIR spectra are analyzed below in detail. Figure 3 illustrates a typical FTIR spectrum of the γ-Fe-Aptes-MNPs and the nanobiocatalyst. The peak observed around 1640/cm is characteristic of the C=O stretching vibrations of amide groups and corresponds to Amide I band and confirms the successful immobilization of enzymes on Fe_3O_4 MNPs [14]. The peak observed at 1542/cm can be attributed to the C-N stretching and N-H bending vibrations corresponding to the Amide II band verifying the successful anchoring of the enzyme on the γ-Fe-Aptes-MNPs [16].

1. Prepare a KBr pellet by firstly grounding the powder sample, as well as KBr, in order to reduce the particle size to less than 5 mm in diameter.
2. Take a small amount of powder sample (enzyme or nanomaterial powder, about 0.1–2% of the KBr amount, or just enough to cover the tip of spatula) and then mix with the KBr powder.
3. Subsequently grind the mixture for 3–5 min.
4. Assemble the die set and then add the powder to the 7 mm collar. Put the die together with the powder into the Pike Hand-Press.
5. Press the powder for 2 min to form a pellet (*see* **Note 14**).
6. Prepare all samples in the form of KBr pellets containing ca. 2 wt % of the enzyme, the nanomaterials, or the nanobiocatalyst.
7. Disassemble the die set and take out the 7 mm collar. Put the collar together with the pellet onto the sample holder.
8. Open the Program "Spectra Manager" on the PC connected to the infrared spectrometer.
9. Record spectra in the range of 400 to 4000/cm and with an average of about 32 scans.

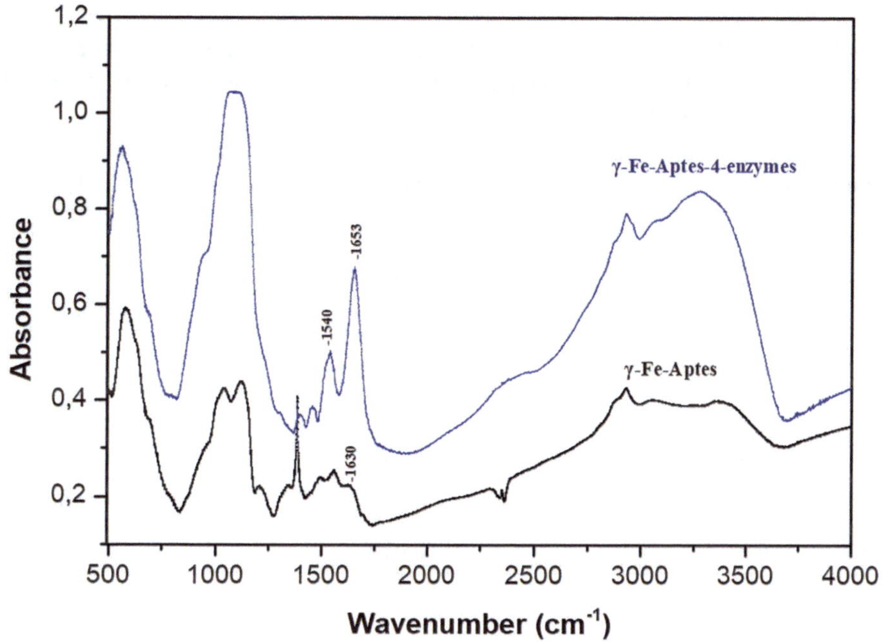

Fig. 3 FTIR spectra of the γ-Fe-Aptes and γ-Fe-Aptes-4enzymes

3.5 Activity and Kinetic Studies of Free and Co-immobilized Cellulase

1. For the measurement of the activity of cellulase (*see* **Note 11**), briefly, 3 μg/mL of native enzyme (or equivalent amount of co-immobilized enzyme) are added to a soluble cellulose solution (1% *w/v* in 100 mM sodium acetate buffer, pH 5.0) in a test tube and the mixture is incubated at 50 °C for 15 min.
2. The released reducing sugar can be determined according to the DNSA method, as previously described in Subheading 3.3.
3. The apparent kinetic parameters (K_m, V_{max}) of the free and co-immobilized form of cellulase were determined by measuring the initial reaction rates for each form by varying concentrations of CMC, following the assay described above.
4. The concentrations of CMC ranged from 0.1 to 3.5% w/v. The Michaelis–Menten constant (K_m) and maximum velocity (V_{max}) were calculated from non-linear regression fitting of the initial reaction rates corresponding to different substrate concentrations by the EnzFitter (Biosoft) software (Table 1).

3.6 Activity and Kinetic Studies of Free and Co-immobilized bgl

1. For the measurement of the activity of bgl (*see* **Note 12**), 0.12 μg/mL of free enzyme solution (or equivalent amount of co-immobilized nanobiocatalyst) is mixed with 2.0 mM *p*-NPG in 100 mM citrate phosphate buffer, pH 6.5 at 50 °C.
2. The reaction is followed according to the methodology described above in Subheading 3.3.

Table 1
Intrinsic kinetic constants of the free form of cellulase, bgl, GOx, and HRP and apparent kinetic constants of the co-immobilized form of cellulase, bgl, GOx, and HRP at 50 °C (the standard deviation was less than 5% in all cases)

Form of the enzyme	K_m	V_{max} (μmol/min)
Free cellulase	6.09	1.41
Co-immobilized cellulase	6.78	0.40
Free bgl	0.25	36.54
Co-immobilized bgl	0.36	9.46
Free GOx	1.72	12.07
Co-immobilized GOx	2.76	5.30
Free HRP	0.014	20.74
Co-immobilized HRP	0.017	5.46

K_m values for cellulase are expressed in terms of mass (mg/mL), while for bgl, GOx, and HRP, K_m is expressed in terms of mM

3. The kinetic parameters (K_m, V_{max}) of free and co-immobilized form of bgl were determined by measuring the initial reaction rates for each form by varying concentrations of *p*-NPG, following the assay described above.
4. The concentrations of *p*-NPG were ranged from 0.05 to 15 mM. The Michaelis–Menten constant (K_m) and maximum velocity (V_{max}) were calculated from non-linear regression fitting of the initial reaction rates corresponding to different substrate concentrations by the EnzFitter (Biosoft) software (Table 1).

3.7 Activity and Kinetic Studies of Free and Co-immobilized GOx

1. In order to measure the activity of GOx (*see* **Note 13**), 0.5 μg/mL of free enzyme solution (or equivalent amount of co-immobilized GOx) are mixed with 250 mM glucose solution, 0.4 mM ABTS and 0.3 μg/mL HRP.
2. The reaction is followed by measuring the increase in absorbance of $ABTS^+$ at 405 nm. Absorption values are taken every 1 min for a total of 5 min according to the methodology described above in Subheading 3.3.
3. The apparent kinetic parameters (K_m, V_{max}) of the free and co-immobilized form of GOx were determined by measuring the initial reaction rates for each form by varying concentrations of glucose, following the assay described above.
4. The concentrations of glucose ranged from 1 to 25 mM. The Michaelis–Menten constant (K_m) and maximum velocity (V_{max}) were calculated from non-linear regression fitting of

the initial reaction rates corresponding to different substrate concentrations by the EnzFitter (Biosoft) software (Table 1).

3.8 Activity and Kinetic Studies of Free and Co-immobilized HRP

1. For the determination of the enzymatic activity of HRP (*see* **Note 14**), 0.5 mM ABTS are mixed with 0.1 mM hydrogen peroxide and 0.4 μg/mL of free enzyme solution (or equivalent amount of co-immobilized HRP) to a final total volume of 1 mL supplemented with 100 mM sodium phosphate buffer, pH 6.0.
2. The reaction is followed by measuring the increase in absorbance of $ABTS^+$ at 405 nm, according to the methodology described above in Subheading 3.3.
3. The apparent kinetic parameters (K_m, V_{max}) of the free and co-immobilized form of HRP can be determined by measuring the initial reaction rates for each form by varying concentrations of H_2O_2, following the assay described above.
4. The concentrations of H_2O_2 ranged from 0.015 to 0.275 mM. The Michaelis–Menten constant (K_m) and maximum velocity (V_{max}) were calculated from non-linear regression fitting of the initial reaction rates corresponding to different substrate concentrations by the EnzFitter (Biosoft) software (Table 1).

3.9 Monitoring the Reaction Progress of the 4-step Cascade Biotransformation of Cellulose

The catalytic performance of this magnetic four-enzyme nanobiocatalyst can be estimated concerning a four-step cascade reaction involving the cellulase-mediated hydrolysis of soluble cellulose (CMC) to cellobiose that can be consecutively hydrolyzed by bgl to D-glucose. D-glucose can be successively oxidized by GOx to gluconic acid and H_2O_2, followed by reduction of H_2O_2 to H_2O by HRP in the presence of an electron donor (ABTS).

1. Briefly, add 0.3 mg/mL of the magnetic nanobiocatalyst into 1% w/v CMC solution containing 3 mM ABTS in 100 mM sodium phosphate buffer, pH 6.0 at 50 °C. The initially colorless substrate solution gradually turns green due to oxidation of ABTS and the formation of the corresponding radical cation. The change in the color of the reaction is shown at Fig. 4.
2. The reaction can be also followed by measuring the increase in absorbance at 730 nm ($ABTS^+$) at a UV-Vis photometer equipped with a Peltier temperature controller.

4 Notes

1. Work in a fume hood and wear a mask while working with the glutaraldehyde solution, as it is considered a hazardous chemical according to its safety data sheet.

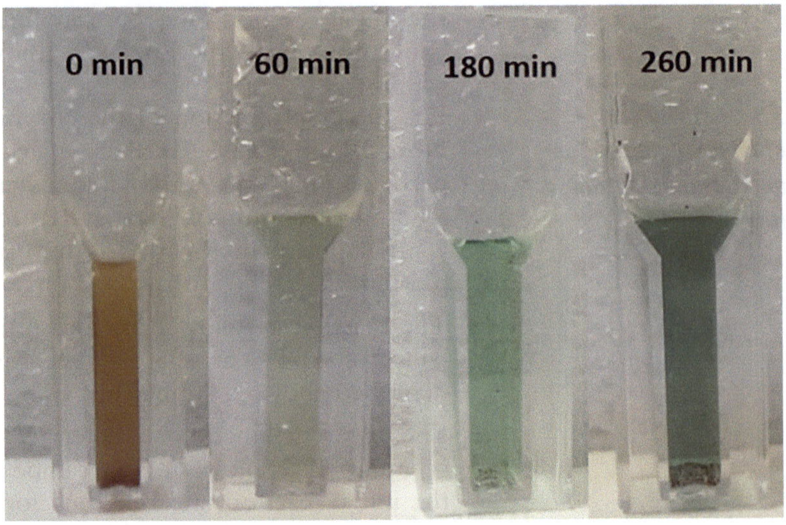

Fig. 4 Representation of the reaction progress at various time intervals ($t = 0$, 60, 180 and 260 min) at 50 °C. The aqueous mixture of CMC and ABTS turns progressively green as oxidized ABTS is released at the final stage of the cascade reaction

2. Keep all enzymes (cellulase, bgl, GOx, and HRP) on ice to ensure no loss in enzymatic activity.
3. Work in a fume hood and wear a mask while transferring the acetic acid, as it is considered a hazardous chemical according to its safety data sheet.
4. To accelerate the dissolution of CMC, place the glass beaker on a magnetic stirrer at low temperature and add slowly the CMC powder at small doses at a time (waiting for it to be dissolved before adding the next dose) as it can easily form aggregates that hinder its uniform solubility.
5. To accelerate the dissolution of sodium hydroxide, place the beaker on a magnetic stirrer at 50 °C for about 10 min.
6. Transfer sodium tartrate very slowly into the solution containing DNSA and NaOH. Keep in mind that adding small doses at a time and waiting for it to dissolve before adding the next dose can avoid the formation of aggregates. Also, place the beaker on a magnetic stirrer with a low temperature to accelerate dissolution and achieve uniform mixing.
7. Through the whole procedure, wrap the beaker with foil as DNSA is considered as a light-sensitive reagent.
8. To achieve effective dissolution of glucose, place the beaker on a magnetic stirrer at 50 °C for about 10 min.

9. ABTS is considered a light-sensitive reagent that can be easily oxidized so it is recommendable to wrap the Eppendorf tube with a foil to prevent oxidation.
10. Enzymatic activities of the free and co-immobilized cellulase are determined by the amount of reducing sugars released from soluble cellulose (CMC) by the 3,5-dinitrosalicylic acid (DNSA) method.
11. The activity of bgl is estimated by using *p*-NPG as a substrate. Under optimum conditions, β-glucosidase hydrolyzes *p*-NPG into *p*-NP (colored product) and glucose. The increase in absorbance of *p*-NP at 410 nm during the reaction can be monitored spectrophotometrically.
12. The study of GOx activity is based on its coupling reaction with horseradish peroxidase. Specifically, GOx utilizes D-glucose as a substrate, converting it to D-glucono-S-lactone and hydrogen peroxide. HRP then uses hydrogen peroxide to convert ABTS to $ABTS^{+}$. The reaction is followed by measuring the increase in absorbance of $ABTS^{+}$ at 405 nm.
13. HRP utilizes hydrogen peroxide to convert chromogenic substrates (e.g., ABTS). Hence, the study of HRP activity is based on the oxidation reaction of ABTS to $ABTS^{+}$. The increase in absorbance of $ABTS^{+}$ is measured at 405 nm.
14. KBr pellets should be thin and transparent. Opaque or blurred pellets give poor spectra because they prevent the infrared beam permeability. Also, the appearance of white spots on a pellet is an indicator that the powder has not been grounded well enough, or it is not dispersed properly in the pellets. In this case, remake the pellet.

Acknowledgements

This work was supported by the project "Synthetic Biology: from omics technologies to genomic engineering (OMIC-ENGINE)" (MIS 5002636), which is implemented under the Action "Reinforcement of the Research and Innovation Infrastructure," funded by the Operational Program "Competitiveness, Entrepreneurship and Innovation" (NSRF 2014–2020) and co-financed by Greece and the European Union (European Regional Development Fund).

References

1. An J, Li G, Zhang Y et al (2020) Recent advances in enzyme-nanostructure biocatalysts with enhanced activity. Catalysts 10:338
2. Ren S, Li C, Jiao X et al (2019) Recent progress in multienzymes co-immobilization and multienzyme system applications. Chem Eng J 373:1254–1278
3. Eshkenazi I, Maltz E, Zion B et al (2000) A three-cascaded-enzymes biosensor to

determine lactose concentration in raw milk. J Dairy Sci 83:1939–1945

4. Giannakopoulou A, Gkantzou E, Polydera A et al (2019) Multienzymatic nanoassemblies: recent progress and applications. Trends Biotechnol 38:202–216
5. Ji Q, Wang B, Tan J et al (2016) Immobilized multienzymatic systems for catalysis of cascade reactions. Process Biochem 51:1193–1203
6. Min K, Yoo YJ (2014) Recent progress in nanobiocatalysis for enzyme immobilization and its application. Biotechnol Bioprocess Eng 19:553–567
7. Yang H, Wei W, Liu S (2014) Monodispersed silica nanoparticles as carrier for co-immobilization of bi-enzyme and its application for glucose biosensing. Spectrochim Acta A Mol Biomol Spectrosc 125:183–188
8. Jia F, Narasimhan B, Mallapragada S (2013) Materials-based strategies for multi-enzyme immobilization and co-localization: a review. Biotechnol Bioeng 111:209–222
9. Schoffelen S, Van Hest JCM (2013) Chemical approaches for the construction of multi-enzyme reaction systems. Curr Opin Struct Biol 23:613–621
10. Liu W, Zhang S, Wang P (2009) Nanoparticle-supported multi-enzyme biocatalysis with in situ cofactor regeneration. J Biotechnol 139:102–107
11. Bilal M, Asgher M, Cheng H et al (2019) Multi-point enzyme immobilization, surface chemistry, and novel platforms: a paradigm shift in biocatalyst design. Crit Rev Biotechnol 39:202–219
12. Hoang DQ, Tran TV, Tran NQ et al (2016) Functionalization of Fe_3O_4 nanoparticles with biodegradable chitosan-grafted-mPEG for paclitaxel delivery. Green Process Synth 5:459–466
13. Akın D, Yakar A, Gündüz U (2015) Synthesis of magnetic Fe_3O_4-chitosan nanoparticles by ionic gelation and their dye removal ability. Water Environ Res 87:425–436
14. Majoul N, Aouida S, Bessais B (2015) Progress of porous silicon APTES-functionalization by FTIR investigations. Appl Surf Sci 331:388–391
15. Giannakopoulou A, Patila M, Spyrou K et al (2019) Development of a four-enzyme magnetic nanobiocatalyst for multi-step cascade reactions. Catalysts 9:995
16. Díaz-Hernández A, Gracida J, García-Almendárez BE et al (2018) Characterization of magnetic nanoparticles coated with chitosan: a potential approach for enzyme immobilization. J Nanomater 2018:9468574

Chapter 17

A Bi-enzymatic Immobilized Nanobiocatalyst for the Biotransformation of Oleuropein to Hydroxytyrosol

Archontoula Giannakopoulou, Alexandra V. Chatzikonstantinou, Aliki Tsakni, Christos L. Chochos, Dimitra Houhoula, and Haralambos Stamatis

Abstract

Multi-enzymatic assemblies offer the opportunity of bringing in proximity several enzymes that are enabled to work together for the catalysis of multi-step reactions. Especially, the development of robust nanobiocatalytic systems comprising of several enzymes has gained considerable attention over the last few years for the catalysis of complex reactions and the production of high added-value products. In the present chapter, we describe the methodology for the development of a bi-enzymatic nanobiocatalyst consisting of the enzymes β-glucosidase from *Thermotoga maritima* and lipase A from *Candida antarctica* (CalA) co-immobilized on chitosan-coated magnetic nanoparticles. This nanobiocatalyst can be efficiently applied for the biotransformation of oleuropein to hydroxytyrosol, a reaction of increased biotechnological interest. Several techniques, as well as methodologies that are required for the characterization of the structure and the activity of such systems are also comprehensively described.

Key words Multi-enzymatic assemblies, Nanobiocatalyst, Chitosan-coated, Magnetic nanoparticles, β-glucosidase, Lipase, Biotransformation, Oleuropein, Hydroxytyrosol

1 Introduction

Enzymes are protein biomolecules that offer numerous advantages over conventional chemical catalysts, such as high stereo- and regioselectivity, high turnover rates under mild reaction conditions as well as biodegradability and non-toxicity. Enzymes, therefore, can be incorporated in many biotransformations in an expanded range of fields, such as bioremediation processes, biofuel production, and pharmaceutical and food industry [1]. However, under extreme temperature or pH conditions that are usually essential for many of those applications, enzymes are unstable and can easily undergo denaturation. For this reason, enzyme immobilization on specific solid supports can provide an efficient solution for the

Haralambos Stamatis (ed.), *Multienzymatic Assemblies: Methods and Protocols*,
Methods in Molecular Biology, vol. 2487, https://doi.org/10.1007/978-1-0716-2269-8_17,
© The Author(s), under exclusive license to Springer Science+Business Media, LLC, part of Springer Nature 2022

development of robust biocatalysts that present enhanced stability over a wide range of operating conditions while ensuring their easy recovery and efficient re-use [2].

Enzyme immobilization, namely bringing an enzyme in close proximity with a suitable support, can be carried out via either physical or chemical methods. Chemical methods are characterized by the formation of stable chemical bonds between the enzyme and the support material [3]. The most widely applied techniques for enzyme immobilization are adsorption, covalent binding, entrapment, encapsulation, and cross-linking [4]. Among them, covalent immobilization has prevailed over the last years in the immobilization field because one of its main benefits, in the majority of the cases studied, is the prolonged biocatalyst's stability compared with the free form of the enzyme [5, 6].

Until now, individually immobilized enzymes have been widely exploited in various biotransformations of industrial importance. However, the synthesis of more complicated bio-products or the catalysis of complex biochemical biotransformations requires the co-localization of multiple enzymes. So, multi-enzyme immobilization is generally applied when several free enzymes need to be brought in proximity to catalyze a multi-step or a cascade reaction. This type of immobilization enables the substrate channeling effect, namely the direct transfer of the product of an enzyme to the active center of the following enzyme offering a plethora of advantages [7]. For example, enhanced reaction rates, shorter cycle times, elimination of reaction by-products and intermediates, or reduction of diffusion losses have been reported to arise from the co-localization of different enzymes on a single matrix [8, 9]. The attachment techniques, as well as the supports used in single enzyme immobilization can be also suitable for multi-enzyme immobilization. However, the properties of both the enzymes and the carrier require careful selection to preserve the overall enzymatic activity of the whole biocatalytic system [10].

The rapid growth of nanotechnology over the last years has led to the development of innovative nanobiocatalysts, defined as nanomaterial-enzyme conjugates, which have been utilized in an expanded range of biotransformations [11, 12]. A special family of nanomaterials consists of the magnetic nanoparticles (MNPs) which are characterized by small size, large surface area, high biocompatibility, and low toxicity, establishing them as suitable nanocarriers both in single and multi-enzyme immobilization. Exceptionally, their magnetic core, providing them with superparamagnetism discriminates them from other nanomaterials due to the advantage of the easy separation from the reaction mixture and subsequently their efficient reusability that lowers the cost of biotransformation [13]. Special treatments, such as their coating with natural biopolymers, e.g., chitosan have been widely applied to reduce their toxicity and provide them with functional chemical

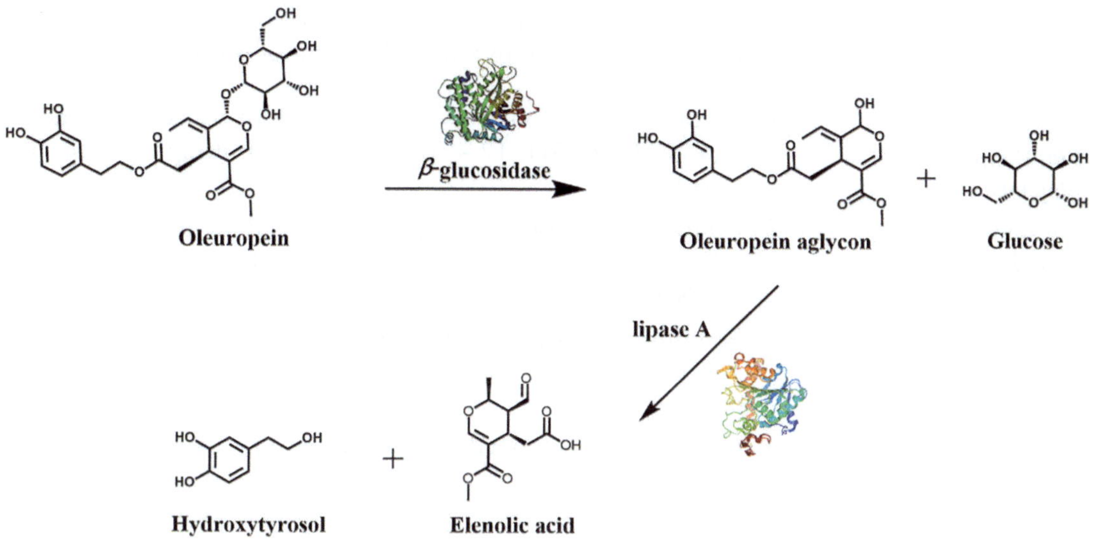

Fig. 1 Schematic representation of the hydrolysis of oleuropein by the synergistic action of β-glucosidase and lipase. Reproduced with permission from [Catalysts], published by [MDPI], [2021] [20]

groups (e.g., amine) to increase their biocompatibility and facilitate the overall immobilization process [14–16].

Nanobiocatalysts have been widely applied for the catalysis of various biotransformations presenting biotechnological interest. A representative example consists of the degradation of oleuropein for the production of hydroxytyrosol, a naturally derived antioxidant [17, 18]. This reaction can be efficiently catalyzed by the synergistic action of two enzymes, namely β-glucosidase and lipase, as shown in Fig. 1. In a first step, glucosidase can cleave the glycosidic bond of oleuropein, producing oleuropein aglycone and glucose. Subsequently, a lipase can catalyze the hydrolysis of oleuropein aglycone to elenolic acid and hydroxytyrosol.

In the present chapter, we present in detail the required steps for the covalent co-immobilization of β-glucosidase from *Thermotoga maritima* and lipase A from *Candida antarctica* onto chitosan-coated magnetic nanoparticles for the development of a bi-enzymatic nanobiocatalyst, able to catalyze the biotransformation of oleuropein to hydroxytyrosol. Several spectroscopic techniques, such as Scanning Electron Microscopy (SEM) and Circular Dichroism (CD) that are used for the structural characterization of the nanobiocatalyst are also being analyzed.

2 Materials

Prepare all solutions using ultrapure water (prepared by purifying deionized water, to attain a sensitivity of 18 MΩ-cm at 25 °C) and analytical grade reagents. Prepare and store all reagents at room temperature (unless indicated otherwise). Diligently follow all waste disposal regulations when disposing of waste materials.

2.1 Synthesis of Chitosan-Coated Magnetic Nanoparticles (CS-MNPs)

1. Acetic acid solution: 1% v/v. Add 1 mL of concentrated acetic acid (CH_3COOH, 17.4 M) to 99 mL of ultrapure water in a glass beaker.
2. Chitosan (CS) solution: 0.2% w/v. Weigh 0.2 g chitosan flakes from crab shells (low viscosity) and transfer to the glass beaker (*see* **Note 1**). Mix at 70 °C for about 60 min.
3. Sodium Tripolyphosphate (TPP) solution: 4 mM. Weigh 1.5 mg TPP in an Eppendorf tube (*see* **Note 2**) and add 1 mL of ultrapure water.
4. Commercially supplied iron oxide (Fe_3O_4) nanoparticles.
5. Magnet.
6. Ethanol.
7. High purity water.

2.2 Co-immobilization of bgl and CalA on Chitosan-Coated Magnetic Nanoparticles

1. 0.1 M solution of citric acid (A): Weigh 19.21 g of citric acid and transfer it to 1000 mL ultrapure water in a glass beaker. Mix at 25 °C for about 10 min.
2. 0.1 M solution of dibasic sodium phosphate (B): Weigh 53.65 g of $Na_2HPO_4 \cdot 7H_2O$ and transfer to 1000 mL ultrapure water in a glass beaker. Mix at 25 °C for about 10 min.
3. 0.1 M citrate phosphate buffer, pH 6.5: Add 13.6 mL of solution A to a 100 mL graduated cylinder and 36.4 mL of solution B. Then, add 50 mL of ultrapure water to reach a final volume of 100 mL. Mix and adjust pH with HCL. Store at 4 °C.
4. Freshly prepared chitosan-coated magnetic nanoparticles.
5. Tween-20 solution: 1% v/v.
6. β-glucosidase (bgl) solution: 0.3 mg/mL. Add 30 μL of commercially supplied β-glucosidase enzyme solution to 970 μL of 0.1 M buffer citrate phosphate, pH 6.5 (*see* **Note 3**).
7. lipase A (CalA) solution: 0.8 mg/mL. Add 50 μL of commercially supplied lipase enzyme solution to 950 μL of 0.1 M buffer citrate phosphate, pH 6.5.
8. Glutaraldehyde solution: 25% v/v (*see* **Note 4**).
9. SpeedVac vacuum concentrator.

2.3 Determination of Enzymes Activity Recovery

1. Spectrophotometer Shimadzu, UV-1601, UV-visible.
2. Dimethyl sulfoxide anhydrous (DMSO) solution, ≥99.9%.
3. Citrate phosphate buffer 0.1 M, pH 6.5. As described in Subheading 2.2.

4. *p*-nitrophenyl-β-D-glucopyranoside (*p*NPG) solution, 5 mM: Weigh 1.5 mg of *p*NPG in an Eppendorf tube and add 1 mL of citrate phosphate buffer 0.1 M, pH 6.5.
5. *p*-nitrophenyl-butyrate (*p*NPB) solution, 50 mM: Add 4.4 μL of concentrated *p*NPB solution into 495.6 mL of dimethyl sulfoxide (DMSO).
6. 0.2 M solution of monobasic sodium phosphate (A): Weigh 27.8 g and dissolve in 1000 mL ultrapure water in a glass beaker.
7. 0.2 M solution of dibasic sodium phosphate (Na_2HPO4) (B): Weigh 53.65 g of Na_2HPO_4 and dissolve in 1000 mL ultrapure water in a glass beaker.
8. 0.05 M phosphate buffer, pH 7.5: Add 8 mL of solution A to a 100 mL graduated cylinder and 42 mL of solution B. Then, add 50 mL of ultrapure water to reach a final volume of 100 mL. Mix and adjust pH with HCL. Store at 4 °C.
9. bgl solution: 0.3 mg/mL. As described in Subheading 2.2.
10. CalA solution: 0.8 mg/mL. As described in Subheading 2.2.
11. Nanobiocatalyst solution: 1 mg/mL. Weigh 1 mg of nanobiocatalyst (co-immobilized bgl and CalA) in an Eppendorf tube and then add 1 mL buffer citrate phosphate 0.1 M, pH 6.5. Disperse the nanobiocatalyst with the help of an ultrasonic bath.

2.4 Characterization of Nanobiocatalysts-Field Emission Scanning Electron Microscopy (FE-SEM)

1. FE-SEM measurements can be performed on the JEOL JSM-7610FPlus that integrates a full set of detectors for secondary electrons, backscattered electrons, energy dispersive X-ray spectroscopy (EDS).
2. Nanobiocatalyst powder (approximately 0.5 mg).

2.5 Characterization of Nanobiocatalysts: Circular Dichroism (CD)

1. Jasco J-1500 spectropolarimeter (Tokyo, Japan), equipped with a Peltier system for temperature control.
2. A quartz cell of 10 mm path length.
3. 0.005 M citrate phosphate buffer, pH 6.5. Prepare the 0.1 M citrate phosphate buffer, pH 6.5, according to Subheading 2.2, and dilute with ultrapure water to reach a final concentration of 0.005 M (*see* **Note 5**).
4. 0.005 M phosphate buffer, pH 7.5. Prepare the 0.05 M phosphate buffer, pH 7.5 according to Subheading 2.3, and dilute with ultrapure water to reach a final concentration of 0.005 M.
5. bgl solution: 6 μg/mL. Prepare the bgl solution (0.3 mg/mL) as described in Subheading 2.2 but add 970 μL of 0.005 M buffer citrate phosphate, pH 6.5. Then, dilute the prepared

solution with 0.005 M buffer citrate phosphate, pH 6.5, to reach a final enzyme concentration of 6 μg/mL (*see* **Note 6**).

6. CalA solution: 5 μg/mL. Prepare the CalA solution (0.8 mg/mL) as described in Subheading 2.2 but add 970 μL of 0.005 M buffer phosphate, pH 7.5. Then, dilute the prepared solution with 0.005 M buffer citrate phosphate, pH 7.5, to reach a final enzyme concentration of 5 μg/mL.
7. CS-MNPs: 0.5 mg/mL. In an Eppendorf tube, weigh 0.5 mg of freshly prepared CS-MNPs as described in Subheading 2.1. Then add 1 mL of ultrapure water and disperse the nanoparticles to form a solution with the help of an ultrasonic bath. Finally, dilute this solution with ultrapure water to attain a final concentration of 1.5 μg/mL of CS-MNPs (*see* **Note 7**).

2.6 Kinetic Studies of Free and Co-immobilized Enzymes

1. Spectrophotometer Shimadzu, UV-1601, UV-visible.
2. 0.05 M phosphate buffer, pH 7.5: As described in Subheading 2.3.
3. 0.1 M citrate phosphate buffer, pH 6.5: As described in Subheading 2.2.
4. *p*NPG solution, 5 mM: As described in Subheading 2.3.
5. *p*NPB solution, 50 mM: As described in Subheading 2.3
6. bgl solution: 0.3 mg/mL. As described in Subheading 2.2.
7. CalA solution: 0.8 mg/mL. As described in Subheading 2.2, but in 0.05 M phosphate buffer, pH 7.5.
8. Nanobiocatalyst solution: 1 mg/mL. As described in Subheading 2.3.

2.7 Reusability Studies of Bi-enzymatic Nanobiocatalyst

1. Oleuropein solution: 3.7 mM. Weigh 2 mg oleuropein and transfer to 1 mL buffer citrate phosphate 0.1 M, pH 6.5 in an Eppendorf tube.
2. Nanobiocatalyst powder: Weigh 1 mg of the nanobiocatalyst powder in an Eppendorf tube.
3. Reaction solution. Transfer the nanobiocatalyst powder to the oleuropein solution and disperse the nanobiocatalyst into the solution using an ultrasonic bath.

2.8 Monitoring the Reaction Progress: High-Pressure Liquid Chromatography (HPLC)

1. The quantitative and qualitative analysis of both oleuropein and hydroxytyrosol can be performed by HPLC (Shimadzu, Tokyo, Japan) equipped with a μBondapack C18 column, particle size 10 μm, length 300 mm, diameter 3.9 mm, and a diode array UV detector.
2. Acetonitrile, HPLC Grade.
3. Methanol, HPLC Grade.
4. Water HPLC grade.

5. Whatman grade filters, 25 mm.
6. Acetic acid solution: 0.1% v/v. Add 0.1 mL of concentrated acetic acid (CH_3COOH, 17.4 M) to 99.9 mL of HPLC grade water.
7. Oleuropein solution: As described in Subheading 2.7.
8. Hydroxytyrosol solution: 1 mg/mL. Weigh 1 mg of hydroxytyrosol in an Eppendorf tube and add 1 mL of 0.1 M citrate phosphate buffer, pH 6.5 (prepared as described in Subheading 2.2).
9. Reaction solution: As described in Subheading 2.7.

2.9 Liquid Chromatography-Mass Spectrometry Analysis (LC-MS)

1. Analyses can be performed on a Mass Spectrometer Advion equipment equipped with an atmospheric pressure chemical ionization (APCI) interface. The mass analyzer is a single quadrupole. The quadrupole is consisted of four parallel rods that use electric fields to filter ions based on the *m/z* ratio. The type of detector which is used is an electron amplifier. Separations are performed with a μBondapack C18 column, particle size 10 μm, length 300 mm, diameter 3.9 mm, and a diode array UV detector.
2. Acetonitrile, HPLC grade.
3. Acetic acid solution: 0.1% v/v. As described in Subheading 2.8.
4. Methanol, HPLC grade.
5. Reaction solution: As described in Subheading 2.7.
6. Oleuropein solution: As described in Subheading 2.7.
7. Hydroxytyrosol solution: As described in Subheading 2.8.
8. Lyophilizer.

3 Methods

3.1 Synthesis of Chitosan-Coated Magnetic Nanoparticles

Iron magnetic nanoparticles (Fe_3O_4) functionalized with chitosan can be synthesized according to the ionic gelation method with sodium tripolyphosphate (TPP) as the cross-linking agent [15, 19]. The protocol below describes in detail the steps followed for the preparation of the chitosan-coated magnetic nanoparticles.

1. Weigh 8 mg of iron oxide magnetic nanoparticles (Fe_3O_4) and disperse them in 0.96 mL chitosan solution (0.2% w/v).
2. Add dropwise 0.32 mL of TPP solution into the dispersion. Mix constantly under magnet stirring (100 × *g*) for 30 min.
3. Collect the magnetic nanoparticles with the use of an external magnet and remove the supernatant.

4. Wash thrice, firstly with water and then with ethanol. After each wash, collect the nanoparticles with a magnet, remove either the water or the ethanol and repeat.
5. Let the chitosan magnetic nanoparticles at 45 °C overnight for drying.

3.2 Co-immobilization of bgl and CalA on Chitosan-Coated Magnetic Nanoparticles

CalA and bgl can be covalently attached onto the surface of chitosan-functionalized magnetic nanoparticles by employing glutaraldehyde. The optimization of several parameters affecting the co-immobilization yield such as the cross-linker concentration, the ratio of CS-MNPs to total enzyme mass, the ratio between bgl and CalA, and the incubation time were examined and optimized in our previous work [20]. Here, based on the optimized preparation conditions (glutaraldehyde concentration, 10% v/v; mass ratio of CS-MNPs to enzyme content, 1:8; mass ratio of bgl to CalA, 1:3 and incubation time, 5 h), we describe in detail the co-immobilization protocol followed for the preparation of the bi-enzymatic magnetic nanobiocatalyst.

1. Weigh 6 mg of CS-MNPs and disperse them in 15 mL citrate phosphate buffer 0.1 M, pH 6.5 using ultrasonic bath.
2. Then, add slowly 200 μL of Tween-20 and leave in ultrasonic bath for about 30 min.
3. Add 5 mL of a stock glutaraldehyde solution (25% v/v) and incubate under constant stirring (100 × *g*) at 30 °C for 30 min.
4. After incubation, centrifugate the solution at 1200 × *g* for 10 min, thrice. After each centrifugation remove the supernatant and then add 4 mL of citrate phosphate buffer.
5. After the last centrifugation, redisperse the nanomaterials in 10 mL buffer for 10 min.
6. Then, transfer the enzymatic solutions of bgl and CalA to the nanomaterial solution.
7. Leave the final solution for incubation at 30 °C for 1 h, under constant stirring (100 × *g*).
8. Centrifugate the final mixture at 15,000 × *g* for 10 min and consequently remove the supernatant.
9. The final precipitate is left to air dry in the SpeedVac vacuum concentrator.

3.3 Determination of Enzymes Activity Recovery

1. *p*NPG is used as a substrate to determine the enzymatic activity of bgl before and after the co-immobilization procedure. A reaction mixture consisting of 2 mM *p*NPG in aqueous citrate phosphate solution 0.1 M, pH 6.5 and bgl either free (with a final enzyme concentration of 0.18×10^{-3} mg/mL) or co-immobilized (with a final biocatalyst concentration of 0.01 mg/mL) is incubated at 50 °C for 10 min and the increase

in absorbance due to the *p*NP released from the reaction is measured at 410 nm. Absorption measurements are performed in triplicate and the mean is subtracted.

2. *p*NPB is used as a substrate for the determination of the enzymatic activity of CalA before and after the co-immobilization procedure. A reaction mixture consisting of 0.5 mM *p*NPB in aqueous phosphate solution 0.05 M, pH 7.5 and CalA either free (with a final enzyme concentration of 0.08×10^{-3} mg/mL) or co-immobilized (with a final biocatalyst concentration of 0.1 mg/mL) is incubated at 50 °C for 10 min and the increase in absorbance due to the *p*NP released from the reaction is measured at 348 nm. Absorption measurements are performed in triplicate and the mean is subtracted (*See* **Note 8**).

3.4 Characterization of CS-MNPs and CS-MNPs-2 Enzymes: Scanning Electron Microscopy (SEM)

Scanning Electron Microscopy (SEM) was applied for the characterization of the chitosan-coated magnetic nanoparticles, regarding their size, shape, and distribution. The main methodology followed for the characterization is described above in detail.

1. Before analysis, the samples are deposited on a precleaned gold (Au) cylindrical surface as powders without any sputtering process. Then, the samples are inserted in a vacuum chamber with 9.6×10^{-5} bar pressure and visualized in SEM and energy dispersive X-ray spectroscopy (EDS) modes. (*see* Figs. 2 and 3 below).
2. From Figs. 1a and 2a, it is shown that the immobilization of the enzyme to the Fe_3O_4-chitosan nanoparticles does not affect the structure of the nanoparticles.
3. From Figs. 1b and 2b, it is shown that the Fe_3O_4-chitosan nanoparticles loaded with enzyme present larger sizes (~80 nm) than the unloaded ones.
4. From the EDS elemental maps of Figs. 1d and 2d it is evident that the Fe_3O_4 nanoparticles are deposited on the surface of the chitosan.
5. The additional two new peaks of sodium (Na) and phosphorus (P) present at the EDS spectrum of Fig. 2e can be attributed to the presence of the enzyme.

3.5 Circular Dichroism Studies

1. Turn on N_2 gas.
2. Turn on the cooling water bath.
3. Wait about 15 min to completely purge the system with N_2 gas.
4. Turn on the power switch.
5. Start the "Spectra Manager" software.
6. Set the sample temperature.

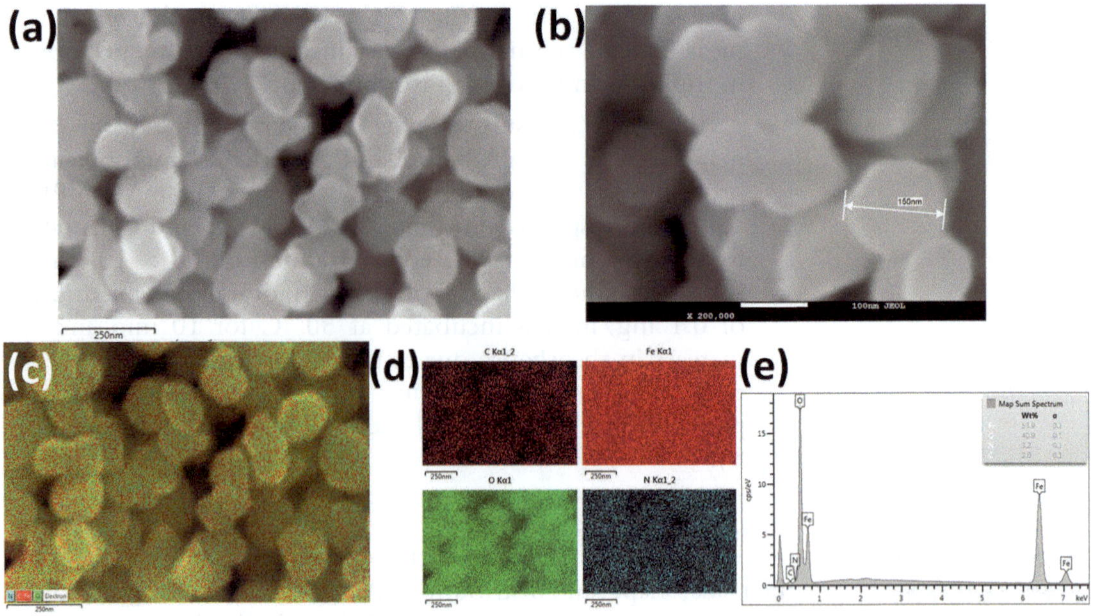

Fig. 2 Fe_3O_4-Chitosan nanoparticles: (**a**) EDS electronic image, (**b**) FE-SEM image, (**c**) EDS elemental map, (**d**) EDS elemental maps of C (red), Fe (red), O (green), N (blue) are shown for each class, and (**e**) EDS spectrum

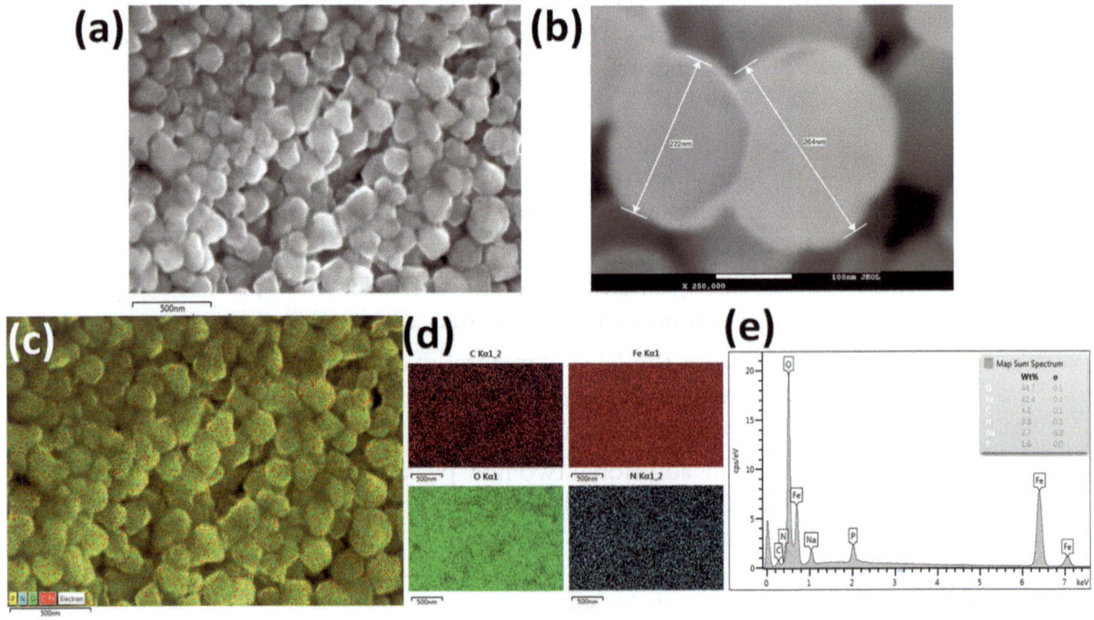

Fig. 3 Fe_3O_4-Chitosan nanoparticles loaded with enzyme: (**a**) EDS electronic image, (**b**) FE-SEM image, (**c**) EDS elemental map, (**d**) EDS elemental maps of C (red), Fe (red), O (green), N (blue) are shown for each class, and (**e**) EDS spectrum

7. Set the parameters. Record all spectra at 25 °C, with a 2 nm bandwidth of the incident light and a scan speed of 50 nm/min (*see* **Note 9**).
8. Record and subtract baseline from the protein spectrum for every sample scanned.
9. Collect and average at least two scans.

3.6 Kinetic Studies of Free and Immobilized Biocatalysts

To study the apparent kinetic constants of the enzymes β-glucosidase and lipase, both free and co-immobilized, one should determine the initial reaction rates for different concentrations (of the corresponding substrate of each enzyme), as calculated from the slope of the linear portion of the absorption increase curve as a function of time for each reaction. The estimation of the apparent constants K_m and V_{max} is calculated by the EnzFitter (Biosoft) software.

1. *p*NPG is used as a substrate to determine the apparent kinetic constants for β-glucosidase. The reaction is carried out at 50 °C in aqueous citrate phosphate solution 0.1 M, pH 6.5. The increase in absorbance due to the *p*NP released from the hydrolysis of *p*NPG is measured at 410 nm with a total reaction time of 10 min. The final concentration of the enzyme in the reaction is 0.18×10^{-3} mg/mL. The range of final concentrations of *p*NPG is from 0.05 to 40 mM. Absorption measurements are performed in triplicate and the mean is subtracted.
2. To study the kinetics of immobilized β-glucosidase, follow the same procedure as described in **step 2**. However, here the final concentration of the immobilized enzyme is 0.01 mg/mL.
3. *p*NPB is used as a substrate for the determination of apparent kinetics constants for lipase. The reaction is carried out at 50 °C in 0.05 mM aqueous phosphate solution pH 7.5. The increase in absorbance due to the *p*NP released from the hydrolysis of *p*NPB is measured at 348 nm with a total reaction time of 5 min. The final concentration of the enzyme in the reaction is 84×10^{-6} mg/mL. The range of final *p*NPB concentrations is from 0.025 to 0.75 mM. Absorption measurements are performed in triplicate and the mean is subtracted.
4. The same procedure as described in **step 4** is used to study the apparent kinetics of immobilized lipase. However, here the final concentration of the immobilized enzyme is 0.1 mg/mL (Table 1).

3.7 Re-use Studies of the Bi-enzymatic Nanobiocatalyst

The immobilization of enzymes on magnetic nanoparticles provides the possibility of isolating and reusing the nanobiocatalyst [21]. Therefore, the reusability (Fig. 4) of the nanobiocatalyst was studied, according to the steps described above.

Table 1
Intrinsic kinetic constants of the free form of bgl and CalA and apparent kinetic constants of the co-immobilized form of bgl and CalA at 60 °C (the standard deviation was less than 5% in all cases)

Biocatalyst	K_m (mM)	V_{max} (μmol/min)
Free β-glucosidase	0.72	9.27
Co-immobilized β-glucosidase	0.85	5.60
Free CalA	0.07	0.07
Co-immobilized CalA	0.17	0.04

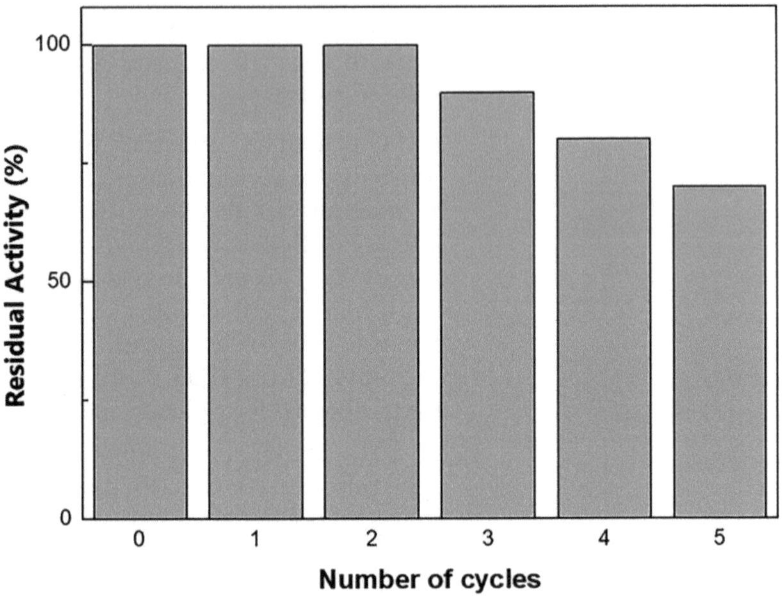

Fig. 4 Re-use of the bi-enzymatic magnetic nanobiocatalyst in the hydrolysis of oleuropein for 5 successive cycles (the standard deviation was less than 5% in all cases)

1. Weigh 2 mg of oleuropein and add 1 mL of 0.1 M buffer citrate phosphate, pH 6.5.
2. Weigh 1 mg of the nanobiocatalyst and add to the oleuropein solution. Disperse using ultrasonic bath.
3. Incubate the mixture with constant stirring at 120 × *g*, at 60 °C for 24 h.
4. After 24 h, isolate the biocatalyst by external magnetic field application. Wash with water in order to remove the oleuropein that might have not reacted and resuspend the nanobiocatalyst in a fresh substrate solution to initiate a new reaction cycle.

5. After each cycle, dilute the sample at 1:1 with methanol, filter, and analyze by HPLC. Figure 4 shows the residual activity of the nanobiocatalyst (%) after five successive reaction cycles.

3.8 Monitoring the Reaction Progress: HPLC

1. Prepare the reaction mixture as described in Subheading 2.8 and incubate under constant stirring at 120 × *g*, at 60 °C.
2. At standard time intervals (0, 1, 3, 6, 8, and 24 h), take a sample of 150 μL from the reaction mixture and dilute it with the addition of 150 μL of methanol.
3. Then, filter the sample and inject 20 μL of the sample in the load position of the HPLC injector.
4. The HPLC analysis conditions for the modification of oleuropein as well as the analysis elution program for the detection of oleuropein and hydroxytyrosol are as follows: the mobile phase consists of acetonitrile (A) and 0.1% acetic acid in water (B). The elution conditions applied for solvent B are as follows: 0–30 min 80–50%, 30–35 min 50%, and 35–40 min 80%. The elution is performed at 27 °C with a flow rate of 1 mL/min, and the samples are detected at 280 nm. The quantification and characterization of oleuropein and hydroxytyrosol are based on standard compounds and calibration curves under the same conditions.
5. Analyze your samples by HPLC. The chromatographs of oleuropein and hydroxytyrosol obtained by HPLC at 280 nm are presented in Fig. 5. The peak of oleuropein is observed at an elution time of 20.19 min, while hydroxytyrosol is characterized by an elution time of 7.45 min.
6. Finally, the percentage of oleuropein enzymatic degradation at standard time intervals is depicted on Fig. 6.

3.9 Liquid Chromatography-Mass Spectrometry Analysis (LC-MS)

1. Lyophilize all samples (oleuropein, hydroxytyrosol, and reaction solutions) and dissolve them in 300 μL methanol HPLC grade.
2. Set the mobile phases (*see* **Note 10**).
3. Open Advion software and wait 30 min until the instrument is stabilized.
4. Use the Atmospheric Pressure Chemical Ionization source with positive polarity.
5. Tune and calibrate the instrument with the following parameters: nebulizing gas (N_2) flow rate 1.5 L/min, Capillary Voltage 180 V, Source Voltage Offset 25 V, Source Voltage Span 20 V, Source Gas Temperature 350 °C, Capillary Temperature 200 °C and APCI Corona Discharge 5 μA.
6. Select full scan MS analysis. Spectra are recorded in the positive mode in the range of *m/z* 50–700.

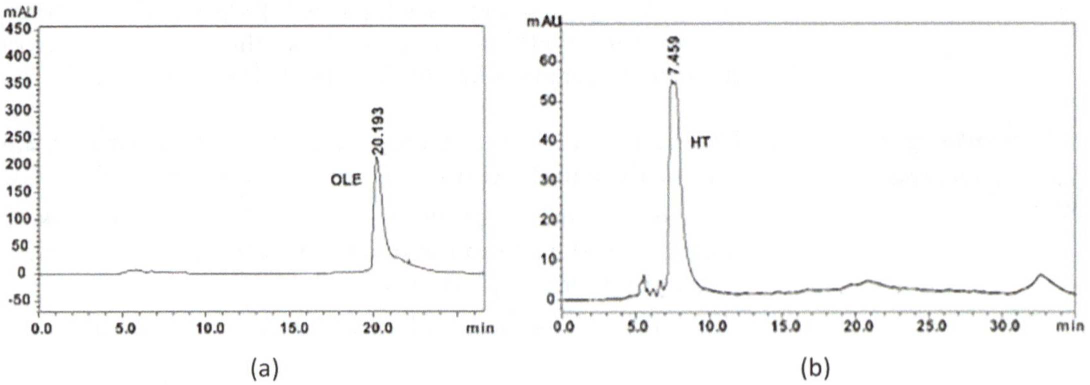

Fig. 5 Oleuropein (OLE) (**a**) and hydroxytyrosol (HT) (**b**) chromatographs at 280 nm. The peak of oleuropein is observed at an elution time of 20.19 min, while hydroxytyrosol is characterized by an elution time of 7.45 min

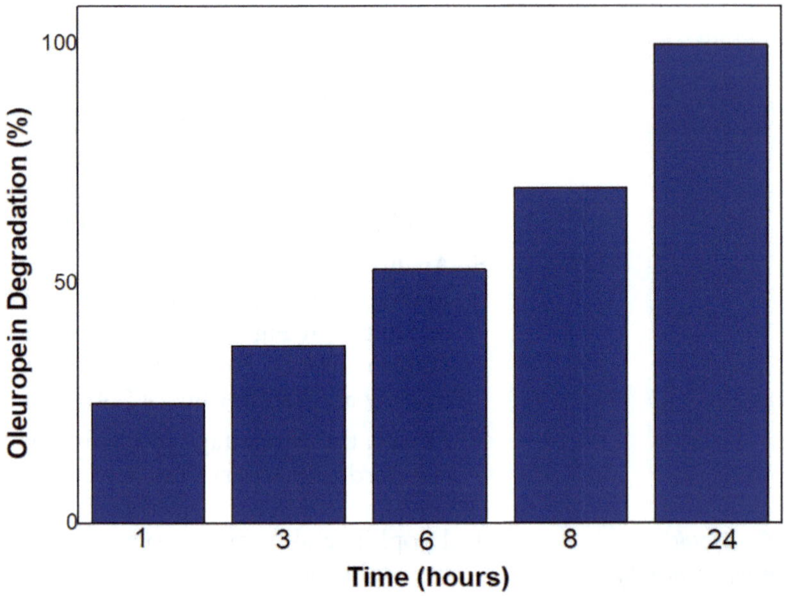

Fig. 6 Percentage (%) of oleuropein degradation by the bi-enzymatic nanobiocatalyst for the intervals 1, 3, 6 and 24 h at 60 °C (the standard deviation was less than 5% in all cases)

7. Inject 200 μL of the sample in load position of the MS injector.
8. Switch to inject position (*see* **Note 11**).
9. Display the mass spectrum (*see* **Note 12** and Fig. 7).

4 Notes

1. When transferring chitosan flakes to the glass beaker containing the acetic acid solution, work in a fume hood and place the beaker on a magnetic stirrer placed inside the hood with low

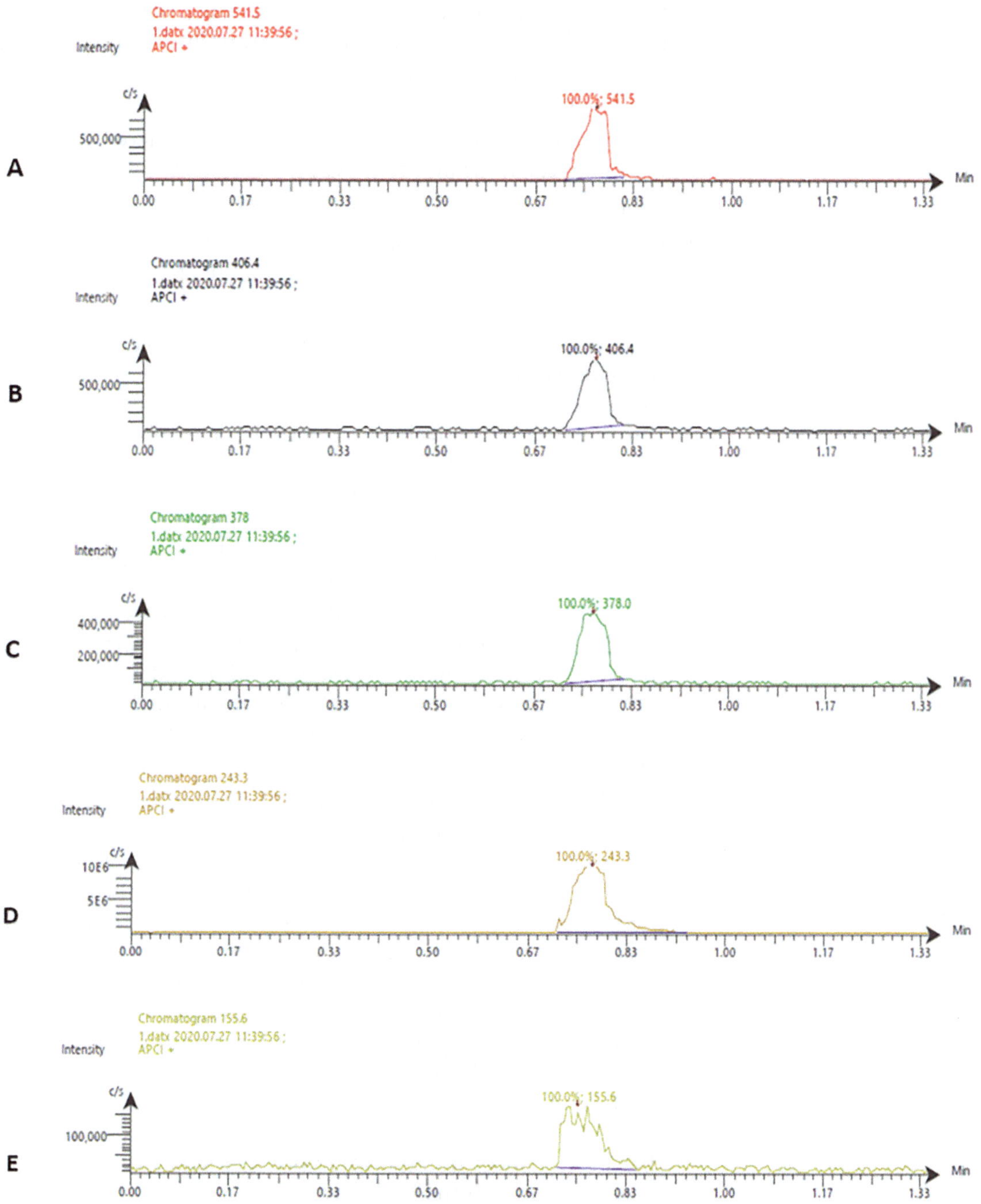

Fig. 7 Positive ion APCI mass spectra analyses of (**a**) oleuropein ([M+H$^+$] = m/z 541.5), (**b**) elenolic acid glucoside ([M+H$^+$] = m/z 406), (**c**) aglycon of oleuropein ([M+H$^+$] = m/z 378), (**d**) elenolic acid ([M+H$^+$] = m/z 243), and (**e**) hydroxytyrosol ([M+H$^+$] = m/z 155.6)

temperature to accelerate dissolution. Also, transfer chitosan powder slowly, adding small doses at a time and waiting for it to dissolve before adding the next dose, to avoid the formation of aggregates.

2. Wear a mask when weighing TPP, as it is considered a hazardous chemical according to its safety data sheet.
3. Keep all enzymes on ice to ensure no loss in enzymatic activity.
4. Work in a fume hood and wear a mask while transferring the glutaraldehyde solution, as it is considered a hazardous chemical according to its safety data sheet.
5. Note that the buffer or the solvent of the sample could affect the circular dichroism spectrum. Especially, for secondary structures studies in the far-UV, an ideal buffer has a low absorbance in the wavelength region being probed. The addition of salt to any buffer will increase absorbance and potentially cause scattering, reducing the signal-to-noise ratio. Phosphate buffers (potassium or sodium, for example) are good to start with. Also, note that any denaturing agent may have high absorbance below 200 nm so many protein denaturation studies must be measured at wavelengths above 200 nm. Finally, buffers consisting of sugars not only absorb highly in the far-UV but also have a CD signal and should be avoided for CD measurements. For the reasons mentioned above, it is advantageous to use low concentrations of buffers and salts (typically 1 to 5 mM).
6. It is important to note that the CD signal arises from the peptide bonds of the enzyme and therefore larger proteins will give a stronger CD signal than a smaller protein at the same concentrations. For this reason, the size of the protein is used when determining molar ellipticity values. It is also the reason that the concentrations of the enzymes for CD measurements are recommended to be expressed in mg/mL rather than μM. For secondary structure studies (180–250 nm), 0.1 mg/mL of sample in a 1 mm path length cell is typically recommended. However, this depends on the specific characteristics of each protein. One should test different enzyme concentrations in order to result in a protein concentration that its CD spectrum and therefore its secondary structure is in agreement with the bibliography.
7. A final concentration of 1.5 μg/mL of CS-MNPs was chosen in order to study the changes in the enzyme's secondary structure upon their interaction with the nanoparticles. The low concentrations of nanoparticles are generally preferred in such studies to minimize any optical phenomena that could lower the signal-to-noise ratio.

8. The activities of the enzymes can be expressed as activity recovery (%), which is determined as the percent activity of each immobilized enzyme divided by the total initial activity of each enzyme.
9. The scanning speed determines how quickly the monochromator scans to acquire data points at the specified data interval. Typical scan speeds for proteins are 20 to 100 nm/min. A good starting scanning speed for biological samples is typically 50 nm/min.
10. Mobile phases consist of (A) acetonitrile and (B) water with 0.1% acetic acid. Isocratic elution is performed consisting of 20% (A) and 80% (B) at a flow rate 0.5 mL/min. The column must be equilibrated for 15 min before each analysis.
11. The experiment is complete when the signal returns to the baseline.
12. The differences between the analysis of the samples are detected on the intensity of the ions of each substance.

Acknowledgements

This research was co-financed by the European Regional Development Fund of the European Union and Greek national funds through the Operational Program Competitiveness, Entrepreneurship and Innovation, under the call 'Aquaculture'—'Industrial Materials'—'Open Innovation in Culture' (project: AntiMicrOxiPac, project code: T6YBΠ-00232).

References

1. Robinson PK (2015) Enzymes: principles and biotechnological applications. Essays Biochem 59:1–41
2. Mohamad NR, Marzuki NH, Buang NA et al (2015) An overview of technologies for immobilization of enzymes and surface analysis techniques for immobilized enzymes. Biotechnol Biotechnol Equip 29:205–220
3. Zdarta J, Meyer AS, Jesionowski T et al (2018) A general overview of support materials for enzyme immobilization: characteristics, properties, practical utility. Catalysts 8:92
4. Nguyen HH, Kim M (2017) An overview of techniques in enzyme immobilization. Appl Sci Converg Technol 26:157–163
5. Zucca P, Sanjust E (2014) Inorganic materials as supports for covalent enzyme immobilization: methods and mechanisms. Molecules 19:14139–14194
6. Wang JH, Tang MZ, Yu XT et al (2019) Site-specific, covalent immobilization of an engineered enterokinase onto magnetic nanoparticles through transglutaminase-catalyzed bioconjugation. Colloids Surf B Biointerfaces 177:506–511
7. Giannakopoulou A, Gkantzou E, Polydera A et al (2019) Multienzymatic nanoassemblies: recent progress and applications. Trends Biotechnol 38:202–216
8. Garcia-Galan C, Berenguer-Murcia Á, Fernandez-Lafuente R et al (2011) Potential of different enzyme immobilization strategies to improve enzyme performance. Adv Synth Catal 353:2885–2904
9. Zhang G, Quin MB, Schmidt-Dannert C (2018) Self-assembling protein scaffold system for easy in vitro coimmobilization of biocatalytic cascade enzymes. ACS Catal 8:5611–5620

10. Wang SZ, Zhang YH, Ren H et al (2017) Strategies and perspectives of assembling multi-enzyme systems. Crit Rev Biotechnol 37:1024–1037
11. Gupta MN, Kaloti M, Kapoor M et al (2011) Nanomaterials as matrices for enzyme immobilization. Artif Cells Blood Substit Immobil Biotechnol 39:98–109
12. Rostro-Alanis MDJ, Mancera-Andrade EI, Patiño MBG et al (2016) Nanobiocatalysis: Nanostructured materials – a minireview. Biocatalysis 2:1–24
13. Bilal M, Zhao Y, Rasheed T et al (2018) Magnetic nanoparticles as versatile carriers for enzymes immobilization: a review. Int J Biol Macromol. 120(Pt B):2530–2544
14. Akın D, Yakar A, Gündüz U (2015) Synthesis of magnetic Fe_3O_4-chitosan nanoparticles by ionic gelation and their dye removal ability. Water Environ Res 87:425–436
15. Nguyen TV, Nguyen TTH, Wang SL et al (2017) Preparation of chitosan nanoparticles by TPP ionic gelation combined with spray drying, and the antibacterial activity of chitosan nanoparticles and a chitosan nanoparticle – amoxicillin complex. Res Chem Intermed 43: 3527–3537
16. Hoang DQ, Tran TV, Tran NQ et al (2016) Functionalization of Fe_3O_4 nanoparticles with biodegradable chitosan-grafted-mPEG for paclitaxel delivery. Green Process Synth 5:459–466
17. Liu M, Yong Q, Yu S (2018) Efficient bioconversion of oleuropein from olive leaf extract to antioxidant hydroxytyrosol by enzymatic hydrolysis and high-temperature degradation. Biotechnol Appl Biochem 65:680–689
18. Chatzikonstantinou AV, Gkantzou E, Thomou E et al (2019) Enzymatic conversion of oleuropein to hydroxytyrosol using immobilized β-glucosidase on porous carbon cuboids. Nano 9:1166
19. Wu Y, Wang Y, Luo G et al (2009) In situ preparation of magnetic Fe3O4-chitosan nanoparticles for lipase immobilization by cross-linking and oxidation in aqueous solution. Bioresour Technol 100:3459–3464
20. Giannakopoulou A, Chatzikonstantinou AV, Chalmpes N et al (2021) Development of a novel bi-enzymatic nanobiocatalyst for the efficient bioconversion of oleuropein to hydroxytyrosol. Catalysts 11:749
21. Orfanakis G, Patila M, Catzikonstantinou AV et al (2018) Hybrid nanomaterials of magnetic iron nanoparticles and graphene oxide as matrices for the immobilization of β-glucosidase: synthesis, characterization, and biocatalytic properties. Front Mater 5:25

Chapter 18

A Simple Recombinant *E. coli* Cell Lysate-Based Biocatalyst for ATP-Dependent Multi-step Reactions

Natalia Bartzoka, Thomas D. Loan, Hideki Onagi, and Apostolos Alissandratos

Abstract

The utility of ATP-dependent multi-enzymatic reactions is limited by their requirement for stoichiometric amounts of this expensive cofactor or additional purified enzymes for its recycling. Here we describe a simple method for the production of recombinant cell-free extracts (or lysates) of *E. coli* that support ATP-dependent biotransformations. The inexpensive preparation described is obtained with modest processing from a single recombinant bacterial culture of *E. coli*. In addition to recombinantly overexpressed enzymes that catalyze the primary ATP-dependent reactions of interest, endogenous kinases that are naturally present in the extract catalyze recycling of the requisite ATP. This means that only catalytic amounts of cofactor are necessary to drive the biotransformation, and without the requirement for additional purified enzymes. This approach has been applied successfully to an array of in vitro enzymatic cascades with multiple ATP-dependent steps.

Key words Cell-free extract, Lysate, Multi-enzymatic cascades, Multi-enzymatic assemblies, ATP recycling, In vitro biocatalysis

1 Introduction

In vitro multi-enzymatic cascades offer attractive green alternatives to multi-step chemical processes [1]. However, their utility is limited by the cost and complexity of assembling the necessary mixtures of purified enzymes, and the requirement for additional enzymes to balance costly cofactors (e.g., adenosine 5′-triphosphate or ATP) [2]. Conversely, lysates (or cell-free extracts) of *E. coli* that over-express combinations of recombinant enzymes may be produced through straightforward and accessible molecular and synthetic biology, and with modest down-stream requirements. Importantly, the cell-free extracts also retain useful endogenous activities, such as those necessary for cofactor recycling. Our group and others have demonstrated the great potential

Haralambos Stamatis (ed.), *Multienzymatic Assemblies: Methods and Protocols*,
Methods in Molecular Biology, vol. 2487, https://doi.org/10.1007/978-1-0716-2269-8_18,
© The Author(s), under exclusive license to Springer Science+Business Media, LLC, part of Springer Nature 2022

of this approach for the recycling of ATP as part of important biotechnological applications (e.g., cell-free protein synthesis, nucleic acid amplification, biochemical syntheses) [3–9].

We have described a number of ATP-dependent multi-enzymatic one-pot syntheses that proceed with high-turnovers and yields, when catalyzed by a cell-free extract obtained from a single recombinant bacterial culture. This has included syntheses of amino acids [4, 5], sugar phosphates [6], nucleosides [6], nucleotides [3, 6, 7], and nucleic acids [3, 7]. In all cases, the multi-step reactions proceed through the combined activities of recombinantly overexpressed and naturally abundant endogenous enzymes, while the necessary ATP is synthesized and recycled from catalytic concentrations of diphosphate (ADP) and/or monophosphate (AMP), through highly active endogenous kinases and with the addition of acetyl phosphate (Fig. 1a). The latter is commercially available, but also easily synthesized from acetic anhydride and inorganic phosphate through a simple procedure [3, 10].

Here, we provide detailed methodology for the production of a recombinant cell-free extract of *E. coli* with the capacity to support ATP recycling as part of in vitro biochemical syntheses. As an example, we present the protocol for the co-expression of carbamate kinase and ornithine transcarbamoylase from hyperthermophilic archaea using a pRSF-Duet vector [5]. This generates a cell-free extract that catalyzes the ATP-dependent synthesis of citrulline via carbamoyl phosphate (Fig. 1b). As highlighted

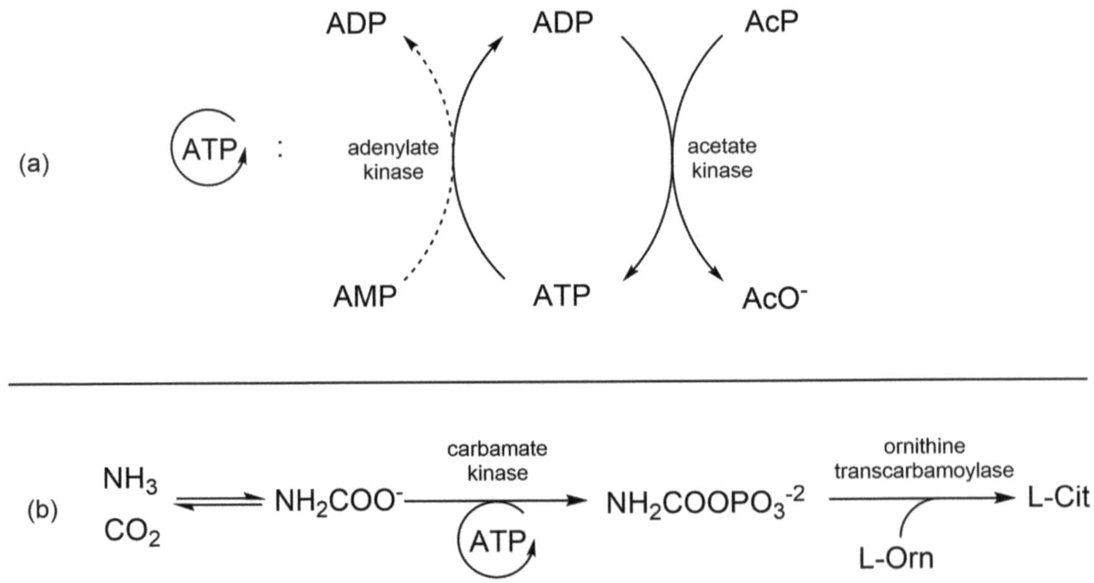

Fig. 1 Reaction catalyzed by the cell-free extract: (**a**) endogenous kinases catalyze the synthesis and recycling of ATP from ADP or AMP with acetyl phosphate (AcP), (**b**) recombinant enzymes catalyze the ATP-dependent reactions of interest, herein, carbamate kinase catalyzed synthesis of carbamoyl phosphate from CO_2, NH_3, and ATP, and subsequent carbamylation of ornithine, catalyzed by ornithine transcarbamoylase

throughout the Subheading 4, the method is easily transferable to other ATP-dependent reactions by replacing this plasmid with one or more encoding enzymes of interest. Regardless, the underlying strategy remains the same and involves an overnight culture of *E. coli* harboring the appropriate plasmid(s). The cells are then harvested and resuspended in a buffer, followed by lysis and recovery of the soluble fraction. This fraction is the cell-free extract biocatalyst which retains its ATP recycling activity for over a year when stored frozen (−80 °C) but may also be lyophilized.

Overall, the straightforward approach presented is highly versatile, and has been applied in our lab without issue and only slight modification (*see* Subheading 4) for ATP-dependent reactions ranging from 2 to 8 steps [2].

2 Materials

2.1 Growth of Recombinant Bacterial Cells

1. Equipment: Waterbath (42 °C, for heat-shock cell transformation), Agar plate incubator (37 °C), Incubated shaker (37 °C, 200 rpm) with 2 L flask holder, Refrigerated centrifuge with rotor to accommodate 1 L bottles at 4000 × *g*, −80 °C Freezer, Autoclave, 2 L Baffled Erlenmeyer flask.
2. H_2O is Milli-Q® ultrapure water (for all Subheading 2).
3. *E. coli* BL21(DE3) (*see* **Note 1**) transformed with pRSFDuet-CK-OTC [5] (*see* **Note 2**): Glycerol stock or fresh transformation. Chemically competent cells are transformed in accordance with the manufacturer's protocol. The plasmid may be obtained by contacting authors or by sub-cloning (e.g., restriction digestion—ligation, Gibson assembly) synthetic genes into the coding regions of the expression sites of pRSFDuet-1 (Novagen) (*see* **Notes 3** and **4**).
4. Lysogeny Broth (LB) (100 mL): Dissolve 1 g tryptone and 0.5 g yeast extract in approx. 75 mL H_2O. Adjust pH to 7.4 with NaOH (2 M). Add H_2O to a final total volume of 100 mL. Autoclave (121 °C, 20 min) and store at 4 °C.
5. ZY medium (1 L) (*see* **Note 5**): Dissolve 10 g tryptone and 5 g yeast extract in approx. 750 mL H_2O. Add H_2O to a final total volume of 1 L. Autoclave (121 °C, 20 min) and store at 4 °C.
6. 5052 solution 50× (100 mL): 25% w/v glycerol, 2.5% w/v glucose, 10% w/v α-lactose in H_2O. Mix in sequence 25 g glycerol, 73 mL H_2O, 2.5 g D-glucose, and 10 g α-lactose. Ensure final volume is 100 L, autoclave and store at room temperature.
7. NPS 20× (100 mL): 2 M Na_2HPO_4, 2 M KH_2PO_4, 500 mM $(NH_4)_2SO_4$ in H_2O. To 90 mL H_2O add in sequence 6.6 g

$(NH_4)_2SO_4$, 13.6 g KH_2PO_4, 14.2 g Na_2HPO_4. Ensure final volume is 100 mL, autoclave and store at room temperature.

8. 1 M $MgSO_4$ in H_2O (10 mL). Dissolve 1.2 g $MgSO_4$ (anhydrous) in 6 mL H_2O. Adjust volume to 10 mL with H_2O. Filter sterilize (0.2 μ) and store at room temperature.
9. 100 mg/mL kanamycin sulfate in H_2O (10 mL). Dissolve 1 g kanamycin sulfate in 8 mL H_2O. Adjust volume to 10 mL with H_2O (*see* **Note 6**)

2.2 Preparation of Cell-Free Extract Biocatalyst

1. Equipment: Omni Sonic Ruptor 400 Ultrasonic homogenizer (*see* **Note 7**), Refrigerated high-speed centrifuge with rotor to accommodate 30 mL centrifuge tubes at 20,000 × *g* (must be resistant to high-speed centrifugation, e.g., Oak Ridge tubes), Labconco FreeZone 4.5 L Benchtop Freeze Dry system, −80 °C Freezer.
2. Lysis buffer 10× (200 mL): 100 mM Tris-acetate pH 8.3, 160 mM potassium acetate (KOAc), 140 mM magnesium acetate (MgOAc). Dissolve 3.6 g Tris-acetate, 3.1 g KOAc and 6.0 g MgOAc in 170 mL H_2O and adjust pH to 8.3 with NaOH (2 M). Adjust volume to 200 mL with H_2O, autoclave and store at 4 °C.
3. 100 mM phenylmethylsulfonyl fluoride (PMSF) in ethanol (*see* **Note 8**): Purchase ready solution or dissolve 1.7 mg PMSF in 1 mL absolute ethanol (exercise caution, *see* **Note 8**). Prepare on the day of the experiment and keep on ice or store at −20 °C until use (*see* **Note 9**).
4. 100 mM DL-dithiothreitol (DTT) in H_2O (1 mL): Prepare on the day of the experiment and keep on ice or store at −20 °C until use (*see* **Note 10**).
5. Phosphate-buffered saline (PBS) 10× (100 mL): 1.37 M NaCl, 27 mM KCl, 100 mM Na_2HPO_4, 18 mM KH_2PO_4. Dissolve 8.0 g NaCl, 0.2 g KCl, 1.4 g Na_2HPO_4 and 0.24 g KH_2PO_4 in 800 mL H_2O. Adjust to pH 7.4 with HCl and final volume to 1 L using H_2O. Use this stock to prepare a 1× solution by mixing 5 mL with 45 mL cold H_2O.
6. Lyoprotectant (*see* **Note 11**): 10% w/v sucrose solution in H_2O (100 mL). Dissolve 10 g sucrose in 75 mL H_2O. Adjust volume to 100 mL with H_2O.

2.3 Cell-Free Extract Catalyzed Synthesis of ATP from AMP/ADP

1. Equipment: Set of calibrated micropipettes for pipetting volumes in the 10–1000 μL range.
2. Assay buffer: 400 mM sodium phosphate pH 7.4. Dissolve 1.24 g NaH_2PO_4 and 4.36 g of Na_2HPO_4 (anhydrous) in 75 mL H_2O. Adjust volume to 100 mL with H_2O.

3. 100 mM adenosine 5′-monophosphate (AMP) in H_2O (1 mL): Dissolve 39.1 mg AMP disodium salt in 1 mL H_2O.
4. 100 mM adenosine 5′-diphosphate (ADP) in H_2O (10 mL): Dissolve 47.1 mg ADP disodium salt in 1 mL H_2O.
5. 1 M acetyl phosphate in H_2O (1 mL) (*see* **Note 12**): Dissolve 184 mg lithium potassium acetyl phosphate in 0.7 mL H_2O. Adjust volume to 1 mL with H_2O.
6. 0.2% (w/v) sodium dodecyl sulfate (SDS) in H_2O (10 mL) (*see* **Note 13**). To 2 mL of a 10% w/v SDS solution (commercially available) add 8 mL H_2O.
7. 1 M $MgCl_2$ in H_2O (1 mL) (*see* **Note 14**). Purchase ready solution or dissolve 203 mg $MgCl_2$ hexahydrate in 6 mL H_2O. Adjust volume to 10 mL with H_2O.

2.4 HPLC Analysis of ATP Synthesis

1. Equipment: Agilent 1100 HPLC system equipped with Alltima HP C18 column (5 μ, 250 × 4.6 mm with a 7.5 × 4.6 mm guard) (Fig. 2) (*see* **Note 15**).
2. HPLC Solvent A - aqueous mobile phase (1 L): 60 mM ammonium dihydrogen phosphate, 5 mM tetrabutylammonium phosphate, pH 5.0 (*see* **Note 16**). In a 1 L volumetric flask, dissolve 6.9 g $(NH_4)(H_2PO_4)$ (ammonium dihydrogen phosphate) in 750 mL H_2O and add 5 mL tetrabutylammonium phosphate monobasic solution (0.1 M in H_2O; available commercially). Adjust the pH to 5.0 using a 28% (w/v) ammonia solution. Adjust the final volume of the solution to 1 L with H_2O.
3. HPLC Solvent B - organic mobile phase (1 L): 5 mM tetrabutylammonium in methanol. Mix 995 mL methanol (HPLC grade) and 5 mL tetrabutylammonium phosphate monobasic solution (0.1 M in H_2O; available commercially).

2.5 Cell-Free Extract Catalyzed Synthesis of Citrulline Coupled to ATP Recycling

1. Synthesis buffer (100 mL): 400 mM $NaHCO_3$ pH 8.5, 400 mM $NH_4(OH)$ in H_2O. Dissolve 3.36 g $NaHCO_3$ in 75 mL H_2O. Slowly add 2.75 mL of $NH_4(OH)$. Adjust to pH 8.5 with HCl. Adjust volume to 100 mL with H_2O.
2. 100 mM L-ornithine in H_2O (1 mL): Dissolve 16.9 mg L-ornithine monohydrochloride in 1 mL H_2O.
3. 100 mM adenosine 5′-triphosphate (ATP) in H_2O (1 mL): Dissolve 55.1 mg ATP disodium salt in 1 mL H_2O.
4. 1 M $MgCl_2$ in H_2O (1 mL) (*see* **Note 14**). Prepared as before (*see* Subheading 2.3, **item 7**).
5. 1 M acetyl phosphate in H_2O (1 mL). Prepared as before (*see* Subheading 2.3, **item 5**).
6. Acetonitrile (ACS grade) to quench enzymatic reaction.

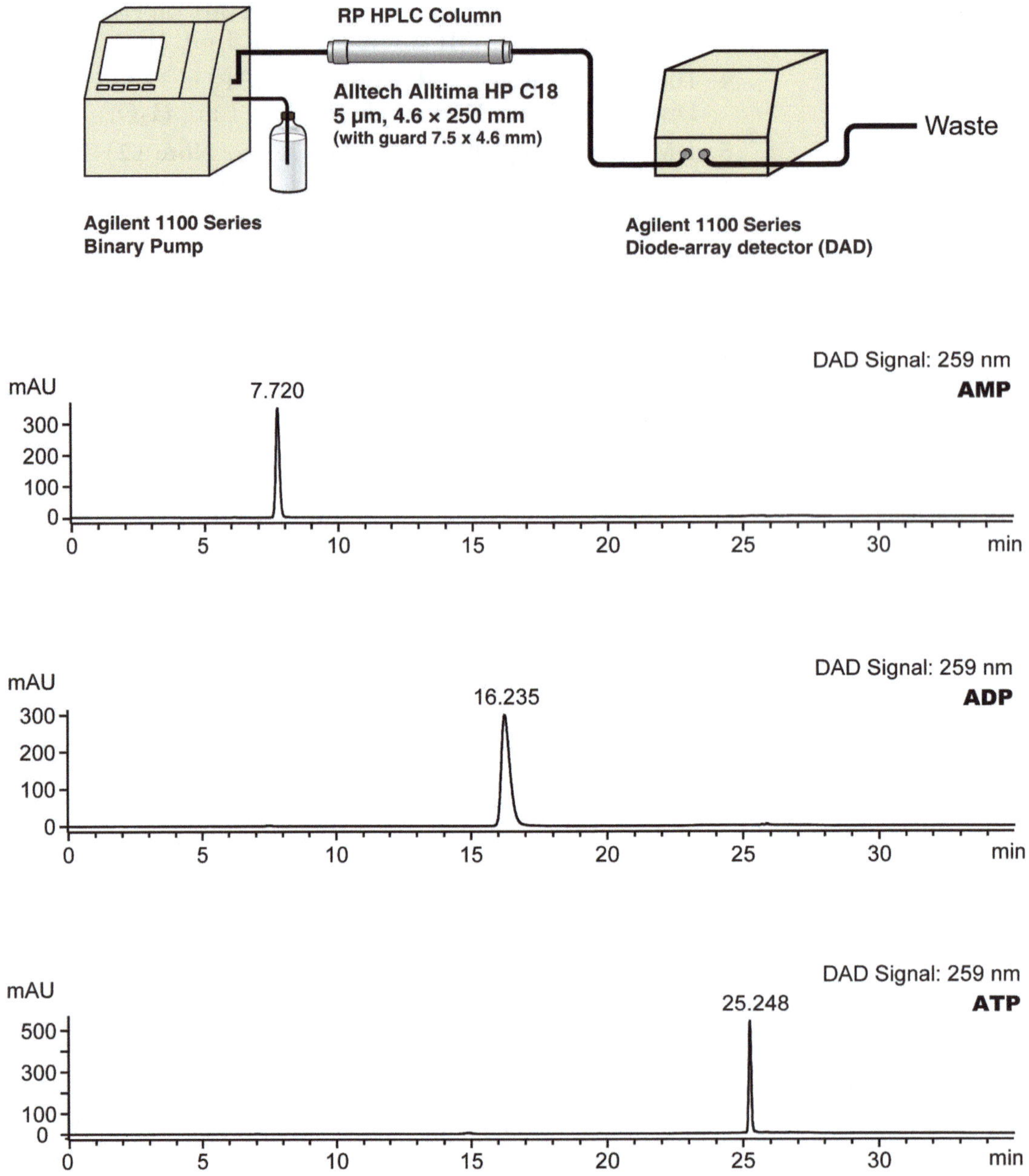

Fig. 2 Configuration of equipment for the analysis of cell-free extract catalyzed synthesis of ATP from AMP and ADP, and chromatograms obtained for authentic commercial AMP, ADP, and ATP

2.6 LC-TQD Analysis of Citrulline

1. Equipment: Waters Alliance 2695 HPLC system coupled to a Waters tandem quadrupole detector (TQD), equipped with a Dionex Acclaim Mixed-mode WCX-1 column (5 μ, 250 × 4.6 mm) (Fig. 3).
2. MS (mass spectrometry) Solvent A - aqueous mobile phase (1 L): H_2O, 0.1% Formic acid. Reagents should be MS grade.

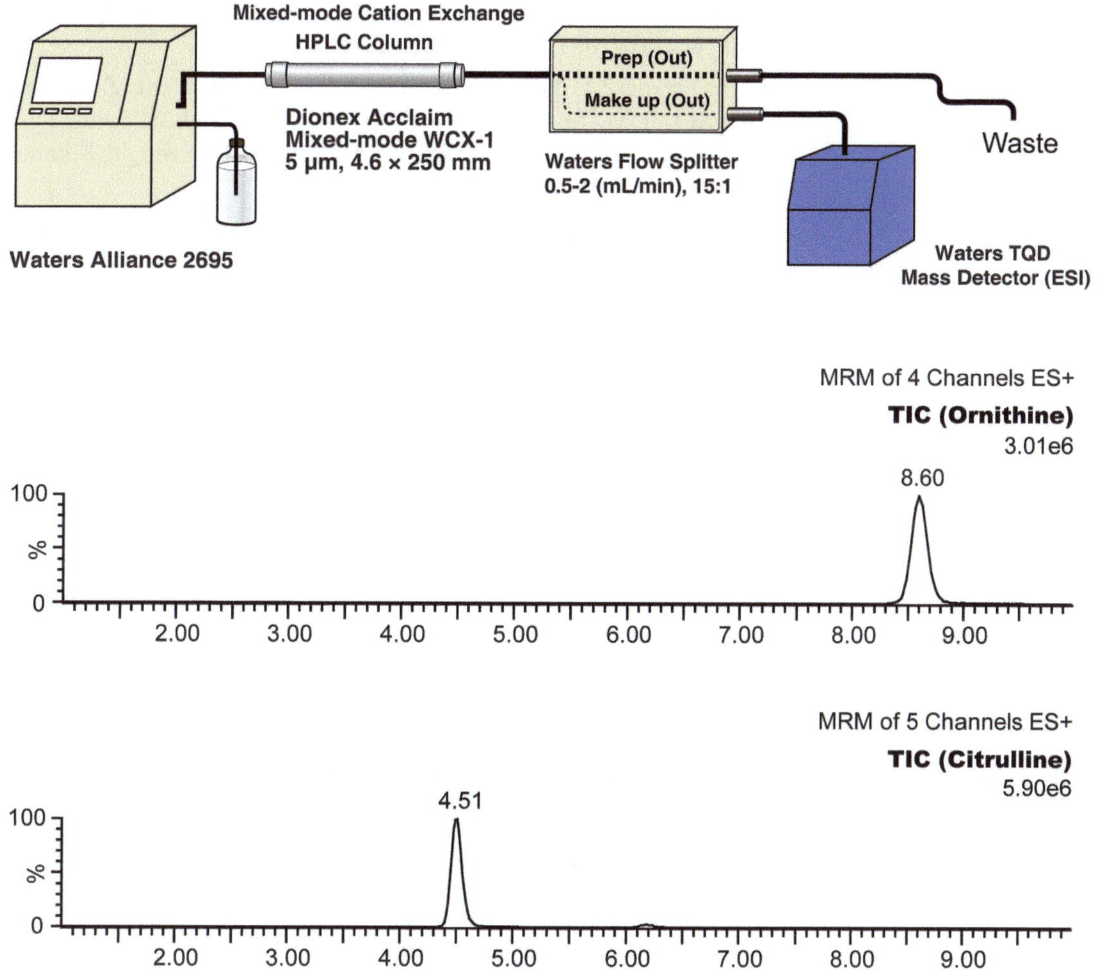

Fig. 3 Configuration of equipment for the analysis of cell-free extract catalyzed synthesis of citrulline, and chromatograms obtained for authentic commercial citrulline and ornithine

3. MS Solvent B - organic mobile phase (1 L): acetonitrile, 0.1% formic acid. Reagents should be MS grade.

3 Methods

3.1 Growth of Recombinant Bacterial Cells

1. Spread and culture *E. coli* BL21(DE3) cells harboring pRSF-Duet-CK-OTC (*see* **Note 2**) on an LB-kanamycin (*see* **Note 6**) agar plate until single colonies can be observed (overnight, 37 °C).
2. Place 5 mL of LB medium containing 100 μg/mL kanamycin into a sterile 50 mL Falcon tube.
3. Inoculate with a single colony from the agar plate, using aseptic technique.

4. Incubate the culture overnight in an incubated shaker set at 30 °C and 180 rpm (*see* **Note 17**).
5. To a sterile 2 L baffled Erlenmeyer flask add 928 mL ZY, 1 mL $MgSO_4$ 1 M (final conc. 1 mM), 20 mL 5052 (final conc. 1×), 50 mL NPS (final conc. 1×), 1 mL of the 100 mg/mL kanamycin stock (final conc. 100 μg/mL).
6. Inoculate the 1 L medium mixture through addition of the entire overnight bacterial culture.
7. Incubate the 1 L culture in an incubated shaker set at 37 °C and 180 rpm for 18 h (*see* **Note 18**).
8. Transfer the culture into a 1 L centrifuge bottle and centrifuge for 15 min at 4000 × *g* and 4 °C (*see* **Note 19**).
9. Discard the supernatant and re-suspend the pellet on ice (*see* **Note 20**) using 40 mL PBS 1×, transfer the suspension to a pre-weighed 50 mL Falcon tube and re-centrifuge using the same conditions as above (**step** 7).
10. Carefully discard the supernatant and note the weight of the wet cell pellet.
11. Place the Falcon tube containing the pellet in a −80 °C freezer and store until lysis (*see* **Note 21**).

3.2 Preparation of Cell-Free Extract

1. Remove the frozen cell pellet from the −80 °C freezer and place on ice to thaw (*see* **Note 22**).
2. Freshly prepare a 10 mL batch of lysis buffer by mixing 8.8 mL H_2O, 1 mL lysis buffer10×, 100 μL each of the 100 mM DTT solution (final conc. 1 mM), and the 100 mM PMSF solution (final conc. 1 mM) (*see* **Note 9**).
3. To the Falcon tube containing the thawed pellet, add 1 mL of fresh lysis buffer for every g wet cell pellet (*see* **Note 23**).
4. Re-suspend the pellet in the lysis buffer (*see* **Note 24**).
5. Lyse cells by sonication of the suspension (50% power, 50% pulse) for 6 min on ice (*see* **Note 25**).
6. Distribute lysate into pre-chilled high-speed centrifuge tubes and place on ice.
7. Centrifuge the lysate for 1 h at 20,000 × *g* and 4 °C (*see* **Note 26**).
8. Carefully collect the soluble supernatant fraction into a pre-chilled 14 mL Falcon tube and place on ice (*see* **Note 27**). This fraction constitutes the recombinant bacterial cell-free extract (*see* **Note 28**).

3.3 Cell-Free Extract Storage

1. Distribute 0.5 mL aliquots of the cell-free extract into pre-chilled 1.5 mL microcentrifuge tubes (*see* **Notes 29** and **30**).

2. Snap-freeze by placing sealed tubes in liquid nitrogen (*see* **Note 31**).
3. Frozen cell-free extract can be kept at −80 °C for long-term storage or at −20 °C for more immediate use (*see* **Note 32**).

3.4 Lyophilization

If the cell-free extract is to be lyophilized, then the following protocol must be followed after **step 8** of Subheading 3.4.

1. Add an equal volume of the lyoprotectant (10% w/v sucrose solution in H_2O) to the cell-free extract and gently mix (final conc. 5% w/v sucrose) (*see* **Notes 11** and **33**).
2. Distribute 0.5 mL aliquots of the cell-free extract into pre-chilled 1.5 mL microcentrifuge tubes (*see* **Notes 29** and **30**).
3. Snap-freeze by placing sealed tubes in liquid nitrogen (*see* **Note 31**).
4. Open microcentrifuge lids, seal with parafilm and carefully pierce 3–4 times using a sterile needle (*see* **Note 34**).
5. Place on a tube rack and transfer to the freeze-dryer, already set to −51 °C, and place under vacuum (*see* **Note 35**).
6. Lyophilize for 24 h, remove from freeze-drying chamber, close tube lids, and store in fridge or at −20 °C (*see* **Note 36**).

3.5 Assay for Cell-Free Extract Catalyzed Synthesis of ATP from AMP/ADP

An assay is carried out to determine the presence of the endogenous kinases that are necessary to catalyze synthesis of ATP in the cell-free extract (Fig. 1a). This demonstrates its capacity to support ATP-dependent cascades through recycling of ATP from ADP or AMP.

1. Thaw the frozen cell-free extract on ice (*see* **Note 37**) and prepare a ten-fold dilution by pipetting 10 μL into 90 μL H_2O.
2. Aliquot 50 μL of the 0.2% SDS solution into 5 microcentrifuge tubes (*see* **Note 38**).
3. In a 1.5 mL microcentrifuge tube mix 500 μL assay buffer (final conc. 200 mM phosphate) (*see* **Note 39**), 420 μL H_2O (to adjust to 1 mL final volume), 10 μL $MgCl_2$ (final conc. 10 mM), 10 μL AMP (or ADP) (final conc. 1 mM), 50 μL acetyl phosphate (final conc. 50 mM), and 10 μL of the cell-free extract dilution (*see* **Note 40**). A negative control in the absence of acetyl phosphate (50 μL H_2O instead) should also be carried out. Time the reaction carefully from the addition of the cell-free extract.
4. Obtain 50 μL aliquots from the reaction mixture and quench enzymatic activity by transferring into to microcentrifuge tubes containing SDS (final conc. 0.1% w/v) at 1, 2.5, 5, 15, and 30 min and place on ice (*see* **Note 41**).

5. The quenched samples are centrifuged on a microcentrifuge (top speed) for 10 min and at 4 °C (*see* **Note 42**) and transferred to HPLC vials (*see* **Note 43**). Samples are then analyzed by HPLC to detect the presence of AMP, ADP, and ATP.

3.6 HPLC Analysis of ATP Synthesis

Quenched samples from ATP formation assays are analyzed through ion-pair reversed-phase HPLC. An autosampler (*see* **Note 44**) is used to deliver 5 μL injections at a flow rate of 1 mL/min and separation is performed with the heated C18 column (37 °C) through gradient elution (0–18 min, 87% HPLC Solvent A: 13% HPLC Solvent B; 19–23.3 min; 70% HPLC Solvent A: 30% HPLC Solvent B; 24–35 min 87% HPLC Solvent A: 13% HPLC Solvent B; linear transition between mixtures). AMP, ADP, and ATP are analyzed through monitoring of absorbance at 259 nm with a diode-array detector (AMP, 7.7 min; ADP, 16.2 min; ATP, 25.2 min) (Fig. 2) (*see* **Note 45**). Quantification is carried out through comparison with calibration curves developed with standard solutions prepared from authentic commercial materials (*see* **Note 46**). Plots for ATP concentration vs time are prepared to evaluate cell-free extract catalyzed formation of ATP (*see* **Note 47**).

3.7 Multi-enzymatic Synthesis of Citrulline

The ATP recycling activity of the cell-free extract is then coupled to the primary reactions of interest (*see* **Note 48**), herein, synthesis of citrulline from NH_3, CO_2, and ornithine (Fig. 1b).

1. Thaw the frozen cell-free extract on ice (*see* **Note 37**).
2. Aliquot 50 μL of quenching acetonitrile into 8 microcentrifuge tubes (*see* **Note 38**).
3. In a 1.5 mL microcentrifuge tube mix 500 μL synthesis buffer (final conc. 200 mM $NaHCO_3$, 200 mM $NH_4(OH)$) (*see* **Note 49**), 399 μL H_2O (to adjust to 1 mL final volume), 40 μL L-ornithine (final conc. 4 mM), 1 μL ATP (final conc. 0.1 mM) (*see* **Note 50**), 50 μL acetyl phosphate (final conc. 50 mM) (*see* **Note 51**), and 10 μL of the cell-free extract. Time the reaction carefully from the addition of the cell-free extract.
4. Obtain 50 μL aliquots from the reaction mixture and quench enzymatic activity by transferring into the microcentrifuge tubes containing acetonitrile at 5, 10, 15, 20, 30, 45, 60, and 90 min.
5. The quenched samples are centrifuged on a benchtop microcentrifuge (top speed) for 10 min (*see* **Note 42**), and the supernatant is transferred into HPLC vials. Samples are then analyzed by LC-TQD to detect the presence of citrulline.

3.8 LC-TQD Analysis of Citrulline

Analysis of the citrulline product in the quenched samples is carried out by LC-TQD (*see* **Note 52**). 5 μL sample injections are delivered by a refrigerated autosampler (4 °C) and separation is performed with column heating (30 °C) and isocratic elution (60% MS Solvent A: 40% MS Solvent B) at a flow rate of 1.0 mL/min (15:1 flow splitter) for 10 min (1.5–10 min retention window). Citrulline is analyzed using integrated multiple reaction monitoring (MRM) and quantified through comparison with a calibration curve developed with authentic commercial material (Fig. 3). The MRM parameters employed are citrulline (parent [*m/z*], 175.9999), channel 1 (daughter [*m/z*], 27.9658; dwell [s], 0.025; cone [V], 20; collision [V], 34), channel 2 (daughter [*m/z*], 43.0312; dwell [s], 0.025; cone [V], 20; collision [V], 32), channel 3 (daughter [*m/z*], 70.053; dwell [s], 0.025; cone [V], 20; collision [V], 26), channel 4 (daughter [*m/z*], 112.7336; dwell [s], 0.025; cone [V], 20; collision [V], 14), channel 5 (daughter [*m/z*], 113.0588; dwell [s], 0.025; cone [V], 20; collision [V], 18); ornithine (parent [*m/z*], 132.9302), channel 1 (daughter [*m/z*], 27.9706; dwell [s], 0.025; cone [V], 18; collision [V], 30), channel 2 (daughter [*m/z*], 42.9769; dwell [s], 0.025; cone [V], 18; collision [V], 24), channel 3 (daughter [*m/z*], 70.0061; dwell [s], 0.025; cone [V], 18; collision [V], 18), channel 4 (daughter [*m/z*], 133.1233; dwell [s], 0.025; cone [V], 18; collision [V], 12).

4 Notes

1. Several strains of *E. coli* have been employed (e.g., BL21(DE3), BL21 Star (DE3), JM109(DE3)). The strain must contain the λ(DE3) lysogen to allow overexpression of recombinant genes under the control of a T7 promoter.
2. The plasmid encodes the enzyme(s) targeted for overexpression, in this example carbamate kinase and ornithine transcarbamoylase. Target enzymes must be suitable for overexpression in *E. coli*, e.g., enzymes that require specialized cofactors not found in prokaryotes may not be suitable. When selecting target enzymes, the user should consider heterologous enzymatic activities required for the reaction of interest, and that are not present in *E. coli* or for which endogenous activities are low. This is usually determined by assaying a native cell-free extract prepared from cells transformed with a blank vector. In addition to carbamate kinase and ornithine transcarbamoylase from *Thermococcus* spp. [5], we have applied this same approach to produce cell-free extracts with overexpressed uridine kinase from *Thermus thermophilus* for the synthesis of uridine monophosphate [3], deoxynucleoside kinase from *Drosophila melanogaster* for the synthesis of deoxynucleoside monophosphates

[7], DNA polymerase I from *Pyrococcus furiosus* for the synthesis of DNA [7] and others. However, the same approach may be applied to enrich the activity of endogenous *E. coli* enzymes. For example, we have overexpressed ribokinase and phosphoribosyl pyrophosphate (PRPP) kinase from *E. coli* to catalyze the production of PRPP from ribose, as part of a multi-enzyme cascade [6]. Though these enzymes are present in *E. coli*, their activity in the native cell-free extract is very low. Lastly, it is noteworthy that the plasmid may be omitted altogether to produce cell-free extract from untransformed cells [3, 4]. This preparation is an effective low-cost biocatalyst for ATP recycling with acetyl phosphate, when combined with other enzymes in vitro. It also catalyzes the production of other natural (deoxy)-nucleoside triphosphates from their corresponding monophosphates, that are necessary as cofactors in enzymatic reactions and substrates in the synthesis of nucleic acids [3].

3. Plasmids are prepared by sub-cloning (e.g., restriction digestion—ligation, Gibson assembly) the genes of interest into the expression site(s) of a commercial vector. Synthetic genes are purchased commercially, codon-optimized for *E. coli* but we have also sub-cloned bacterial genes from source. Several different vectors have been utilized to construct plasmids (pRSF-Duet, pET-Duet, pACYC-Duet, pETMCSIII). Duet vectors are particularly convenient as they enable co-expression of two genes from different T7 promoters, and there are four compatible Duet vectors (i.e., different origins of replication and antibiotic resistance), that allow co-expression of up to eight genes. In addition, we have co-expressed enzymes using single plasmids with multiple genes sub-cloned into the coding region of a single T7 promoter (polycistronic transcripts).

4. An affinity-tag (e.g., polyhistidine-tag) may be included in the synthetic gene or incorporated using an appropriate vector. This facilitates recovery of the recombinant protein for characterization but is not necessary for application of the cell-free extract.

5. We have found the ZY-5052 auto-induction medium to be convenient for protein expression with vectors requiring lactose induction, such as the Duet vectors that contain lactose operators. This strategy leads to good yields, following simple overnight expression and without need for careful monitoring of growth. IPTG induction at exponential growth has also been employed but without observed advantage. Finally, when employing a plasmid with a constitutive T7 promoter and no operator (e.g., pETMCSIII; can be obtained by contacting the authors), the auto-induction protocol may be replaced by overnight expression in antibiotic-supplemented LB medium.

6. Kanamycin is employed here for the selection of cells containing pRSF-Duet-CK-OTC. For other expression vectors, kanamycin should be replaced with the appropriate antibiotic.
7. We have prepared cell-free extracts following lysis with the listed sonicator, a French pressure cell, and a high-pressure homogenizer (Avestin Emulsiflex B15) without significant variation in the observed ATP recycling activity. We do not recommend chemical lysis methods as the chemical agents are not removed from the cell-free extract, and could inhibit application as a biocatalyst.
8. PMSF is a suicide protease inhibitor that attacks catalytic serines and is acutely toxic and corrosive. It must therefore be handled with caution. Commercial solutions are available to minimize handling. It is commonly used in lysis buffers, including in the preparation of S30 extracts for cell-free protein synthesis [3], which we have adapted for our application. If enzymes with catalytic serines are involved in the reaction of interest, PMSF should be omitted from the lysis buffer. We have determined that the presence of PMSF is not necessary, however, stability is expected to diminish in its absence.
9. PMSF has low solubility in H_2O and must be dissolved in anhydrous ethanol before dilution in the lysis buffer. When diluting, ensure the PMSF ethanol solution is added to the buffer, and not vice versa, to avoid it crashing out. PMSF is unstable in H_2O and therefore is added to the lysis buffer before the cells are resuspended, and the buffer is kept cold before and during lysis.
10. DTT is commonly employed in lysis buffers to inhibit non-specific disulfide formation, leading to protein aggregation. As with PMSF, we have not determined its presence to be necessary but omission would limit the cell-free extract's shelf life.
11. May be replaced by other common lyoprotectants (e.g., trehalose).
12. The acetyl phosphate stock can be prepared from commercial acetyl phosphate or synthesized in-house. We have routinely used acetyl phosphate prepared through the synthesis described by Whitesides and coworkers [10], as it is easy to produce from cheap components (inorganic phosphate and acetic anhydride). Following reaction, the only necessary work-up is repeated extraction with organic solvent to remove side-products. This produces an aqueous acetyl phosphate solution (~1 M) that may be added directly to enzymatic reactions. This method does not require complex or specialized techniques, and may be easily carried out in most chemistry labs.

13. SDS is used to quench enzymatic activities but may be replaced by other appropriate chemical or physical methods (e.g., addition of organic solvent). It should be noted that SDS shortens the life of the HPLC column. For this reason, use of a guard cartridge is highly recommended.
14. A source of Mg^{2+} is necessary for enzymatic binding of ATP. $MgSO_4$ may be used instead.
15. Other reversed-phase HPLC systems equipped with a diode-array detector may be employed following appropriate optimization. We have obtained similar results using a Waters Alliance 2695 system. We have not tested other stationary phases.
16. Tetrabutylammonium is present in both aqueous and organic mobile phases and acts as the ion-pairing agent.
17. Typically, 18 h at which point the cells are chilled on ice until inoculation of the 1 L culture.
18. Cell growth is typically carried out overnight. The length of incubation should be optimized in each case to increase the concentration of target recombinant enzymes, as it does not strongly influence ATP recycling activity in the cell-free extract (8–24 h have been employed).
19. If 1 L centrifuge bottles are not available, the culture can be split and centrifuged in smaller bottles or tubes (e.g., 2×0.5 L bottles), using the same conditions. Each pellet is then resuspended sequentially in PBS 1× and combined with the rest.
20. The procedure may be carried out on ice; however, a cold room is preferable. Pellet resuspension is carried out by gentle manual agitation or low-speed (<100 rpm) magnetic stirring or vortexing.
21. This step can be omitted if cells are to be lysed on the same day, however, a freeze-thaw cycle increases lysis efficiency.
22. Inspect visually until the cell pellet no longer appears frozen. Resuspension can be initiated with the still frozen pellet but this will require longer agitation.
23. Typical yields for a 1 L overnight culture are 3–5 g of wet cell pellet, requiring 3–5 mL lysis buffer.
24. As above, pellet resuspension should be carried out on ice (or in a cold room) and may be achieved by gentle manual agitation or low-speed magnetic stirring or vortexing. This procedure will result in a concentrated thick suspension. Care must be taken to ensure all cell clumps have been dispersed and cells are properly resuspended. Avoid excessive agitation to prevent foaming.
25. To prevent overheating during sonication, the cell suspension is kept cold, usually by resting the tube in a container filled with

ice. As the ice will melt during sonication, take care to secure the tube (e.g., using a retort stand) to avoid accidental submersion of the suspension mixture.

26. Setting the high-speed centrifuge to cool to 4 °C before the cell lysis ensures it is already at temperature at the beginning of the centrifugation.
27. For the conditions described here, this should yield approximately 5–10 mL of cell-free extract. Lower or higher biomass yields will result in variations to the final volume of the cell-free extract.
28. Total protein concentration of the cell-free extract is typically around 150 mg/mL as determined by Bradford assay (e.g., using Biorad Quick-Start Bradford 1× dye). Absorbance at 280 nm is less reliable, mainly due to the presence of nucleic acids absorbing at 260 nm. SDS-PAGE electrophoresis may also be carried out to detect the presence of overexpressed recombinant proteins. These should appear in the stained gel as bold prominent bands of the expected size. However, the absence of visible bands above the protein background does not necessarily indicate absence of recombinant enzymatic activity and this should be tested through standard enzyme activity assays.
29. To avoid unnecessary handling, any tube labeling should be made prior to aliquoting lysate and snap-freezing.
30. This will require about 20 tubes. Though smaller volumes may be aliquoted, volumes above 0.5 mL will delay freezing.
31. Dry ice may also be used to snap-freeze. To ensure rapid freezing, place the microcentrifuge tubes on the dry ice before aliquoting the lysate.
32. The cell-free extract's shelf life is mainly determined by the stability of the recombinant enzymes. In our experience, ATP recycling activity is observed even after several years of storage at −80 °C.
33. Sucrose acts as a lyoprotectant and stabilizes the proteins in the cell-free extract during the freeze-drying process.
34. The tube lids should be left open to enable H_2O sublimation. The pierced parafilm (aluminum foil may also be used) cover minimizes loss of the lyophilized cell-free extract due to bumping.
35. A pressure controller is not necessary but the vacuum should be low enough to ensure H_2O sublimation (below triple point). Using the described setup and with good system maintenance, vacuum <0.1 mbar is typically achieved.

36. We have limited data on the long-term stability of the ATP recycling enzymes in the lyophilized cell-free extract, however, we have observed significant improvement of the stability of some recombinant enzymes in lyophilized cell-free extract stored in the fridge.
37. For lyophilized cell-free extract, reconstitute with H_2O. The volume used should be half the volume of cell-free extract aliquoted into the microcentrifuge tube prior to freeze-drying (e.g., 250 μL if 500 μL aliquots were used). This is to account for the cell-free extract's dilution with the sucrose lyoprotectant solution and restores the cell-free extract's original protein concentration.
38. The number of tubes should match the number of intended time-points to be analyzed.
39. The volume of cell-free extract is indicative and may vary depending on the application. We recommend carrying out assays for a range of diluted and undiluted cell-free extract volumes.
40. When preparing the assay mixtures, keep all reagent stocks on ice except for the assay buffer and H_2O, which should be incubated at the assay temperature. Add the acetyl phosphate second last and the cell-free extract last to initiate the enzymatic reaction.
41. To quench, withdraw the sample approximately 10 s prior and mix with the SDS at the desired quench time. Take care so that the entire volume of the sample is mixed with the entire volume of SDS. Cooling the sample protects the synthesized ATP from non-enzymatic hydrolysis.
42. Centrifugation is to remove any insoluble aggregates that have formed after quenching, prior to HPLC analysis.
43. HPLC vials must suit the autosampler and accommodate injections from reduced sample volumes (100 μL). On the Agilent 1100 system, we commonly employ 2 mL Agilent vials fitted with 400 μL flat-bottom glass inserts.
44. A refrigerated autosampler will minimize non-enzymatic decomposition of ATP before analysis. Nevertheless, cold quenched samples may be left at room temperature for several hours without significant effect on ATP concentrations.
45. All adenine-containing species absorb at this wavelength and thus peaks are observed for AMP, ADP, and ATP. Retention increases with negative charge and therefore with the number of phosphate groups in each species. The retention times should be confirmed through comparison with authentic commercial standards.

46. Calibration curves are developed by linear regression analysis of HPLC peak integration data from standard solutions (0.1–10 mM range). Standard solutions are prepared using authentic commercial material. Modern HPLC software packages include user-friendly tools for straightforward automated peak integration. Care should be taken so that integration parameters are consistent for the calibration standards and reaction samples. Linear regression may be carried out using standard statistical analysis software.
47. The progress curve should demonstrate accumulation of ATP with time as evidence of endogenous kinase activity in the cell-free extract. When coupled to the activity of recombinant enzymes, the rapid accumulation and maintenance of ATP levels demonstrate cofactor recycling. Assay conditions can be adapted to suit the reaction of interest. For example, Fig. 4 demonstrates cell-free extract catalyzed synthesis and recycling of ATP from AMP, coupled to the synthesis of a product (uridine triphosphate) [6].

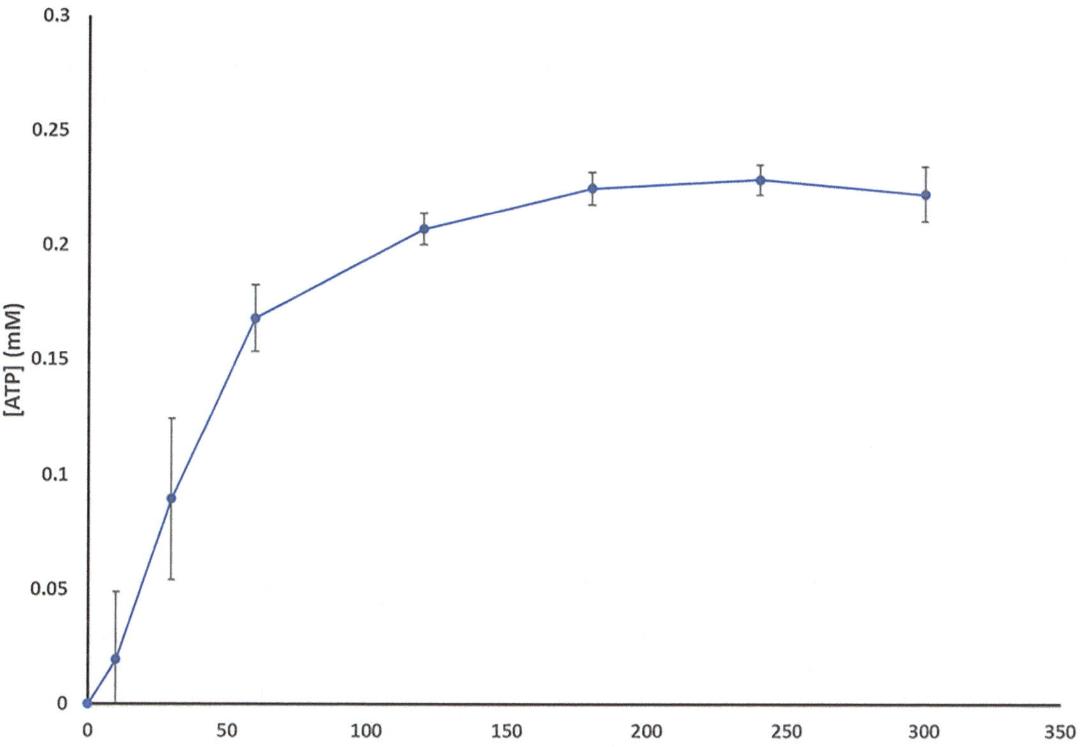

Fig. 4 Example of cell-free extract catalyzed synthesis (from AMP) and recycling of ATP coupled to ATP-dependent synthesis of uridine triphosphate from ribose and orotic acid

48. Further analysis of the ATP-dependent reaction cascade will depend on the product of interest. Citrulline analysis is presented as an example.
49. The $NaHCO_3$ and $NH_4(OH)$ in the synthesis buffer dissociate to produce CO_2 and NH_3 which spontaneously form the necessary carbamate kinase substrate NH_2COO^- (carbamate).
50. As ATP is recycled, only sub-stoichiometric amounts of cofactor are necessary. Note that the reaction may be initiated with AMP or ADP instead. In the presence of cell-free extract and acetyl phosphate, ADP is rapidly transformed to ATP, which is subsequently recycled. The initial transformation of AMP to ADP and then ATP is accompanied by a short lag (<5 min for the conditions described here) as this system relies on traces of ADP/ATP in the extract to initiate the cascade (Fig. 1a). It should also be noted that at higher cell-free extract loadings (10% v/v), catalytic ATP amounts are spontaneously formed from acetyl phosphate and endogenous components [6], and exogenous cofactor addition may be omitted.
51. In the synthesis of citrulline, acetyl phosphate is added portionwise [5] due to the promiscuity of ornithine transcarbamoylase but note that this is not typically required. For other reactions, all acetyl phosphate is added at the beginning.
52. Alternatively, citrulline analysis may be carried out by HPLC and detection of absorbance by employing post-reaction amino acid tagging with Waters AccQ-Fluor reagent, as previously reported [5].

Acknowledgments

Apostolos Alissandratos is the recipient of a CSIRO Future Science Fellowship in Synthetic Biology.

References

1. Bornscheuer UT (2018) The fourth wave of biocatalysis is approaching. Philos Trans R Soc A 376:20170063
2. Alissandratos A (2020) In vitro multi-enzymatic cascades using recombinant lysates of *E. coli*: an emerging biocatalysis platform. Biophys Rev 12:175–182
3. Alissandratos A, Caron K, Loan TD et al (2016) ATP recycling with cell lysate for enzyme-catalyzed chemical synthesis, protein expression and PCR. ACS Chem Biol 11: 3289–3293
4. Hennessy JE, Latter MJ, Fazelinejad S et al (2018) Hyperthermophilic carbamate kinase stability and anabolic in vitro activity at alkaline pH. Appl Environ Microbiol 84: e02250–e02217
5. Alissandratos A, Hartley CJ, French NG et al (2019) One-pot multienzymatic transformation of NH_3, CO_2, and ornithine into the organic nitrogen plant fertilizer citrulline using a single recombinant lysate of *E. coli*. ACS Sustain Chem Eng 7:8522–8529
6. Loan TD, Easton CJ, Alissandratos A (2019) Recombinant cell-lysate-catalysed synthesis of uridine-5′-triphosphate from nucleobase and ribose, and without addition of ATP. New Biotechnol 49:104–111

7. Loan TD, Easton CJ, Alissandratos A (2019) DNA amplification with in situ nucleoside to dNTP synthesis, using a single recombinant cell lysate of *E. coli*. Sci Rep 9:15621
8. Calhoun KA, Swartz JR (2005) An economical method for cell-free protein synthesis using glucose and nucleoside monophosphates. Biotechnol Prog 21:1146–1153
9. Caschera F, Karim AS, Gazzola G et al (2018) High-throughput optimization cycle of a cell-free ribosome assembly and protein synthesis system. ACS Synth Biol 7:2841–2853
10. Crans DC, Whitesides GM (1983) A convenient synthesis of disodium acetyl phosphate for use in insitu ATP cofactor regeneration. J Org Chem 48:3130–3132

Chapter 19

CO_2 to Methanol: A Highly Efficient Enzyme Cascade

Io Antonopoulou, Ulrika Rova, and Paul Christakopoulos

Abstract

Carbon dioxide (CO_2) has been increasingly regarded not only as a greenhouse gas but also as a valuable feedstock for carbon-based chemicals. In particular, biological approaches have drawn attention as models for the production of value-added products, as CO_2 conversion serves many natural processes. Enzymatic CO_2 reduction in vitro is a very promising route to produce fossil free and bio-based fuel alternatives, such as methanol. In this chapter, the advances in constructing competitive multi-enzymatic systems for the reduction of CO_2 to methanol are discussed. Different integrated methods are presented, aiming to address technological challenges, such as the cost effectiveness, need for material regeneration and reuse and improving product yields of the process.

Key words Methanol, Formate dehydrogenase, Formaldehyde dehydrogenase, Alcohol dehydrogenase, Enzyme cascade, Carbon dioxide, Reduction, Cofactor regeneration, Oxidoreductase, Clean fuel

1 Introduction

Climate change and resource scarcity are two of the most pressing challenges that the human kind has faced. Although many strategies have been developed for replacing fossil fuels with renewable energy, a major percentage of the global energy requirements is still met from burning fossil fuels, a process that emits greenhouse gases (GHGs) to the atmosphere [1]. Among them, carbon dioxide (CO_2) is considered the main culprit behind climate change and global warming [2]. Deeper cuts in GHG emissions have been proposed in order to meet the goal of becoming a carbon neutral society, such as use of biomass feedstock [3], electrification [4], and carbon capture utilization and storage [5]. In particular, carbon capture and utilization (CCU) has a great potential of substituting the conventional production of various chemicals by capturing CO_2 from industrial point sources or ambient air and use the captured CO_2 as alternative carbon source for chemical production.

Enzymatic conversion of CO_2 to chemicals has been intensively explored during the past decade, bringing highly attractive and

Haralambos Stamatis (ed.), *Multienzymatic Assemblies: Methods and Protocols*,
Methods in Molecular Biology, vol. 2487, https://doi.org/10.1007/978-1-0716-2269-8_19,
© The Author(s), under exclusive license to Springer Science+Business Media, LLC, part of Springer Nature 2022

advantageous characteristics in CCU, such as high selectivity, operation under mild conditions (low temperature), and avoiding use of harsh chemicals and toxic catalysts. Different enzymatic activities have been identified in vitro, acting as monoenzymes or synergistically as a cascade, for the biocatalytic conversion of CO_2 to an extended portfolio of intermediate or final products, such as bicarbonate [6–8], carbon monoxide [9, 10], formic acid/formate [11–14], formaldehyde [15–17], methane [18], and methanol [15].

Methanol is a good fuel alternative due to its high cetane number, efficient combustion, wide availability, and easy distribution. It is used as a transportation fuel in different manners: directly as fuel, as part of the biodiesel production process, blended with gasoline or converted to dimethyl ether (DME) as replacement for diesel [19]. Conventional methanol production entails the reforming of methane to synthesis gas followed by high-pressure catalytic conversion of synthesis gas to methanol [20]. Application of enzymes for converting CO_2 to methanol is one of the most promising possible routes for green and sustainable production, by avoiding use of fossil resources and harvesting anthropogenic CO_2 [21]. An overview of the potential routes for CO_2 conversion to methanol involving enzymes is presented in (Fig. 1). This chapter will focus on the application of a multi-enzyme cascade for the direct conversion of CO_2 to methanol.

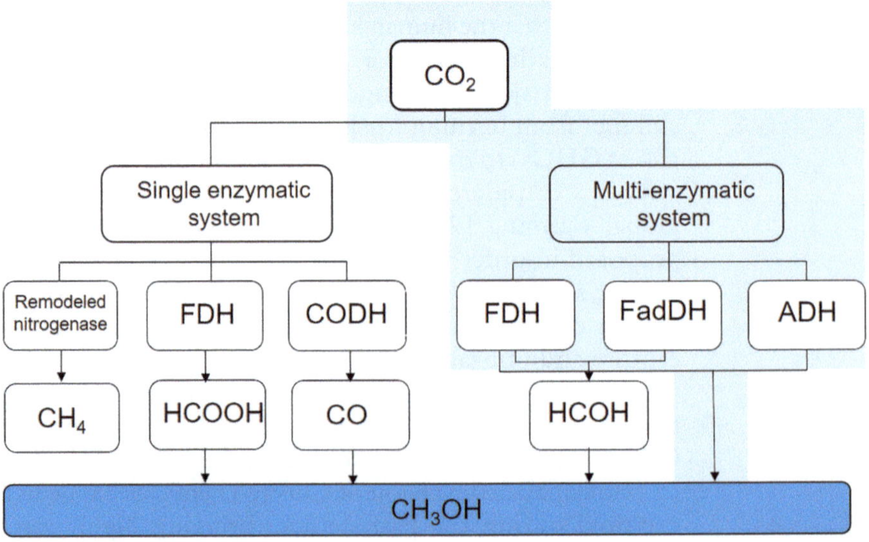

Fig. 1 Different routes for the enzymatic CO_2 reduction to methanol. *FDH*: Formate dehydrogenase; *CODH*: Monoxide dehydrogenase; *FaldDH*: Formaldehyde dehydrogenase; *ADH*: Alcohol dehydrogenase. Adapted from Sultana et al. [19]

2 Enzymes Involved in the Multi-enzymatic CO_2 Reduction to Methanol

The most prevalent enzyme cascade for the CO_2 reduction to methanol is based on three oxidoreductases. The process involves the initial reduction of CO_2 to formate by formate dehydrogenase (FDH), followed by reduction of formate to formaldehyde by formaldehyde dehydrogenase (FaldDH) and the final reduction of formaldehyde to methanol by alcohol dehydrogenase (ADH), via equimolar cofactor oxidation of NADH to NAD^+ (Fig. 2). As NADH acts as a terminal electron donor for each dehydrogenase-catalyzed reaction, three moles of NADH should be consumed to produce one mol of methanol. This reaction sequence is highly appealing, as it involves a double gain: (1) the potential utilization of CO_2 from industrial emissions battling climate change and (2) the production of a high-added value product as an alternative to fossil fuel, methanol.

2.1 Formate Dehydrogenase (FDH)

FDH (EC 1.17.1.9, formerly EC 1.2.1.2) constitutes a diverse group of enzymes found in both eukaryotes and prokaryotes and catalyzes the oxidation of formic acid with the concomitant reduction of NAD^+. FDHs are active over a wide pH range (pH 5.5–11.0) and thus they can be coupled with almost any other dehydrogenase. When FDH is exposed to CO_2 it catalyzes the reverse reaction, being the initiator enzyme in the multi-enzymatic conversion of CO_2 (Eq. 1).

$$CO_2 + NADH + H^+ \leftrightarrow HCOOH + NAD^+ \quad (1)$$

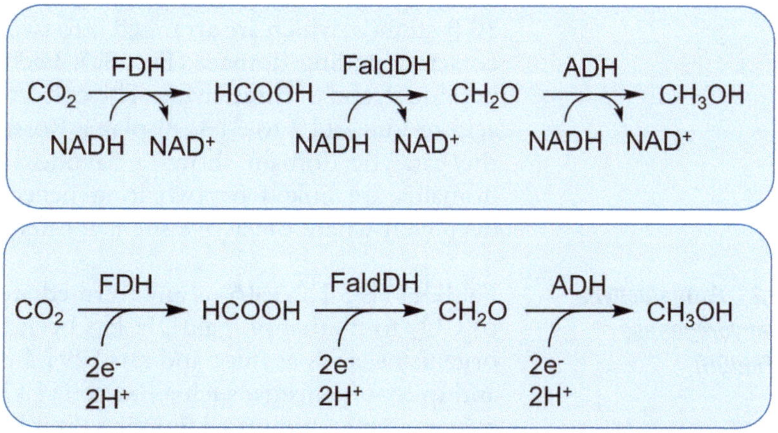

Fig. 2 Enzymatic CO_2 reduction to methanol via a three-dehydrogenase cascade comprising of formate dehydrogenase (FDH), formaldehyde dehydrogenase (FaldDH), and alcohol dehydrogenase (ADH). NADH/NAD^+: cofactor; nicotinamide adenine dinucleotide

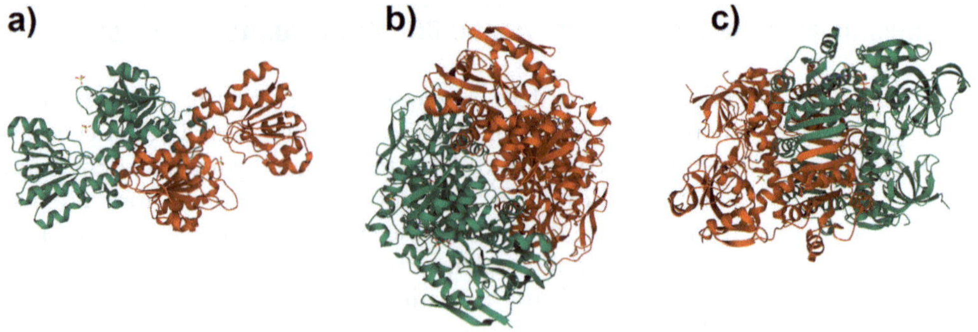

Fig. 3 Structure of (**a**) FDH from *C. boidinii* (Two homodimers; PDB: 5DNA) (**b**) FaldDH from *P. putida* (Homotetramer; PDB: 1KOL) and (**c**) ADH from *S. cerevisiae* (Homotetramer; PDB: 4W6Z)

FDHs are divided into two distinct classes, having distinctly different structure and catalytic mechanisms: to metal-independent and metal-dependent FDHs. This categorization is based on the metal content (e.g., molybdenum, tungsten) and thus the related mechanism of the reaction in the active site [22]. In contrast to most FDHs from aerobic bacteria, FDHs from anaerobes contain metal cofactors [23]. Another categorization divides FDHs to NAD(H)-dependent and NAD(H)-independent, however this division is less distinct as NADH-independent FDHs may still utilize NADH [24]. Metal-independent FDH, which are NAD (H)-dependent, belongs to the super family of D-specific-2-hydroxy acid dehydrogenases [25].

FDH from the yeast *Candida boidinii* is an oxygen-stable and a metal-independent and NAD(H)-dependent enzyme that has been predominantly used for the conversion of CO_2 to formate [26]. As a NAD(H)-dependent FDH, it forms homodimers. Each subunit has a molecular weight of 40 kDa and contains 15 α-helices and 13 β-strands, which are arranged into two domains: a catalytic and a cofactor-binding domain (Fig. 3a). Each domain contains a polypeptide strand of three layers. The cofactor-binding domain (amino acid residues 119 to 313) display a Rossmann fold structure while the catalytic domain shows a flavodoxin-like topology. The two domains are linked by two long helices and are separated by a deep cleft where the active site is located [27, 28].

2.2 Formaldehyde Dehydrogenase (FaldDH)

FaldDH (EC 1.2.1.46) is an intermediate enzyme in the reduction of CO_2 to methanol. FaldDH has been found in a wide variety of organisms, such as mice and rats [29], humans [30], bovines [31], and microorganisms, such as bacteria [32, 33] and yeast [34]. The enzyme has a ubiquitous function, as it has been described to be an important enzyme for detoxification in metabolism catalyzing the oxidation of formaldehyde, which is a highly toxic compound [35]. Almost all FaldDHs require NAD^+ and/or glutathione to activate the reaction. Under availability of formate, the enzyme catalyzes the reverse reaction, reducing formate to formaldehyde via oxidation of the cofactor. FaldDH from bacterium *Pseudomonas putida* is a NAD(H)-dependent and structurally distinguished enzyme that can catalyze the reaction without presence of glutathione [36] (Eq. 2).

$$HCOOH + NADH + H^+ \leftrightarrow HCOH + NAD^+ + H_2O \quad (2)$$

FaldDH from *P. putida* has been predominantly used in enzyme cascades for CO_2 reduction to methanol. Its molecular mass is approximately 42 kDa. In its active form it forms an homotetramer, where each subunit contains two zinc ions and two domains, the catalytic and the cofactor-binding domain. The two domains are separated by a cleft containing a deep pocket, and one of the zinc ions is located at the bottom of the cleft [37]. The catalytic domain is larger and contains 231 amino acid residues while the smaller cofactor-binding domain contains 167 out of the total 398 residues. The secondary structure of the cofactor-binding domain consists of a six stranded β-sheet surrounded by 4 α-helices which shows a Rossmann fold comprised of a $(\beta\alpha\beta\alpha\beta)_2$ unit (Fig. 3b).

2.3 Alcohol Dehydrogenase (ADH)

ADH (EC 1.1.1.1) is the final enzyme in the proposed three-dehydrogenase CO_2-reducing cascade. There are four different ADH families identified in vertebrates, plants, yeast, and bacteria [38] . In humans and many other animals, ADH serves to break down alcohols that are otherwise toxic, and participates in the generation of useful aldehyde, ketone, or alcohol groups during biosynthesis of various metabolites. In yeast, plants, and many bacteria, some ADHs catalyze the opposite reaction as part of fermentation to ensure a constant supply of NAD^+ [39] (Eq. 3).

$$HCOH + NADH + H^+ \leftrightarrow CH_3OH + NAD^+ \quad (3)$$

Yeast ADH is a zinc-containing medium chain dehydrogenase and a member of the third family [40]. It is one of the first enzymes to be ever isolated and purified and is found into three isoenzyme forms (I, II, and III) found in *Saccharomyces cerevisiae* [41]. ADH-1 is generally expressed during anaerobic fermentation, while ADH-2 is another cytoplasmic form repressed by glucose and ADH-3 is found in the mitochondria [42]. ADH-1 plays a fundamental role in the growth of *S. cerevisiae*. ADH-1 is a homo tetramer where each subunit consists of a single polypeptide chain with 347 amino acid residues and with a molecular mass of 36 kDa [43]. The subunit is divided into two domains: the catalytic domain containing a zinc ion and the coenzyme-binding domain (Fig. 3c). Yeast ADH is a very important enzyme that is exclusively used in the development of CO_2 reducing multi-enzymatic systems.

3 Immobilization as a Strategy to Develop Robust Multi-enzymatic CO_2 Reducing Systems

A first priority for developing enzyme cascades for CO_2 reduction is the establishment of robust multi-enzymatic systems. To achieve this, immobilization methods were applied in order to enhance enzyme stability and reaction yield but also benefit the economics of the process by enabling reuse of the biocatalyst and product recovery [44]. Immobilization could also aid in the efficient communication and synergism between enzymes and cofactors, which is major requirement for the potential implementation of such processes. Immobilization approaches for multi-enzyme conversion of CO_2 to methanol have been focused on co-immobilization or sequential immobilization, often coupling the technique of encapsulation with other methods.

3.1 Co-immobilization

Obert and Dave [45] reported for the first time an exclusively enzymatic system, where CO_2 gas was bubbled into a solution containing NADH and three co-encapsulated enzymes, FDH, Fald, and ADH, for the formation of methanol. Co-immobilization of the three dehydrogenases in a silica sol gel proved to enhance the enzyme stability and the probability of reaction, due to the local concentration of substrates within the nanopores of the matrix. Indeed, CO_2 reduction employing co-encapsulated enzymes resulted in a four-fold increased methanol yield requiring half concentration of NADH as cofactor. Potential limitations, such as the instability of many hydrogenases in oxygen environment and enzyme inactivation due to vigorous stirring, were overcome by immobilization.

Xu et al. [46] suggested another immobilization method, where the three dehydrogenases were encapsulated in a hybrid alginate-silica gel. The hybrid gel was prepared through in situ growth of the silica precursor in an alginate-silica solution, followed by Ca^{2+} crosslinking. It was demonstrated that methanol yield was superior (98.1%) using the hybrid alginate-silica gel compared alginate as enzyme support (70%). Moreover, it was proved that the stable cross-linked support could aid in diminishing enzyme leakage and increase storage and reuse stability of the immobilized enzymes.

Sun et al. [47] proposed a green and efficient way to encapsulate the three dehydrogenases in protamine-templated titania (TiO_2) particles, via a simple and mild biomimetic mineralization process. Compared to the free enzyme system, higher reaction yield in a wider pH and temperature range was obtained by the closed-style co-immobilization multi-enzyme system. In particular, the methanol yield was approximately six-fold higher than the free system, requiring half concentration of NADH. The immobilized enzymes could maintain >90% of their initial activity after 21 days of storage and approximately 90% of the initial methanol yield after 8 cycles of reaction.

Wang et al. [48] proposed a bioinspired approach for the immobilization of the three dehydrogenases coupling two platform technologies: biomimetic mineralization and bioadhesion. The microcapsules were prepared via the deposition of catechol-modified gelatin (GelC) following by in situ growth of silica nanoparticles on the GelC layer. Immobilization included three sequential steps: FDH-containing $CaCO_3$ templates were synthesized via co-precipitation and were coated with a GelC layer via entrapment, FaldDH was covalently attached on GelC via Michaelis addition and Schiff base reactions, and ADH was entrapped accompanying the formation of silica particles due to GelC-induced hydrolysis and condensation of silicate. After removal of the $CaCO_3$ templates, the GelC-Si based multi-enzyme system was applied in CO_2 conversion to methanol, displaying higher methanol yield and selectivity (71.6% and 86.7%) than the free system (35.5% and 47.3%).

Although co-immobilization by encapsulation provides high stability and improved product yields, potential limitations might arise due to the unnecessary collision of enzymes causing a loss of their full potential [19]. In principle, co-immobilization could accomplish catalysis in a highly efficient manner, as the reaction intermediates are consumed in situ by the enzymes, causing low product inhibition. However, enzyme co-encapsulation within only one support may suffer from inherent drawbacks, such as the undesired interaction between different enzymes, lowering the catalytic activity of certain enzymes, and the potential manipulation of immobilization process or catalytic property of an individual enzyme, by integrating different components in the reaction mixture. Therefore, an integrated, spatially separated system could aid in allowing sequential reactions to happen under well-defined reaction zones.

3.2 Sequential Immobilization

Jiang et al. [49] proposed a novel kind of capsules-in-bead scaffold in order to build a spatially separated multi-enzyme cascade for CO_2 reduction. Each enzyme was encapsulated in a guest microcapsule based on protamine-silica by coupling two approaches: biomimetic mineralization and layer-by-layer assembly. Compared to the co-encapsulation of the three dehydrogenases in a single bead gel, the novel approach offered a four-fold increase in methanol production, and substantially improved the activity, recyclability, and storage stability of the enzymes.

Luo et al. [50] sequentially immobilized the three dehydrogenases in a polymeric membrane by the fouling formation technique. The membrane consisted of regenerated cellulose skin layer and a polypropylene woven support. However, when CO_2 was bubbled through, methanol yields were similar to the respective free enzymatic and co-immobilized enzymatic system. According to the authors, two main reasons could lead to this unexpected result under these conditions. One reason could be that the immediate consumption of the intermediate products by the subsequent enzymes results in a constant low substrate concentration and the trade-off between addressing product inhibition and low substrate concentration for the adjacent enzyme makes co-immobilization meaningless. Another reason could be that the pressure-driven mass transfer overcomes the diffusion barrier and the intermediates rapidly reach the targeted enzymes in the sequentially immobilized system. Nevertheless, sequential immobilization holds a great potential for cascade reactions, since the operational conditions of each single step can be individually optimized exploiting the optimal operating conditions of each enzyme, not only during enzyme immobilization but also during the CO_2 reduction reaction. Moreover, assembly on a membrane could overcome the diffusion resistance between the enzymes, due to the pressure-driven mass transfer during the flow through mode. A summary of the reported studies is presented in Table 1.

4 Addressing the Need for Cofactor Regeneration

Regeneration and reutilization of cofactors remain the biggest challenge for the industrial application of cofactor-dependent redox reactions [51]. During the past years, different approaches have been developed for integrating cofactor regeneration with robust multi-enzyme systems for enzymatic CO_2 reduction, in order to increase the methanol/cofactor ratio and provide an economically feasible scenario. Reduction of the enzymatically produced NAD^+ to NADH can be done applying a diverse set of approaches: (1) by conventional catalysis using chemicals that act as reducing agents or using transition metal complexes, coupled with various reducing agents including H_2, (2) by enzyme catalysis using an oxidoreductase, (3) by photocatalysis using semiconductors or metal systems that may use light as a source of energy, (4) by electrocatalysis that may reduce NAD^+ under electrochemical conditions. The following paragraph will focus on the integration of multi-enzymatic CO_2 reduction with green and sustainable approaches for cofactor regeneration, employing enzymes, light and/or electricity.

5 Integrated Approaches for CO_2 Reduction to Methanol with In Situ Cofactor Regeneration

5.1 Enzyme Catalysis

Cofactor regeneration by enzyme catalysis has been proposed by incorporating an additional oxidoreductase to the three-dehydrogenase cascade, which reduces the enzymatically produced

Table 1
Reduction of CO_2 to methanol via enzyme catalysis

Enzyme	Enzyme source	Form	Cofactor	Reaction conditions	Maximum methanol yield/ concentration	Residual activity of immobilized enzymes: After storage	Residual activity of immobilized enzymes: After reuse	Ref.
FDH FaldDH ADH	n.m.	Free in solution	25–100 mM NADH	50 mM phosphate pH 7.0, medium bubbled with CO_2 during reaction, 2 mL, 3 h	21% (at 50 mM NADH) 5.6 mM (at 100 mM NADH)	n.a.	n.a.	[45]
		Co-encapsulated in alginate-silica hybrid gel			91.2% (at 25 mM NADH) 14.6 mM (at 50 mM NADH)			
FDH FaldDH ADH	*Candida boidinii* *Pseudomonas* sp. Yeast	Free in solution Co-encapsulated in alginate-silica hybrid gel	940 μM NADH	50 mM Tris–HCl pH 7.0, medium bubbled with CO_2 during reaction, 2 mL, 8 h, 37 °C	98.8% 98.1%	76.2% (60 days)	78.5% (10 cycles)	[46]
FDH FaldDH ADH	*Candida boidinii* *Pseudomonas* sp. Yeast	Free in solution	25–100 mM NADH	50 mM Tris–HCl pH 7.0, medium presaturated with CO_2 for 1 h, 2 mL, 8 h, 37 °C	~60% (at 25 mM NADH) ~12.5 mM (at 100 mM NADH)	>90% (21 days)	~90% (8 cycles)	[47]
		Co-encapsulated in titania particles			~10% (at 50 mM NADH) ~2.5 mM (at 100 mM NADH)			

(continued)

Table 1
(continued)

Enzyme	Enzyme source	Form	Cofactor	Reaction conditions	Maximum methanol yield/ concentration	Residual activity of immobilized enzymes: After storage	Residual activity of immobilized enzymes: After reuse	Ref.
FDH FaldDH ADH	*Candida boidinii* *Pseudomonas* sp. Yeast	Free in solution Co-encapsulated in catechol-modified gelatin	50 mM NADH	50 mM PBS pH 7.0, medium presaturated with CO_2 for 0.5 h, 2 mL, 6 h, 37 °C, 0.3 MPa	35.5% 71.6%	~60% (20 days)	52.6% (9 cycles)	[48]
FDH FaldDH ADH	*Candida boidinii* *Pseudomonas* sp. Yeast	Free in solution Co-encapsulated in protamine-silica gel beads	25 100 mM NADH	Medium bubbled with CO_2 during reaction, 2 mL, 8 h, 0.5 MPa	4 μmol (at 200 μmol NADH) 15 μmol (at 200 μmol NADH)	~70% (24 days)	~80% (8 cycles)	[49]
FDH FaldDH ADH	*Candida boidinii* *Pseudomonas sp.* *Saccharomyces cerevisiae*	Free in solution Co-immobilized on flat-sheet polymeric membranes Sequentially immobilized on stacked membranes	5–50 mM NADH	100 mM Tris–HCl, medium presaturated with CO_2 for 0.5 h, 4 mL, 0.5 h	0.45 mM (at 50 mM NADH) 0.50 mM (at 50 mM NADH) 0.7 mM (at 50 mM NADH)	n.a.	~100% (10 cycles)	[50]

FDH: Formate dehydrogenase; FaldDH: Formaldehyde dehydrogenase; ADH: Alcohol dehydrogenase; NADH: Reduced form of nicotinamide adenine dinucleotide; n.a: not assessed; n.m: not mentioned. Methanol yield: based on amount of cofactor (NADH)

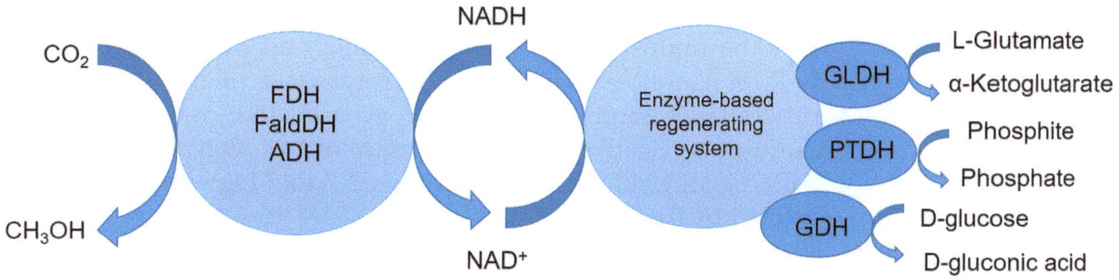

Fig. 4 CO_2 reduction with NADH regeneration via enzyme catalysis. *GLDH*: L-glutamate dehydrogenase; *PTDH*: phosphite dehydrogenase; *GDH*: glucose dehydrogenase; *NADH/NAD*$^+$: cofactor; nicotinamide adenine dinucleotide. The substrate of each regenerative oxidoreductase serves as electron donor

NAD^+ to NADH by oxidation of a substrate that in turn acts as electron donor (Fig. 4). The first integrated approach coupling enzymatic CO_2 reduction to methanol with enzymatic NADH regeneration was proposed by El-Zahab et al. [52]. In this system, glutamate dehydrogenase (GLDH; EC 1.4.1.2) was added to the three-dehydrogenase cascade FDH, FaldDH, and ADH, catalyzing the reduction of NAD^+ to NADH by oxidation of L-glutamate to α-ketoglutarate. The four enzymes were co-immobilized onto polystyrene particles, while NADH was attached separately to the particles in the same way as in enzyme immobilization. By testing different combinations of different forms of enzymes and cofactor (immobilized and free), it was concluded that reactions with free cofactor and immobilized enzymes produced up to approximately 5 times more methanol than other combinations, reaching a methanol yield of approximately 50% at 50 μM NADH. Interestingly, it appeared that immobilized enzymes were more active than free enzymes when free cofactor was used. Although immobilized enzymes are expected to show reduced rates due to potential enzyme conformation changes and substrate mass transfer resistance, this might not be the case for cascade reactions. As co-immobilization concentrates different enzymes to close vicinity, intermediate products may not have to travel distances to reach the active site of the enzymes, as compared to a free enzyme system.

Ji et al. [53] used the same regeneration approach by co-encapsulating the three dehydrogenases and GLDH in situ inside the lumen of hollow nanofibers and incorporating them in the core-phase solution in co-axial electrospinning, in which a cationic polyelectrolyte was dissolved. The polyelectrolyte, penetrating the shell of the hollow nanofibers, could enable efficient tethering and retention of NADH inside the lumen via ion-exchange interactions. Although the reaction was successful, the system could face limitations by introducing extra mass transfer resistance. Thus, another approach was developed by the same team towards constructing the multi-enzyme complex on the outer surface of the hollow nanofibers, assembling a negative charged polyelectrolyte poly(styrene sulfonate) (PSS) [51]. It was shown that the multi-enzyme system assembled on the outer surface of hollow nanofibers offered a methanol yield of 64.70%. On the other hand, sandwiched dehydrogenases and cofactor between double polyelectrolytes on the surface of hollow nanofiber offered a methanol yield 98.27%, which was 2.7 fold higher than the yield offered by free enzymes [51].

Marques Netto et al. [54] immobilized each one of the four enzymes (FDH, FaldDH, ADH, and GLDH) on functionalized magnetite nanoparticles and applied two different methodologies for methanol production: sequential reaction and batch multi-enzymatic reaction. Sequential reaction had an overall methanol yield of only 2.3%, while the multi-enzymatic cascade in a batch pressurized reactor yielded 5.2 nmol of methanol which corresponds to 147% based on NADH consumption. Zhu et al. [55] embedded the four enzymes on ZIF-8 metal-organic network nanoparticles, that were subsequently placed in a membrane in an ordered or disordered fashion. It was revealed that ordered enzymes/ZIF-8 resulted in a production of 5.8 μmol methanol in 6 h, while introduction of NADH in the solution increased the production two-fold, to 12.6 μmol. Immobilization of NADH on the membrane increased further the production to approximately 15 μmol at only 4 h. Using a pervaporation membrane for dead-end filtration, methanol could be timely separated from the alcohol/water mixture, which could efficiently control the reaction direction and enhance catalytic efficiency.

Cazelles et al. [56] demonstrated that NADH regeneration via phosphite dehydrogenase (PTDH; EC 1.20.1.1) could be integrated with the three-dehydrogenase cascade, by catalyzing the reduction of NAD^+ during oxidation of phosphite to phosphate. Although three different NAD^+ regenerating systems were initially evaluated; PTDH, glycerol dehydrogenase, and a natural photosystem extracted from spinach leaves, the first system was proven to be more efficient at neutral pH, and thus was selected for integration to multi-enzymatic CO_2 reduction. The four-enzyme system was encapsulated in phospholipids-silica and the immobilized system showed 55 times higher activity that the free enzymes, with a productivity of 4.3 mmol per g enzyme under pressurized conditions (5 bar). Singh et al. [57] applied the same regeneration approach using free hydrogenases and employing novel FDH and FaldDH for efficient methanol production (3.28 mM).

Zhang et al. [58] proposed that NADH regeneration could be done by employing a glucose dehydrogenase (GDH; EC 1.1.1.118), catalyzing the reduction of NAD^+ during oxidation of D-glucose to gluconic acid. Initially, the three dehydrogenases were immobilized in a membrane assembled in a sandwich mode (polypropylene layer-skin layer-polypropylene support) in such way that the support layer was placed to face the feed. An extra support was positioned under the skin layer in order to act as a cushion to alleviate membrane compression and peeling of the skin layer. Immobilization of the four enzymes resulted in a methanol yield of 100% in only 120 h. The immobilized three-enzyme cascade, without GDH, offered a methanol yield of 75%. The reported approaches integrating CO_2 conversion to methanol with cofactor regeneration, employing enzymes are presented in Table 2.

5.2 Enzymatic Photocatalysis

Photoenzymatic CO_2 reduction is a hybrid approach developed to address the need for cofactor regeneration. In such system, semiconductors or metal systems that may use (solar) light as a source of energy act as photocatalysts for the effective regeneration of NADH [59]. After the photocatalyst is irradiated with light, electron-hole couples (exciton) are generated. The excited electrons are then transferred to reaction centers at the surface of the photo-materials and the conversion of NAD^+ to NADH occurs, involving the oxidation of a sacrificial donor and mediation in electron transfer (Fig. 5).

Dibenedetto et al. [60] proposed for the first time that NADH regeneration during enzyme-catalyzed CO_2 reduction could be performed either chemically by sodium dithionite (SDT) or photochemically using the semiconductor zinc-sulfide (ZnS). The photochemical reaction involved glycerol as hydride donor and required working at the border of the visible light domain. Exploring further this approach, Aresta et al. [61] proposed regeneration of NADH by using visible-light-active heterogeneous TiO_2 photocatalysts in water/bioglycerol mixtures. After irradiation with visible light, $[CrF_5(H2O)]^{2-}$@TiO_2 generated excited electrons that were then transferred and the conversion of NAD^+ to NADH occurred, involving the Rh-complex as electron mediator and the oxidation of glycerol. Integration of this approach with three co-encapsulated hydrogenases FDH, FaldDH, and ADH, in a continuous flow reactor, resulted in the production of 100 to 1000 μmol methanol from one mol of NADH. The reactor consisted of two compartments, one for selective NAD^+ regeneration to1,4-NADH and one for enzymatic CO_2-reduction, and was further coupled with in situ methanol recovery by stripping.

Yadav et al. [62] screened different graphene-based photocatalysts for NADH regeneration. Integrated of the three-dehydrogenase cascade with the covalently attached isatin-porphyrin chromophore to chemically converted graphene (CCG-IP), offered two-fold higher methanol production than with the chemically converted graphene coupled multi-anthraquinone-substituted porphyrin (CCGCMAQSP). The reaction involved selective NAD^+ regeneration to 1,4-NADH, involving use of Rh-complex as mediator and oxidation of triethanolamine (TEOA). Liu et al. [63] reported NADH

Table 2
Reduction of CO_2 to methanol via enzyme catalysis with in situ cofactor regeneration

Enzyme	Enzyme source	Enzyme form	Cofactor regeneration	Cofactor	Reaction conditions	Maximum methanol yield/ concentration	Residual activity of co-immobilized enzymes After reuse	Ref.
CA FDH FaldDH ADH GLDH	*Candida boidinii* *Pseudomonas sp.* Yeast Bovine liver *n.m.*	Free in solution Sandwiched onto the surface of hollow nanofibers Assembled on the surface of hollow nanofibers	GLDH, 10 mM L-glutamate	0.2 mM NADH	50 mM PBS pH 7.0 medium presaturated with CO_2 for 0.5 h, 3 mL, 0.3 MPa, 10 h	36.13% 98.27% 64.70%	90% (10 cycles)	[51]
FDH FaldDH ADH GLDH	*Candida boidinii* *Pseudomonas sp.* Yeast Bovine liver	Free in solution Co-Immobilized on polystyrene particles via covalent binding	GLDH, 1 mM L-glutamate	50–500 μM NADH	100 mM phosphate buffer H 7.0, medium presaturated with 8 mL/min for 30 min, 4 mL, 30 min 0.3 MPa	~20% (at 50 μM NADH) ~5 μM (200 μM NADH) 0.04 μmol/ h g enzyme ~50% (at 50 μM NADH) ~20 μM (500 μM NADH) 0.02 μmol/ h g enzyme	~80% (11 cycles)	[52]

FDH FaldDH ADH GLDH	*Candida boidinii* *Pseudomonas sp.* Yeast Bovine liver	Free in solution Co-encapsulated inside the lumen of hollow nanofibers	GLDH, 10 mM L-glutamate	0.2 mM NADH	50 mM PBS pH 7.0, medium presaturated for 0.5 h, 2 mL, 0.3 MPa	33.1% 95.3%	n.a.	[53]
CA FDH FaldDH ADH GLDH	*Candida boidinii* *Pseudomonas sp.* Yeast Bovine liver *n.m.*	Free in solution Co-encapsulated inside the lumen of hollow nanofibers	GLDH, 10 mM L-glutamate	0.2 mM NADH	50 mM PBS pH 7.0, medium presaturated with CO_2 for 0.5 h, 2 mL, 0.3 MPa, 10 h	36.7% 103.2%	80% (10 cycles)	[53]
FDH FaldDH ADH GLDH	*Candida boidinii* *Pseudomonas putida* *Saccharomyces cerevisiae* Bovine liver	Immobilized individually on super-paramagnetic nanoparticles via crosslinking	GLDH, 4 mM L-glutamate	11 mM NADH	100 mM Tris–HCl pH 7.5, medium supplied with CO_2, 25 °C, 126 psi, 30 min	5.2 mmol	n.a.	[54]
FDH FaldDH ADH GLDH	*Candida boidinii* *Pseudomonas sp.* *Saccharomyces cerevisiae* Bovine liver	Free in solution Co-ordered on metal-organic framework ZIF-8 and placed in a membrane	GLDH, 4 mM L-glutamate	– 10 mM NADH Immobilized on ZIF-8	10 mL, medium presaturated with for 30 min	5.8 µmol (6 h) 12.6 µmol (6 h) 13.4 µmol (6 h) ~15 µmol (4 h)	50% (12 h use)	[55]
FDH FaldDH ADH PTDH	*Candida boidinii* *Pseudomonas putida* *Saccharomyces cerevisiae*	Free in solution Co-encapsulated in phospholipids-silica nanocapsules	PTDH, Na_2HPO_3 40 mM	0.1 M NADH	0.05 M phosphate pH 6.5 containing $KHCO_3$ 0.05 M, medium pre-flushed with CO_2 for 65 h, room temperature, 5 bar, 3 h	0.16 mmol/g enzyme 0.15 mM 4.3 mmol/g enzyme 42 µmol/g_{NPS} 18.96 mM	n.a.	[56]

(continued)

Table 2
(continued)

Enzyme	Enzyme source	Enzyme form	Cofactor regeneration	Cofactor	Reaction conditions	Maximum methanol yield/ concentration	Residual activity of co-immobilized enzymes After reuse	Ref.
FDH FaldDH ADH PTDH	*Clostridium ljungdahlii* *Burkholderia multivorans* *Saccharomyces cerevisiae*	Free in solution	PTDH, 50 mM Na_2HPO_3	1.5 mM NADH	100 mM phosphate pH 7.0, 50 mL/min CO_2 during reaction, 2 bar, 100 rpm, EMIM-ac 1%, 30 °C 6 h	1.31 mM/h 6.75 mM	n.a.	[57]
FDH FaldDH ADH GDH	*Candida boidinii* *Pseudomonas sp.* *Saccharomyces cerevisiae* *Pseudomonas* sp.	Co-immobilized on regenerated cellulose membrane	GDH, 50 mM D-glucose	5 mM NADH	4 mL Pre-Bubbled with CO_2 for 30 min 100 mM Tris–HCl pH 6.5 with 20% IL (CH)(Glu)	100% (120 min)	~85% (6 cycles)	[58]

CA: Carbonic anhydrase; *FDH*: Formate dehydrogenase; *FaldDH*: Formaldehyde dehydrogenase; *ADH*: Alcohol dehydrogenase; *GLDH*: Glutamate dehydrogenase; *PTDH*: Phosphite dehydrogenase; *GDH*: Glucose dehydrogenase; *NADH*: Reduced form of nicotinamide adenine dinucleotide; *EMIM –ac*: 1-ethyl-3-methylimidazolium acetate; n.a: not assessed. Methanol yield: based on amount of cofactor (NADH). Cofactor regeneration reactions: *GLDH*: L-glutamate + H_2O + NAD^+ ↔ α-ketoglutarate + NH_3 + NADH + H^+; *PTDH*: phosphite + NAD^+ + H_2O ↔ phosphate + NADH + H^+; *GDH*: D-glucose + NAD^+ ↔ D-glucono-1,5-lactone + NADH + H^+

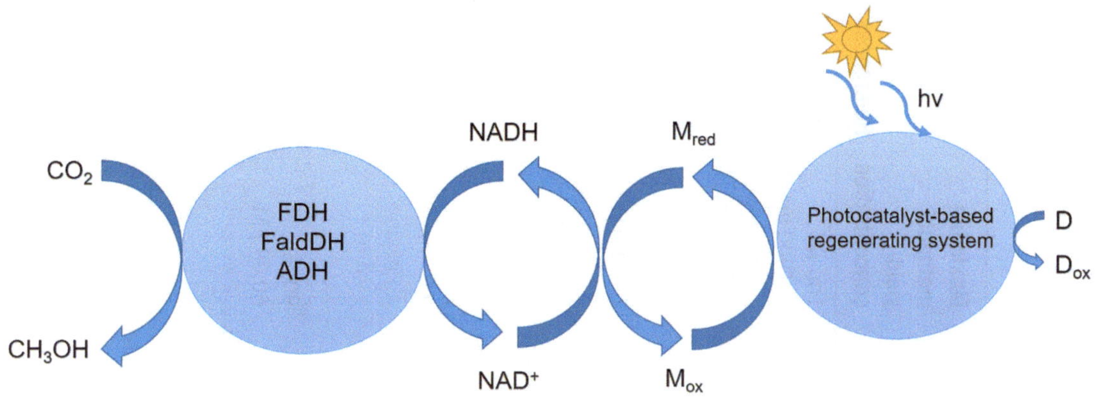

Fig. 5 CO_2 reduction with NADH regeneration via enzymatic photocatalysis. M: Mediator; D: electron donor

regeneration based on carbon nitride array (CNA) material in the presence of Rh-complex as mediator with simultaneous oxidation of TEOA. When combined with a three hydrogenase system, 0.07 mmol methanol per g enzymes were produced in only 3 h. The capability of CNA for NADH regeneration was comparable to PTDH-based regeneration, which obtained a productivity of 0.16 mmol methanol per g enzymes in 3 h [56].

Ji et al. [64] described a completely integrated artificial photosynthesis system for enhanced electronic energy-transfer efficacy, by combining co-axial electrospinning/electrospray and layer-by-layer self-assembly. In particular, the three dehydrogenases were encapsulated in situ inside lumen polyelectrolyte-doped hollow nanofibers or microcapsules, fabricated via co-axial electrospinning/electrospray. NADH regeneration was evaluated by using Eocin Y (EY) as photosensitizer and TEOA as electron donor involving Rh-complex as mediator, resulting in a regeneration yield of 84.06%. The integrated artificial photosynthesis system provided 90.6% methanol yield, where the dehydrogenases immobilized in hollow nanofibers could maintain more than 60% of their initial activity after 10 rounds of operation. A summary of the reported photoenzymatic methods for CO_2 reduction to methanol is shown in Table 3.

5.3 Enzymatic Electrocatalysis

Enzymatic electrocatalysis is an emerging platform that could play a significant role in the industrial application of cofactor-dependent redox by enabling cofactor regeneration or even eliminating its use during enzyme-catalyzed CO_2 reduction [65]. A classic enzymatic electrosynthesis cell (EEC) consists of an enzymatic biocathode and an anode, which are separated by a polymeric membrane that allows ion exchange, and an external circuit that provides power supply and causes water oxidation in the anode (Fig. 6). Water due to its oxidation acts as electron donor; the released electrons travel from the anode to the cathode and are transferred to the enzyme active

Table 3
Reduction of CO_2 to methanol via enzymatic photocatalysis

Enzyme	Enzyme source	Enzyme form	Photocatalyst/ Electron donor	Cofactor	Mediator	Reaction conditions	Maximum methanol yield/ concentration	References
FDH FaldDH ADH	*Candida boidinii* *Pseudomonas sp.* Yeast	Co-encapsulated in Ca-alginate and tetraeth oxysilanes (TEOS)	$[CrF_5(H_2O)]^{2-}$@TiO_2 /glycerol	NAD^+	Rh-complex	pH 7.0, Continuous flow reactor-two compartments	100 to 1000 µmol methanol from one mol of NADH	[61]
FDH FaldDH ADH	n.m.	Free in solution	0.5 mg CCG-IP/ 1.24 mmol TEOA 0.5 mg CCGCMAQSP/ 1.24 mmol TEOA	1.24 µmol NAD^+	0.62 µmol Rh-complex	100 mM sodium phosphate buffer pH 7.0, medium bubbled with 0.5 mL/ min CO_2, 3.1 mL	11.21 µM 5.62 µM	[62]
FDH FaldDH ADH	*Candida boidinii* *Pseudomonas sp.* Yeast	Free in solution	3 mg CNA / 15% w/v TEOA	1 mM NADH	0.125 mM Rh-complex	100 mM PBS pH 7.0, 100 mM $KHCO_3$ presaturated with CO_2, 2 mL	1.8 µmol 0.07 mmol/ g enzymes (3 h)	[63]
FDH FaldDH ADH	*Candida boidinii* *Pseudomonas sp.* Yeast	Co-encapsulated in PAH-doped hollow nanofibers or microcapsules	50 µM EY (50 µM)/ 15% w/v TEOA	1 mM NADH	0.25 mM Rh-complex	50 mM PBS pH 7.0, medium presaturated with N_2/CO_2 for 0.5 h, 10 mL, 0.3 MPa	90.6% 60.39 µM	[64]

FDH: Formate dehydrogenase; *FaldDH*: Formaldehyde dehydrogenase; *ADH*: Alcohol dehydrogenase; *CCG-IP*: Covalently attached isatin-porphyrin (IP) chromophore to chemically converted graphene (CCG); *CCGCMAQSP*: chemically converted graphene coupled multi-anthraquinone-substituted porphyrin; *EY*: Eocin Y; *TEOA*: triethanolamine; *NAD^+/NADH*: Nicotinamide adenine dinucleotide. *n.m*: not mentioned. Methanol yield: based on amount of cofactor (NADH)

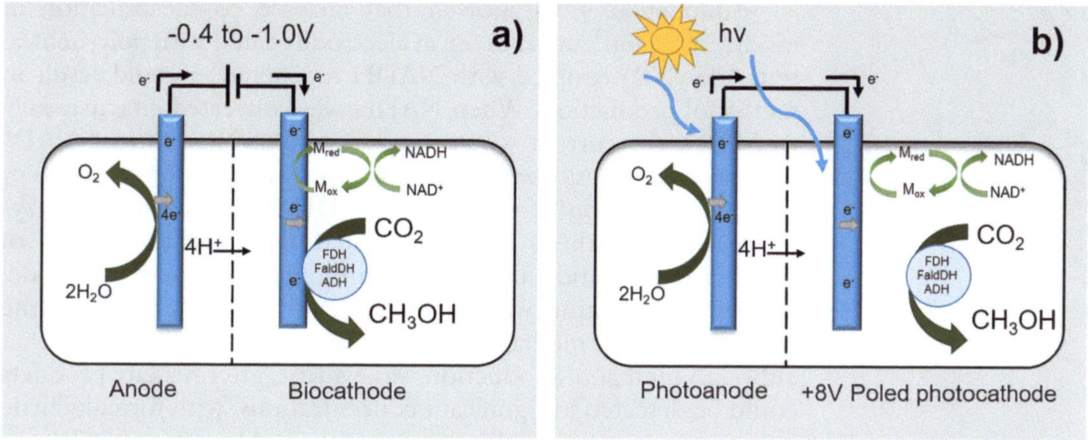

Fig. 6 CO_2 reduction in (**a**) enzymatic electrosynthesis cell (EEC) and (**b**) a photoelectrocatalytic cell (PEC). The scenarios include the use of a mediator (M) and a cofactor (NADH) regeneration. Water serves as electron donor via oxidation occurring at the anode

site, which in turns catalyzes CO_2 reduction to the desired product. The electrons from the cathode may follow a direct path to the catalytic center of the oxidoreductase (direct electron transfer) or an indirect path mediated by electron shuttles (indirect electron transfer) [66–68]. Solid electrodes act as conductors and the cathode can support enzyme and/or mediator immobilization. Since the dehydrogenases applied in CO_2 reduction are usually cofactor dependent (e.g., NADH), the cofactor may be introduced in the catholyte, where it can be continuously regenerated via the occurring redox reaction. A significant advantage of this approach is that the cofactor can be continuously regenerated with beneficial potential effects on the economics of the process.

The first enzyme cascade ever reported for the conversion of CO_2 to methanol was demonstrated in an EEC by Kuwabata et al. [69]. This approach included the use of only two enzymes: FDH from *Pseudomonas oxalaticus* for CO_2 reduction to formate and MDH from *Methylophilus methylotrophus* for the production of methanol from formate via formaldehyde. Although either methyl viologen (MV^{2+}) or pyrroloquinolinequinone (PQQ) was evaluated as mediators for the enzyme cascade, use of PQQ in place of MV^{2+} led to the exclusive production of methanol after some induction period, while formate accumulation was completely suppressed. It was remarkably demonstrated that PQQ could act as a mediator for both FDH and MDH, which was validated by the enhancement of electrolysis in the presence of PQQ during mono-enzymatic reactions. This pioneer report provided a basis for further investigation of enzyme cascades for CO_2 reduction. Although limited, CO_2 reduction to methanol via enzymatic electrocatalysis has been done thereafter by use of the classic cascade: FDH from *C. boidinii*, FaldDH from *Pseudomonas* sp., and yeast ADH.

Addo et al. [70] showed that enzyme co-encapsulation in modified Nafion® micelles on an electrode coated with poly(neutral red) (PolyNR) coupled with NADH regeneration could result in methanol production. When NADH was converted enzymatically to NAD^+, the current would increase as PolyNR reduced NAD^+ back to NADH. Although this report provided a proof of concept for the application of the FDH-FaldDH-ADH cascade in enzymatic electrosynthesis and highlighted the beneficial effects of cofactor regeneration and enzyme immobilization on the cathode, further optimization was proved necessary in order to shift the reaction towards methanol production. In fact, it was shown that although methanol production was evident, intermediate products could be detected in significant concentrations, with formaldehyde being present in the highest concentration. More recently, Zhang et al. [71] showed that enzymes co-encapsulated in ZIF-8 metal-organic network on a cathode that had been previously grafted with Rh-complex as mediator, offered a five-fold increase in methanol production compared to a free enzyme system. When NADH was added, a 12-fold increase was observed compared to the free enzyme system without NADH.

Schlager et al. [72] provided a proof of concept for the reduction of CO_2 without use of any mediator or cofactor. Enzymes were co-encapsulated in an alginate-silicate hybrid gel on the cathode, where the direct injection of electrons into the immobilized dehydrogenases delivered a methanol production of around 0.15 ppm. Although a conservative estimate of coulombic efficiency was around 10%, this value was in the same order of magnitude of efficiencies for methanol production using pyridinium as homogeneous electrocatalyst (~22%) [73]. A summary of CO_2 reduction to methanol via enzymatic electrocatalysis in presented in Table 4.

5.4 Enzymatic Photoelectrocatalysis

Kuk et al. [74] developed a tandem photoelectrochemical cell (PEC) with a three-dehydrogenase cascade, mimicking the natural photosynthetic Z-scheme by acquiring photogenerated electrons from the oxidation of water on illumination with visible light. The PEC cell comprised a cobalt phosphate/hematite (Co-Pi/α-Fe_2O_3), photoanode, and a bismuth ferrite ($BiFeO_3$) photocathode that were separated by an ion-exchange salt bridge and were connected by a conducting wire in order to prevent product mixing and side reactions. This feature is highly advantageous as it allow biocatalysts to maintain their optimal activities. The photoanode takes up electrons from water oxidation after light illumination, and the photocathode transports the electrons via an indirect path to an electron mediator, Rh-complex, and NAD^+ through an internal electric field induced by remnant polarization to regenerate 1,4-NADH (Fig. 5b). As FaldDH is, a rate-limiting

Table 4
Reduction of CO_2 to methanol via enzymatic electrocatalysis/photoelectrocatalysis

Enzyme	Enzyme source	Enzyme form	Cathode electrode	System	Cofactor	Mediator	Reaction conditions	Maximum methanol concentration/ efficiency	Ref.
FDH MDH	*Pseudomonas oxalaticus* *Methylophilus methylotrophus*	Free in solution	Glassy carbon	EEC	–	5 mM PQQ adsorbed on electrode	500 mM phosphate buffer containing 300 mM $NaHCO_3$ pH 7.0, medium presaturated with CO_2 for 1 h, −0.8 V vs. SCE, 5 mL	~0.16 mM 89% coulombic efficiency (70 h)	[69]
FDH FaldDH ADH CA FDH FaldDH ADH	*Candida boidinii* *Pseudomonas sp.* Yeast Human *Candida boidinii* *Pseudomonas sp.* Yeast	Immobilized on electrode via encapsulation in modified Nafion® micelles	Glassy carbon	EEC	1 mM NADH	PolyNR coated on electrode	10 mM phosphate buffer pH 8.0, medium presaturated with CO_2 for 200 s, −0.8 V vs. SCE	0.7 mM formic acid 1.3 mM formaldehyde 0.6 mM methanol 0.2 mM formic acid 0.9 mM formaldehyde 0.9 mM methanol	[70]
FDH FaldDH ADH	*Candida boidinii* *Pseudomonas sp.* *Yeast*	Free in solution Immobilized on electrode via encapsulation in ZIF-8 metal-organic framework	Carbon-felt	EEC	– 1 mM NADH – 1 mM NADH	Rh-complex grafted on electrode	50 mM Tris–HCl pH 7.4, medium presaturated with CO_2 for 30 min, −0.7 V vs. Ag/AgCl, 10 mL	0.061 mM (3 h) 0.320 mM (3 h) ~0.3 mM (3 h) 0.742 mM (3 h)	[71]

(continued)

Table 4 (continued)

Enzyme	Enzyme source	Enzyme form	Cathode electrode	System	Cofactor	Mediator	Reaction conditions	Maximum methanol concentration/ efficiency	Ref.
FDH FaldDH ADH	*Candida boidinii* *Pseudomonas putida* *Saccharomyces cerevisiae*	Immobilized on electrode via encapsulation in alginate–silicate hybrid gel	Carbon-felt	EEC	–	–	50 mM phosphate buffer pH 7.6, medium presaturated with CO_2 for 30 min, −1.2 V vs. Ag/AgCl, 25 mL	~ 0.15 ppm ~10% coulombic efficiency (4 h)	[72]
FDH FaldDH ADH	*Clostridium carboxi divorans* *Pseudomonas cepacia* Yeast	Free in solution	$BiFeO_3$	PEC	NADH 5 mM	Rh-complex	External voltage 0.8 V, 6 h	1.31 mM 220 μM/h 1280 μmol /g enzyme h Coulombic efficiency 31.9%	[74]

CA: Carbonic anhydrase; *FDH*: Formate dehydrogenase; *FaldDH*: Formaldehyde dehydrogenase; *ADH*: Alcohol dehydrogenase; *MDH*: Methanol dehydrogenase. *PQQ*: Pyrroloquinoline quinone; *PolyNR*: Polymerized neutral red; *NADH*: Reduced form of nicotinamide adenine dinucleotide; *EEC*: Enzymatic electrosynthesis cell; *PEC*: photoelectrocatalytic cell enzyme during CO_2 reduction to methanol, a novel FaldDH from *Pseudomonas cepacia* was identified showing a high binding affinity to formate, which significantly enhanced methanol production to 220 μM/h.

6 Addressing CO_2 Solubilization as a Limiting Factor

An identified factor limiting yields in enzymatic CO_2 reduction is the low CO_2 solubility in water (33 mM) [75]. Two main approaches have been applied to address CO_2 solubility in multi-enzymatic methanol production, although in a very limited extent: use of enzyme or co-solvent.

6.1 Use of Carbonic Anhydrase

Integrating carbonic anhydrase (CA; EC 4.2.1.1) in multi-enzyme cascades for CO_2 reduction has been shown to be a promising strategy for enhancing substrate solubilization and thus promoting substrate availability for the CO_2-converting enzyme, shifting the reaction direction towards the desired product [76]. CA catalyzes the reversible hydration of CO_2 to bicarbonate ($CO_2 + H_2O \leftrightarrow HCO_3^- + H^+$) and is one of the fastest enzymes in nature with turnover number up to 10^7/s [77–80]. In a three-dehydrogenase cascade coupled with GLDH-catalyzed cofactor regeneration, addition of CA resulted in an increase in methanol yield from 95.4% to 103.2% for hollow nanofiber-supported enzymes, and from 33.1% to 36.2% for free enzymes [53]. In an enzymatic electrocatalysis concept, addition of CA in the three-dehydrogenase cascade accelerated significantly the process and further improved product selectivity [70]. In particular, a 50% increase in methanol production and a significant decrease in intermediate products; 70% for formic acid and 30% for formaldehyde, was observed.

6.2 Use of Co-solvents

Different ionic liquids have been screened for improving the methanol yield in multi-enzyme cascade for CO_2 reduction [58]. In particular, addition of ionic liquid (CH)(Glu) as co-solvent significantly enhanced the methanol yield (75%). Comparatively, performing the reaction in buffer, allowed a methanol yield of only 20–30%. The reason why methanol yields were improved in the presence of the ionic liquid were investigated. It was revealed that the slightly acidic pH, offered by the ionic liquid addition, favored the reaction. Most importantly, it was shown by molecular dynamics simulations that FDH acquired such conformation in the presence of (CH)(Glu) that CO_2 stayed for longer time in the vicinity of its active site. 1-ethyl-3-methylimidazolium acetate (EMIM-ac) has been also incorporated to multi-enzymatic cascade as co-solvent, offering a 2.4-fold increase in methanol production

(7.86 mM) compared to using only enzymes (3.28 mM) [57]. It was observed that the addition of co-solvent increased CO_2 solubilization, as CO_2 interacted with the electronegative nitrogen and oxygen atoms of the green EMIM-ac solvent.

7 Conclusions

The development of clean technologies for the sustainable production of alternative fuels is anticipated. Dehydrogenases have a key role in the road to a sustainable bio-economy, where anthropogenic CO_2 is used as raw material and is converted to high-added value products. Existing industrial methods involving chemical catalysts often lack in selectivity and yield of the obtained products, requiring high temperature and/or pressure. Direct application of biocatalysts can offer mild operational conditions and become more and more industrially attractive. For this purpose, novel pathways to tackle limitations, such as the cost effectiveness, the need for regeneration and reuse of materials are required. Enzyme cascades can offer significant advantages into the process of clean fuel production, such as environmental protection, high efficiency, and no need for intermediate purification.

References

1. Kober T, Schiffer HW, Densing M, Panos E (2020) Global energy perspectives to 2060 – WEC's World Energy Scenarios 2019. Energy Strategy Rev. 31(December 2019):100523. https://doi.org/10.1016/j.esr.2020.100523
2. Letcher TM (2021) Global warming—a complex situation. Clim Chang:3–17. https://doi.org/10.1016/b978-0-12-821575-3.00001-3
3. Hiloidhari M, Baruah DC, Kumari M, Kumari S, Thakur IS (2019) Prospect and potential of biomass power to mitigate climate change: a case study in India. J Clean Prod 220:931–944. https://doi.org/10.1016/j.jclepro.2019.02.194
4. Schiffer ZJ, Manthiram K (2017) Electrification and decarbonization of the chemical industry. Joule 1(1):10–14. https://doi.org/10.1016/j.joule.2017.07.008
5. Zhang Z et al (2020) Advances in carbon capture, utilization and storage. Appl Energy 278:115627. https://doi.org/10.1016/j.apenergy.2020.115627
6. Bond GM, Stringer J, Brandvold DK, Simsek FA, Medina MG, Egeland G (2000) Development of integrated system for biomimetic CO_2 sequestration using the enzyme carbonic anhydrase. ACS Div Fuel Chem Prepr 45(4):713–717
7. Sivanesan D et al (2017) Enhanced CO_2 absorption and desorption in a tertiary amine medium with a carbonic anhydrase mimic. J Ind Eng Chem 52:287–294. https://doi.org/10.1016/j.jiec.2017.03.058
8. Sjöblom M et al (2020) Enzyme-assisted CO_2 absorption in aqueous amino acid ionic liquid amine blends. ACS Sustain Chem Eng 8(36):13672–13682. https://doi.org/10.1021/acssuschemeng.0c03497
9. Shin W, Lee SH, Shin JW, Lee SP, Kim Y (2003) Highly selective electrocatalytic conversion of CO_2 to CO at −0.57 V (NHE) by carbon monoxide dehydrogenase from moorella thermoacetica. J Am Chem Soc 125(48):14688–14689. https://doi.org/10.1021/ja037370i
10. Woolerton TW, Sheard S, Reisner E, Pierce E, Ragsdale SW, Armstrong FA (2010) Efficient and clean photoreduction of CO_2 to CO by enzyme-modified TiO2 nanoparticles using visible light. J Am Chem Soc 132(7):2132–2133. https://doi.org/10.1021/ja910091z

11. Yu S et al (2020) Light-driven enzymatic nanosystem for highly selective production of formic acid from CO_2. Chem Eng J (August):127649. https://doi.org/10.1016/j.cej.2020.127649
12. Gu FJ, Wang YZ, Meng ZH, Liu WF, Qiu LY (2020) A coupled photocatalytic/enzymatic system for sustainable conversion of CO_2 to formate. Catal Commun 136(December 2019):105903. https://doi.org/10.1016/j.catcom.2019.105903
13. Ren S et al (2020) Co-immobilization multienzyme nanoreactor with co-factor regeneration for conversion of CO_2. Int J Biol Macromol 155:110–118. https://doi.org/10.1016/j.ijbiomac.2020.03.177
14. Li Y, Wen L, Tan T, Lv Y (2019) Sequential co-immobilization of enzymes in metal-organic frameworks for efficient biocatalytic conversion of adsorbed CO_2 to formate. Front Bioeng Biotechnol 7(December):1–13. https://doi.org/10.3389/fbioe.2019.00394
15. Shi J, Wang X, Jiang Z, Liang Y, Zhu Y, Zhang C (2012) Constructing spatially separated multienzyme system through bioadhesion-assisted bio-inspired mineralization for efficient carbon dioxide conversion. Bioresour Technol 118: 359–366. https://doi.org/10.1016/j.biortech.2012.04.099
16. Liu W, Hou Y, Hou B, Zhao Z (2014) Enzyme-catalyzed sequential reduction of carbon dioxide to formaldehyde. Chin J Chem Eng 22(11):1328–1332. https://doi.org/10.1016/j.cjche.2014.09.026
17. Nabavi Zadeh PS, Gomes MZDV, Åkerman B, Palmqvist AEC (2018) Förster resonance energy transfer study of the improved biocatalytic conversion of CO_2 to formaldehyde by coimmobilization of enzymes in siliceous mesostructured cellular foams. ACS Catal 8(8):7251–7260. https://doi.org/10.1021/acscatal.8b01806
18. Yang ZY, Moure VR, Dean DR, Seefeldt LC (2012) Carbon dioxide reduction to methane and coupling with acetylene to form propylene catalyzed by remodeled nitrogenase. Proc Natl Acad Sci U S A 109(48):19644–19648. https://doi.org/10.1073/pnas.1213159109
19. Sultana S, Chandra Sahoo P, Martha S, Parida K (2016) A review of harvesting clean fuels from enzymatic CO_2 reduction. RSC Adv 6(50):44170–44194. https://doi.org/10.1039/c6ra05472b
20. Dalena F, Senatore A, Marino A, Gordano A, Basile M, Basile A (2018) Methanol production and applications: an overview. Elsevier B. V., Amsterdam
21. Zhang N, Long R, Gao C, Xiong Y (2018) Recent progress on advanced design for photoelectrochemical reduction of CO_2 to fuels. Sci China Mater 61(6):771–805. https://doi.org/10.1007/s40843-017-9151-y
22. Maia LB, Moura I, Moura JJG (2017) Molybdenum and tungsten-containing formate dehydrogenases: aiming to inspire a catalyst for carbon dioxide utilization. Inorganica Chim Acta 455:350–363. https://doi.org/10.1016/j.ica.2016.07.010
23. Fauque GD, Barton LL (2012) Hemoproteins in dissimilatory sulfate- and sulfur-reducing prokaryotes, vol 60, 1st edn. Elsevier, Amsterdam
24. Choe H et al (2014) Efficient CO_2-reducing activity of NAD-dependent formate dehydrogenase from Thiobacillus sp. KNK65MA for formate production from CO 2 gas. PLoS One 9(7):14–16. https://doi.org/10.1371/journal.pone.0103111
25. Nielsen CF, Lange L, Meyer AS (2019) Classification and enzyme kinetics of formate dehydrogenases for biomanufacturing via CO_2 utilization. Biotechnol Adv 37(7):107408. https://doi.org/10.1016/j.biotechadv.2019.06.007
26. Schütte H, Flossdorf J, Sahm H, Kula M-R (1976) Purification and properties of formaldehyde dehydrogenase and formate dehydrogenase from candida boidinii. Eur J Biochem 62(1):151–160. https://doi.org/10.1111/j.1432-1033.1976.tb10108.x
27. Schirwitz K, Schmidt A, Lamzin VS (2007) High-resolution structures of formate dehydrogenase from Candida boidinii. Protein Sci 16(6):1146–1156. https://doi.org/10.1110/ps.062741707
28. Guo Q et al (2016) Structural and kinetic studies of formate dehydrogenase from Candida boidinii. Biochemistry 55(19):2760–2771. https://doi.org/10.1021/acs.biochem.6b00181
29. Harris C, Wang SW, Lauchu JJ, Hansen JM (2003) Methanol metabolism and embryotoxicity in rat and mouse conceptuses: comparisons of alcohol dehydrogenase (ADH1), formaldehyde dehydrogenase (ADH3), and catalase. Reprod Toxicol 17(3):349–357. https://doi.org/10.1016/S0890-6238(03)00013-3
30. Yang ZN, Bosron WF, Hurley TD (1997) Structure of human χχ alcohol dehydrogenase: a glutathione-dependent formaldehyde dehydrogenase. J Mol Biol 265(3):330–343. https://doi.org/10.1006/jmbi.1996.0731
31. Pourmotabbed T, Creighton DJ (1986) Substrate specificity of bovine liver formaldehyde

dehydrogenase. J Biol Chem 261(30): 14240–14244. https://doi.org/10.1016/s0021-9258(18)67010-3
32. Marx CJ, Miller JA, Chistoserdova L, Lidstrom ME (2004) Multiple formaldehyde oxidation/detoxification pathways in burkholderia fungorum LB400. J Bacteriol 186(7): 2173–2178. https://doi.org/10.1128/JB.186.7.2173-2178.2004
33. Gutheil WG, Kasimoglu E, Nicholson PC (1997) Induction of glutathione-dependent formaldehyde dehydrogenase activity in Escherichia coli and Hemophilus influenza. Biochem Biophys Res Commun 238(3): 693–696. https://doi.org/10.1016/S0006-291X(00)90000-7
34. Kato N, Sahm H, Wagner F (1979) Steady-state kinetics of formaldehyde dehydrogenase and formate dehydrogenase from a methanol-utilizing yeast, Candida boidinii. BBA Enzymol 566(1):12–20. https://doi.org/10.1016/0005-2744(79)90243-2
35. Thompson CM, Ceder R, Grafström RC (2010) Formaldehyde dehydrogenase: beyond phase I metabolism. Toxicol Lett 193(1):1–3. https://doi.org/10.1016/j.toxlet.2009.11.023
36. Ando M, Yoshtmoto T, Ogushl S, Rikitake K, Shibata S, Tsuru D (1979) Formaldehyde dehydrogenase from Pseudomonas putida: purification and some properties. J Biochem 85(5):1165–1172. https://doi.org/10.1093/oxfordjournals.jbchem.a132440
37. Tanaka N, Kusakabe Y, Ito K, Yoshimoto T, Nakamura KT (2002) Crystal structure of formaldehyde dehydrogenase from Pseudomonas putida: the structural origin of the tightly bound cofactor in nicotinoprotein dehydrogenases. J Mol Biol 324(3):519–533. https://doi.org/10.1016/S0022-2836(02)01066-5
38. Persson B, Hedlund J, Jörnvall H (2008) Medium- and short-chain dehydrogenase/reductase gene and protein families: the MDR superfamily. Cell Mol Life Sci 65(24): 3879–3894. https://doi.org/10.1007/s00018-008-8587-z
39. Hurley TD, Edenberg H (2012) Gene encoding enzymes in alcohol metabolism. Alcohol Res Curr Rev 34(3):339–344. [Online]. Available: https://www.ncbi.nlm.nih.gov/pmc/articles/PMC3756590/pdf/arcr-34-3-339.pdf
40. Vallee BL, Hoch FL (1955) Zinc, A component of yeast alcohol dehydrogenase. Proceedings of the National Academy of Sciences 41 (6):327–338
41. Wills C (1976) Production of yeast alcohol dehydrogenase isoenzymes by selection. Nature 261(5555):26–29. https://doi.org/10.1038/261026a0
42. Leskovac V, Trivić S, Peričin D (2002) The three zinc-containing alcohol dehydrogenases from baker's yeast, Saccharomyces cerevisiae. FEMS Yeast Res 2(4):481–494. https://doi.org/10.1016/S1567-1356(02)00157-5
43. Raj SB, Ramaswamy S, Plapp BV (2014) Yeast alcohol dehydrogenase structure and catalysis. Biochemistry 53(36):5791–5803. https://doi.org/10.1021/bi5006442
44. Zdarta J, Meyer AS, Jesionowski T, Pinelo M (2018) A general overview of support materials for enzyme immobilization: characteristics, properties, practical utility. Catalysts 8(2). https://doi.org/10.3390/catal8020092
45. Obert R, Dave BC (1999) Enzymatic conversion of carbon dioxide to methanol: enhanced methanol production in silica sol-gel matrices [5]. J Am Chem Soc 121(51):12192–12193. https://doi.org/10.1021/ja991899r
46. Xu SW, Lu Y, Li J, Jiang ZY, Wu H (2006) Efficient conversion of CO_2 to methanol catalyzed by three dehydrogenases co-encapsulated in an alginate-silica (ALG-SiO2) hybrid gel. Ind Eng Chem Res 45(13):4567–4573. https://doi.org/10.1021/ie051407l
47. Sun Q, Jiang Y, Jiang Z, Zhang L, Sun X, Li J (2009) Green and efficient conversion of CO_2 to methanol by biomimetic coimmobilization of three dehydrogenases in protamine-templated titania. Ind Eng Chem Res 48(9): 4210–4215. https://doi.org/10.1021/ie801931j
48. Wang X et al (2014) Bioinspired approach to multienzyme cascade system construction for efficient carbon dioxide reduction. ACS Catal 4(3):962–972. https://doi.org/10.1021/cs401096c
49. Jiang Y, Sun Q, Zhang L, Jiang Z (2009) Capsules-in-bead scaffold: a rational architecture for spatially separated multienzyme cascade system. J Mater Chem 19(47): 9068–9074. https://doi.org/10.1039/b914268a
50. Luo J, Meyer AS, Mateiu RV, Pinelo M (2015) Cascade catalysis in membranes with enzyme immobilization for multi-enzymatic conversion of CO_2 to methanol. New Biotechnol 32(3):319–327. https://doi.org/10.1016/j.nbt.2015.02.006
51. Ji X, Su Z, Ma G, Zhang S (2018) Sandwiching multiple dehydrogenases and shared cofactor between double polyelectrolytes for enhanced communication of cofactor and enzymes. Biochem Eng J 137:40–49. https://doi.org/10.1016/j.bej.2018.05.017

52. El-Zahab B, Donnelly D, Wang P (2008) Particle-tethered NADH for production of methanol from CO_2 catalyzed by coimmobilized enzymes. Biotechnol Bioeng 99(3): 508–514. https://doi.org/10.1002/bit.21584
53. Ji X, Su Z, Wang P, Ma G, Zhang S (2015) Tethering of nicotinamide adenine dinucleotide inside hollow nanofibers for high-yield synthesis of methanol from carbon dioxide catalyzed by coencapsulated multienzymes. ACS Nano 9(4):4600–4610. https://doi.org/10.1021/acsnano.5b01278
54. Marques Netto CGC, Andrade LH, Toma HE (2018) Carbon dioxide/methanol conversion cycle based on cascade enzymatic reactions supported on superparamagnetic nanoparticles. An Acad Bras Cienc 90(1):593–606. https://doi.org/10.1590/0001-3765201720170330
55. Zhu D et al (2019) Ordered coimmobilization of a multienzyme cascade system with a metal organic framework in a membrane: reduction of CO_2 to methanol. ACS Appl Mater Interfaces 11(37):33581–33588. https://doi.org/10.1021/acsami.9b09811
56. Cazelles R, Drone J, Fajula F, Ersen O, Moldovan S, Galarneau A (2013) Reduction of CO_2 to methanol by a polyenzymatic system encapsulated in phospholipids-silica nanocapsules. New J Chem 37(11):3721–3730. https://doi.org/10.1039/c3nj00688c
57. Singh RK et al (2018) Insights into cell-free conversion of CO_2 to chemicals by a multienzyme cascade reaction. ACS Catal 8(12): 11085–11093. https://doi.org/10.1021/acscatal.8b02646
58. Zhang Z et al (2018) Efficient ionic liquid-based platform for multi-enzymatic conversion of carbon dioxide to methanol. Green Chem 20(18):4339–4348. https://doi.org/10.1039/c8gc02230e
59. Lee SH, Choi DS, Kuk SK, Park CB (2018) Photobiocatalysis: activating redox enzymes by direct or indirect transfer of photoinduced electrons. Angew Chem Int Ed. 57(27): 7958–7985. https://doi.org/10.1002/anie.201710070
60. Dibenedetto A et al (2012) Hybrid technologies for an enhanced carbon recycling based on the enzymatic reduction of CO_2 to methanol in water: chemical and photochemical NADH regeneration. ChemSusChem 5(2):373–378. https://doi.org/10.1002/cssc.201100484
61. Aresta M, Dibenedetto A, Baran T, Angelini A, Łabuz P, Macyk W (2014) An integrated photocatalytic/enzymatic system for the reduction of CO_2 to methanol in bioglycerol-water. Beilstein J Org Chem 10:2556–2565. https://doi.org/10.3762/bjoc.10.267
62. Yadav RK, Oh GH, Park NJ, Kumar A, Kong KJ, Baeg JO (2014) Highly selective solar-driven methanol from CO_2 by a photocatalyst/biocatalyst integrated system. J Am Chem Soc 136(48):16728–16731. https://doi.org/10.1021/ja509650r
63. Liu J, Cazelles R, Chen ZP, Zhou H, Galarneau A, Antonietti M (2014) The bioinspired construction of an ordered carbon nitride array for photocatalytic mediated enzymatic reduction. Phys Chem Chem Phys 16(28):14699–14705. https://doi.org/10.1039/c4cp01348d
64. Ji X, Su Z, Wang P, Ma G, Zhang S (2016) Integration of artificial photosynthesis system for enhanced electronic energy-transfer efficacy: a case study for solar-energy driven bioconversion of carbon dioxide to methanol. Small 12(34):4753–4762. https://doi.org/10.1002/smll.201600707
65. Wu R, Ma C, Zhu Z (2020) Enzymatic electrosynthesis as an emerging electrochemical synthesis platform. Curr Opin Electrochem 19:1–7. https://doi.org/10.1016/j.coelec.2019.08.004
66. Spinner NS, Vega JA, Mustain WE (2012) Recent progress in the electrochemical conversion and utilization of CO 2. Cat Sci Technol 2(1):19–28. https://doi.org/10.1039/c1cy00314c
67. Schlager S et al (2017) Biocatalytic and bioelectrocatalytic approaches for the reduction of carbon dioxide using enzymes. Energy Technol 5(6):812–821. https://doi.org/10.1002/ente.201600610
68. Schlager S, Fuchsbauer A, Haberbauer M, Neugebauer H, Sariciftci NS (2017) Carbon dioxide conversion to synthetic fuels using biocatalytic electrodes. J Mater Chem A 5(6): 2429–2443. https://doi.org/10.1039/c6ta07571a
69. Kuwabata S, Tsuda R, Yoneyama H (1994) Electrochemical conversion of carbon dioxide to methanol with the assistance of formate dehydrogenase and methanol dehydrogenase as biocatalysts. J Am Chem Soc 116(12): 5437–5443. https://doi.org/10.1021/ja00091a056
70. Addo PK, Arechederra RL, Waheed A, Shoemaker JD, Sly WS, Minteer SD (2011) Methanol production via bioelectrocatalytic reduction of carbon dioxide: role of carbonic anhydrase in improving electrode performance. Electrochem Solid-State Lett 14(4):5–10. https://doi.org/10.1149/1.3537463

71. Zhang Z et al (2021) Encapsulation of multiple enzymes in a metal-organic framework with enhanced electro-enzymatic reduction of CO_2 to methanol. Green Chem 23(6):2362–2371. https://doi.org/10.1039/d1gc00241d
72. Schlager S et al (2016) Electrochemical reduction of carbon dioxide to methanol by direct injection of electrons into immobilized enzymes on a modified electrode. ChemSusChem 9(6):631–635. https://doi.org/10.1002/cssc.201501496
73. Barton Cole E, Lakkaraju PS, Rampulla DM, Morris AJ, Abelev E, Bocarsly AB (2010) Using a one-electron shuttle for the multielectron reduction of CO_2 to methanol: kinetic, mechanistic, and structural insights. J Am Chem Soc 132(33):11539–11551. https://doi.org/10.1021/ja1023496
74. Kuk SK, Singh RK, Nam DH, Singh R, Lee JK, Park CB (2017) Photoelectrochemical reduction of carbon dioxide to methanol through a highly efficient enzyme cascade. Angew Chem Int Ed 56(14):3827–3832. https://doi.org/10.1002/anie.201611379
75. Windle CD, Perutz RN (2012) Advances in molecular photocatalytic and electrocatalytic CO 2 reduction. Coord Chem Rev 256(21–22):2562–2570. https://doi.org/10.1016/j.ccr.2012.03.010
76. Marpani F, Pinelo M, Meyer AS (2017) Enzymatic conversion of CO_2 to CH3OH via reverse dehydrogenase cascade biocatalysis: quantitative comparison of efficiencies of immobilized enzyme systems. Biochem Eng J 127:217–228. https://doi.org/10.1016/j.bej.2017.08.011
77. Effendi SSW, Ng IS (2019) The prospective and potential of carbonic anhydrase for carbon dioxide sequestration: a critical review. Process Biochem. 87(July):55–65. https://doi.org/10.1016/j.procbio.2019.08.018
78. Kanth BK, Lee J, Pack SP (2013) Carbonic anhydrase: its biocatalytic mechanisms and functional properties for efficient CO_2 capture process development. Eng Life Sci 13(5): 422–431. https://doi.org/10.1002/elsc.201200157
79. Russo ME, Olivieri G, Marzocchella A, Salatino P, Caramuscio P, Cavaleiro C (2013) Post-combustion carbon capture mediated by carbonic anhydrase. Sep Purif Technol 107: 331–339. https://doi.org/10.1016/j.seppur.2012.06.022
80. Kupriyanova E, Pronina N, Los D (2017) Carbonic anhydrase — a universal enzyme of the carbon-based life. Photosynthetica 55(1): 3–19. https://doi.org/10.1007/s11099-017-0685-4

Chapter 20

Multienzyme Catalysis in Phase-Separated Protein Condensates

Miao Liu, Xi Chen, and Jiang Xia

Abstract

Liquid–liquid phase separation forms condensates that feature a highly concentrated liquid phase, a defined yet dynamic boundary, and dynamic exchange at and across the boundary. Phase transition drives the formation of dynamic multienzyme complexes in cells, and understanding how phase separation regulates multienzyme catalysis may need the help of in vitro investigations. Recently we have constructed synthetic versions of multienzyme biosynthetic systems by assembling enzymes in protein condensates. Here, we describe the methods for checking the enzyme assembly using fluorescent microscopy and centrifugation assay. We further provide steps for analysis of the cascade enzyme catalytic efficiencies inside the condensates, using enzymes from terpene biosynthesis pathway.

Key words Liquid–liquid phase separation, Enzyme, Cascade catalysis, Terpene biosynthesis, Protein–protein interaction, Protein condensate

1 Introduction

Liquid–liquid phase separation (LLPS) is prevalent in cells in the modern world now. Biomacromolecules including proteins or RNA spontaneously condense and form self-assembled liquid condensates, a.k.a. membraneless organelles, which serve as micro-compartments to selectively concentrate certain molecules while excluding others [1–4]. The liquid condensates are highly mobile and dynamic: they can move around, fuse together, and exchange materials with the surrounding environment.

How LLPS impacts cascade biocatalysis has been an appealing question [5–7]. Metabolic enzymes can form condensates in bacteria, yeast, and mammalian cells under certain conditions, which will regulate enzyme catalysis inside cells [8–11]. Studies of these multienzyme condensates have demonstrated the effects of LLPS on biochemical reactions in different directions, and more insights will be complemented by in vitro investigations. Assembly of enzymes

Haralambos Stamatis (ed.), *Multienzymatic Assemblies: Methods and Protocols*,
Methods in Molecular Biology, vol. 2487, https://doi.org/10.1007/978-1-0716-2269-8_20,
© The Author(s), under exclusive license to Springer Science+Business Media, LLC, part of Springer Nature 2022

in LLPS in vitro was first proved in synthetic polymer condensates [12], then also in RNA or protein-based condensates, which were better mimics of the natural dynamic multienzyme complexes [13–15]. In vitro constitution of LLPS system may give us more tunable properties, while the continuous investigation of different parameters is important for understanding more about the limits and benefits of regulating enzyme activity via LLPS [7].

Notwithstanding, previous enzyme assembly strategies mostly rely on direct participation of guest enzymes or fusion with phase-forming scaffold proteins with the enzymes, which may affect the phase-forming property or the enzymatic property. Interaction-driven assembly, instead, is a more efficient and general method to incorporate guest proteins into phase-separated condensates [16, 17]. We therefore reason that interaction-driven self-assembly will be a feasible method to construct phase-separated multienzyme catalytic systems. Previously, we utilized a high-affinity peptide–peptide interacting pair to assemble multiple biosynthetic enzymes into a synthetic protein phase separation system [18]. RIDD and RIAD are two peptide motifs derived from the natural proteins cAMP-dependent protein kinase and the A kinase-anchoring proteins with high-affinity [19]. Through this interaction, we assemble biosynthetic enzymes to a three-component phase separation system from postsynaptic density in neurological synapses, including GKAP, Shank and Homer, which can self-assemble to form protein phase droplets in vitro [20, 21]. The critical enzymes in terpene biosynthesis pathway, isopentenyl-diphosphate delta-isomerase (Idi), and farnesyl diphosphate synthase (IspA) were chosen because Idi was known to catalyze the rate-limiting step of the cascade reaction [22]. Phase-separated protein coacervation enriched both Idi and IspA in the protein condensate by 50 folds as compared to the bulk solution, such an enrichment increased the production rate of farnesyl pyrophosphate (FPP) by more than 50%. Here, we describe the general method to monitor the phase-separated protein assembly using fluorescently labeled enzymes and confocal microscopy, a centrifugation assay to quantify phase separation based on the distribution of enzymes in the condensate and the bulk solution, and the protocol to evaluate the catalytic efficiencies of enzymes inside the condensates, using enzymes from terpene biosynthesis pathway as an example.

2 Materials

2.1 Recombinant Proteins Expression and Purification

Prepare all media and buffers in deionized water.

1. *E. coli* expression strain such as BL21 (DE3).
2. LB media: 10 g/L tryptone, 5 g/L yeast extract, 5 g/L NaCl, autoclaved.

3. Antibiotics: 100 μg/mL ampicillin or 50 μg/mL kanamycin in LB media.
4. 1 M isopropyl-β-D-thiogalactopyranoside (IPTG) stock solution: 238 mg IPTG in 1 mL water, sterile-filtered.
5. Binding buffer: 50 mM Tris, pH 8.0, 500 mM NaCl, 2 mM β-mercaptoethanol, 10 mM imidazole.
6. Elution buffer: 50 mM Tris, pH 8.0, 500 mM NaCl, 500 mM imidazole.
7. General buffer: 50 mM Tris, pH 8.0, 100 mM NaCl, 1 mM EDTA.
8. Human rhinovirus (HRV) 3C protease.
9. Temperature-controlled centrifuge.
10. FPLC with HisTrap™ HP column, HiTrap® Desalting column, Superdex 200, or 75 column (Cytiva).
11. SDS-PAGE equipment and gels.

2.2 Phase Separation, Confocal Imaging, and Centrifugation/SDS-PAGE Assay

Prepare all the buffers in deionized water.

1. Labeling buffer: 0.1 M sodium bicarbonate, pH 8.3, 100 mM NaCl.
2. Sulfo-Cy5 NHS ester or other amine-reactive dye (*see* **Note 2**).
3. Dimethyl sulfoxide (DMSO).
4. Termination buffer: 200 mM Tris, pH 8.3.
5. General buffer: 50 mM Tris, pH 8.0, 100 mM NaCl, 1 mM EDTA.
6. FPLC with HiTrap® Desalting Column (Cytiva).
7. Nanodrop spectrophotometer.
8. Confocal microscope.
9. High-speed temperature-controlled centrifuge.
10. Glass slide and coverslip.
11. SDS-PAGE equipment and gels.
12. SDS-PAGE loading buffer (5×): 0.25 M Tris, pH 6.8, 10% SDS, 0.5 M dithiothreitol, 0.1% bromophenol blue, 50% glycerol.

2.3 Enzymatic Reaction of Terpene Biosynthesis Enzymes

Prepare all the buffers in deionized water.

1. Reaction buffer: 50 mM Tris, pH 7.6, 100 mM NaCl, 1 mM DTT, 2 mM Mg^{2+}.
2. Purified enzymes: 0.5 μM idi-RIDD and 1 μM RIDD-IspA used in this chapter.
3. Purified phase scaffold proteins: 5 μM each of GKAP, Homer and Shank in this chapter.

4. Isopentenyl pyrophosphate (IPP) and Farnesyl pyrophosphate (FPP) (stock solution available from Sigma-Aldrich).
5. Mobile phase A: 0.01% triethylamine in water.
6. Mobile phase B: acetonitrile.
7. Reverse-phase HPLC with C18 column and MS detector: Waters ACQUITY UPLC System with Thermo Scientific Accucore™ C18 HPLC Column.

3 Methods

3.1 Cloning, Recombinant Protein Expression, and Purification

First, the phase-forming scaffold proteins and enzymes of interest can be obtained by recombinant protein production using *E. coli*. Plasmids will be constructed through restriction cloning methods using pET vectors. To fuse with RI or RIAD tag, two synthesized primers are annealed and inserted into respective plasmids. Detailed sequence information can be found in ref. 18 and 21. The protein expression and purification steps are described here.

3.1.1 Protein Expression

1. Transform the plasmid into an *E. coli* expression strain, e.g., BL21 (DE3).
2. Grow an overnight culture from a single colony at 37 °C in LB media supplemented with ampicillin or kanamycin.
3. Inoculate the overnight culture to LB media supplemented with antibiotics at 1:100 ratio.
4. Grow the culture at 37 °C to reach OD_{600} ~0.6, add 0.25 mM IPTG to induce protein expression, and continue to grow at 16 °C overnight.
5. Harvest the cells by centrifugation at 4000 × *g* for 10 min, store the pellets for protein purification.

3.1.2 Protein Purification

All proteins are purified using Nickel-NTA affinity chromatography through their His tags. Proteins can be further purified using size-exclusion chromatography (**step 6**), and when needed, recombinant protein tags can be cleaved by proteases and further purified by size-exclusion chromatography (**step** 7) (*see* **Note 1**).

1. Resuspend the harvested cells with binding buffer, lyse the cells by sonication.
2. Centrifuge the cell lysate at 13,000 × *g* for 30 min at 4 °C. Load the filtered supernatant into HisTrap column equilibrated by binding buffer.
3. Connect the column to FPLC system. Protein is eluted by imidazole gradient from 10 mM to 500 mM, using binding buffer and elution buffer.

4. Analyze the collected fractions by SDS-PAGE gel.
5. Pure protein fractions are collected and exchanged with general buffer.
6. (Optional, *see* **Note 1**) Further purify the eluted protein using Superdex 200 column equilibrated with general buffer.
7. (Optional, tag removal, *see* **Note 1**) Eluted protein is exchanged with general buffer, tag is removed by HRV 3C protease incubating at 4 °C overnight, protein without tag is purified by Superdex 75 column equilibrated with general buffer.
8. Proteins were flash frozen under liquid nitrogen and stored at −80 °C.

3.2 Fluorescent Protein Labeling

Microscopy techniques are used to observe LLPS condensates [4]. Confocal microscope imaging is a convenient way to examine the assembly of enzymes inside the condensed phase, which need the proteins of interested fluorescent labeled. Here the Cyanine 5 (Cy5) is chosen as the fluorescent dye to label enzymes (*see* **Note 2**).

1. Dilute or dialyze the protein of interest against the labeling buffer at concentration of 2 mg/mL.
2. Dissolve Sulfo-Cy5 NHS ester in DMSO. The concentration of Cy5 can be measured by absorbance at 646 nm (ε_{646} = 271,000 L/mol/cm, according to Lumiprobe [23]).
3. Slowly add Sulfo-Cy5 NHS ester into the protein solution under stirring until the molar ratio of protein to Cy5 reaches 1:1 to 1:2 (*see* **Note 3**).
4. Incubate the reaction for 1 h at room temperature.
5. Stop the reaction by adding termination buffer.
6. Separate the labeled proteins from unreacted labeling reagent using a desalting column equilibrated with the general buffer.
7. Determine the protein concentration and labeling efficiency by absorbance at 280 nm and 646 nm.

3.3 Phase Separation and Confocal Imaging Assay

The labeled enzymes can then be used for assembly, by mixing with scaffold proteins. By using confocal microscope, formation of LLPS condensates as well as the assembly and enrichment of labeled enzymes can be easily detected (Fig. 1a). Steps for confocal imaging assay are listed here, according to ref. 18 and 21.

1. The scaffold proteins and enzymes are prepared in the general buffer. Centrifuge all protein solutions to remove any protein aggregates.

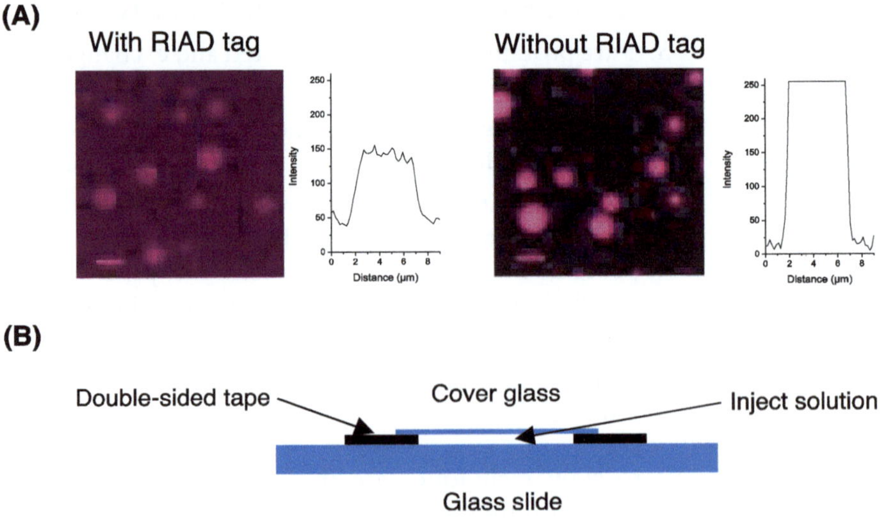

Fig. 1 Interaction-driven enzyme assembly in the phase-separated protein droplets. (**a**) Confocal images showing significant enrichment of RIDD tagged enzyme in the RIAD-tagged protein condensates, but not untagged enzyme. The quantification of Cy5 fluorescence intensities is shown on the right. Scale bar, 5 μm. Protein scaffolds, 5 μM; enzyme, 1 μM with 5% labeled with Cy5. (**b**) Preparation of the homemade flow chamber. Solution can be injected in the space between two glasses

2. Dilute and mix proteins at the designated combinations and concentrations in the general buffer. The solution will turn opaque, which signifies phase separation.
3. Prepare a homemade flow chamber comprised a glass slide sandwiched by a coverslip with one layer of double-sided tape as a spacer (Fig. 1b, *see* **Note 4**).
4. Inject the protein mixture prepared in **step 2** into the flow chamber (*see* **Note 5**). Use confocal microscope for observation at room temperature.
5. Analyze the enrichment of the labeled enzymes according to the fluorescence intensity.

3.4 Analyze Protein Distribution in Condensate by Centrifugation/SDS-PAGE Assay

Besides analyzing the enzyme enrichment inside the condensate using imaging tools, the enrichment of enzymes in the condensate droplets can be judged by a sedimentation assay, which is described in this section.

1. Dilute and mix proteins at the designated combinations and concentrations in the general buffer. The total volume for each reaction is typically 100 μL (*see* **Note 6**).
2. Incubate the mixtures at room temperature for 20 min to allow full formation of LLPS.
3. Centrifuge the mixture at 16,000 × *g* for 10 min at 20 °C on a temperature-controlled microcentrifuge.

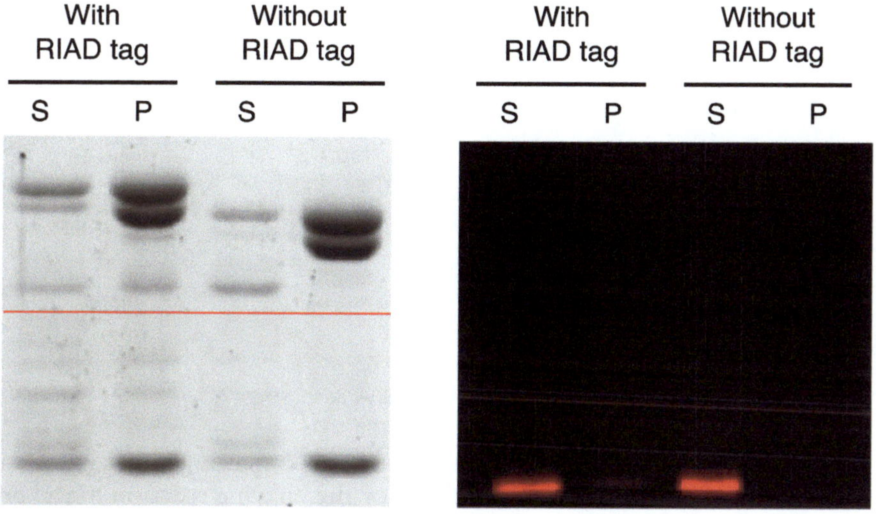

Fig. 2 Centrifugation/SDS-PAGE analysis showing the protein distributions of enzymes in the protein condensate (P) and bulk solution (S). The PAGE can be quantified by either Coomassie blue staining (left) or fluorescent imager (right) if fluorescent labeled enzymes were used. When RIDD tagged enzyme was mixed with untagged phase scaffolds, most of the enzyme was found in supernatant. About 50% of RIDD tagged enzyme was found in pellet when assembled in RIAD-tagged phase scaffolds. Protein scaffold, 5 μM; enzyme, 1 μM with 5% Cy5 labeled

4. After centrifugation, transfer the supernatant into a new tube immediately.
5. Thoroughly resuspend the pellet in the same buffer (i.e., the general buffer) to the same volume (typically 100 μL).
6. Add SDS-PAGE loading buffer to both supernatant and resuspended pellet fractions. Run SDS-PAGE.
7. The amount of enzymes inside the phase droplet can be characterized using pellet percentage, which is calculated according to the following formula by analyzing the band intensities of supernatant and pellet fractions (Fig. 2):

$$\text{Pellet percentage} = \frac{\text{Band intensity of pellet}}{\text{Band intensity of pellet} + \text{Band intensity of supernatant}}$$

3.5 Enzymatic Reaction of Terpene Biosynthesis Inside Phase Condensates

Next the catalytic efficiency of the enzymes in the protein condensates will be analyzed. Here we use enzymes from terpene biosynthesis pathway, isopentenyl-diphosphate delta-isomerase (Idi) and farnesyl diphosphate synthase (IspA) as an example. Idi and IspA can convert the five-carbon precursor isopentenyl pyrophosphate (IPP) to sesquiterpene farnesyl pyrophosphate (FPP) (Fig. 3) [22, 24, 25]. Both IPP and FPP can be detected using reverse-phase HPLC [26, 27].

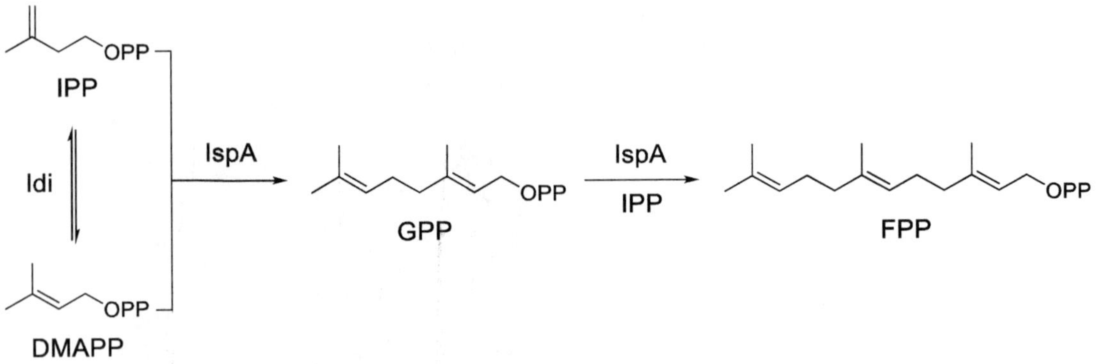

Fig. 3 Terpene biosynthetic pathway catalyzed by Idi and IspA

1. Prepare a reaction mixture containing the tagged enzymes in the reaction buffer. For the enzyme reaction in the presence of phase condensates, scaffold proteins are also added. The total volume for each reaction is typically 100 μL.
2. Incubate the mixtures at room temperature for 10 min to allow full formation of LLPS.
3. Add the reactant IPP to a final concentration of 67 μM (20 μl/mL) to initiate the reaction. Mix the reaction solution thoroughly.
4. At every 15 min, withdraw 0.5 to 2 μL of reaction mixture. Terminate the reaction by diluting with the mobile phase A (*see* **Note 7**).
5. Analyze the product FPP by LC-MS. Samples diluted in mobile phase A are injected into LC-MS and eluted with the following gradient: from 5% B to 99% mobile phase B in 5 min. Column temperature is set at 35 °C. FPP can be monitored by the negative ion mode at *m/z* value of 381.
6. Generate a standard curve of FPP by injecting FPP solutions of different concentrations using the same column and HPLC condition. Integrate the areas of specific peaks to draw the standard curve. Deduce the FPP concentration of the samples.
7. The apparent reaction rates (k_{app}) can be calculated by linear curve fitting of FPP concentration increase within the first hour.

4 Notes

1. Most of the proteins eluted from affinity chromatography should be at least 90% purity according to SDS-PAGE. For some proteins like the scaffold protein Homer, besides affinity purification, size-exclusion chromatography may be utilized to

acquire samples of high purity. Protein tags, e.g., Trx tag used to increase the expression and solubility of GKAP protein, may be removed to avoid any negative effects on phase formation.

2. Besides Cy5, other fluorescent dyes such as iFluor 405, Alexa Fluor 488, Cy3, and Alexa Fluor 647 can also be used for labeling [21].
3. In this ratio, not all the proteins can be labeled with fluorescent dyes. The label efficiency is about 30–60%. Stoichiometric labeling reaction is not a must.
4. Usually, the size of phase droplet is in the range of micrometers. The homemade flow chamber will leave a space of 50–100 μm between the two glasses, which will not disrupt the phase droplets.
5. Typically, 20 μL solution can fill the whole chamber. To prevent evaporation in long-term imaging, the chamber can be sealed by nail polish.
6. 100 μL can ensure all the condensates are sedimented.
7. The pH of 0.01% triethylamine in water is around 10. Adding mobile phase A to the reaction solution will break the phase and terminate the enzymatic reaction.

References

1. Shin Y, Brangwynne CP (2017) Liquid phase condensation in cell physiology and disease. Science 357(6357):eaaf4382
2. Banani SF, Lee HO, Hyman AA, Rosen MK (2017) Biomolecular condensates: organizers of cellular biochemistry. Nat Rev Mol Cell Biol 18(5):285–298
3. Boeynaems S, Alberti S, Fawzi NL et al (2018) Protein phase separation: a new phase in cell biology. Trends Cell Biol 28(6):420–435
4. Bracha D, Walls MT, Brangwynne CP (2019) Probing and engineering liquid-phase organelles. Nat Biotechnol 37(12):1435–1445
5. Prouteau M, Loewith R (2018) Regulation of cellular metabolism through phase separation of enzymes. Biomol Ther 8(4):160
6. Zhang Y, Narlikar GJ, Kutateladze TG (2020) Enzymatic reactions inside biological condensates. J Mol Biol. https://doi.org/10.1016/j.jmb.2020.08.009
7. O'Flynn BG, Mittag T (2021) The role of liquid-liquid phase separation in regulating enzyme activity. Curr Opin Cell Biol 69:70–79
8. O'Connell JD, Zhao A, Ellington AD, Marcotte EM (2012) Dynamic reorganization of metabolic enzymes into intracellular bodies. Annu Rev Cell Dev Biol 28:89–111
9. Noree C, Monfort E, Shiau AK, Wilhelm JE (2014) Common regulatory control of CTP synthase enzyme activity and filament formation. Mol Biol Cell 25(15):2282–2290
10. Pedley AM, Benkovic SJ (2017) A new view into the regulation of purine metabolism: the purinosome. Trends Biochem Sci 42(2): 141–154
11. Kohnhorst CL, Kyoung M, Jeon M et al (2017) Identification of a multienzyme complex for glucose metabolism in living cells. J Biol Chem 292(22):9191–9203
12. Strulson CA, Molden RC, Keating CD, Bevilacqua PC (2012) RNA catalysis through compartmentalization. Nat Chem 4(11):941–946
13. Drobot B, Iglesias-Artola JM, Le Vay K et al (2018) Compartmentalised RNA catalysis in membrane-free coacervate protocells. Nat Commun 9(1):3643
14. Küffner AM, Prodan M, Zuccarini R et al (2020) Acceleration of an enzymatic reaction in liquid phase separated compartments based on intrinsically disordered protein domains. ChemSystemsChem 2:e2000001
15. Peeples W, Rosen MK. (2020) Phase separation can increase enzyme activity by concentration and molecular organization. bioRxiv.

https://doi.org/10.1101/2020.09.15.299115

16. Banani SF, Rice AM, Peeples WB et al (2016) Compositional control of phase-separated cellular bodies. Cell 166(3):651–663
17. Schuster BS, Reed EH, Parthasarathy R et al (2018) Controllable protein phase separation and modular recruitment to form responsive membraneless organelles. Nat Commun 9(1): 2985
18. Liu M, He S, Cheng L, Qu J, Xia J (2020) Phase-separated multienzyme biosynthesis. Biomacromolecules 21(6):2391–2399
19. Carlson CR, Lygren B, Berge T et al (2006) Delineation of type I protein kinase A-selective signaling events using an RI anchoring disruptor. J Biol Chem 281(30):21535–21545
20. Zeng M, Shang Y, Araki Y, Guo T, Huganir RL, Zhang M (2016) Phase transition in postsynaptic densities underlies formation of synaptic complexes and synaptic plasticity. Cell 166(5):1163–1175
21. Zeng M, Chen X, Guan D et al (2018) Reconstituted postsynaptic density as a molecular platform for understanding synapse formation and plasticity. Cell 174(5):1172–1187
22. Zhu F, Zhong X, Hu M, Lu L, Deng Z, Liu T (2012) In vitro reconstitution of mevalonate pathway and targeted engineering of farnesene overproduction in Escherichia coli. Biotechnol Bioeng 111(7):1396–1405
23. Lumiprobe Corporation, Sulfo-Cyanine5 NHS ester. https://www.lumiprobe.com/p/sulfo-cy5-nhs-ester
24. Hahn FM, Hurlburt AP, Poulter CD (1999) Escherichia coli open reading frame 696 is idi, a nonessential gene encoding isopentenyl diphosphate isomerase. J Bacteriol 181(15): 4499–4504
25. Hosfield DJ, Zhang Y, Dougan DR et al (2004) Structural basis for bisphosphonate-mediated inhibition of isoprenoid biosynthesis. J Biol Chem 279(10):8526–8529
26. Zhang D, Poulter CD (1993) Analysis and purification of phosphorylated isoprenoids by reversed-phase HPLC. Anal Biochem 213(2): 356–361
27. Ku B, Jeong JC, Mijts BN, Schmidt-Dannert C, Dordick JS (2005) Preparation, characterization, and optimization of an in vitro C30 carotenoid pathway. Appl Environ Microbiol 71(11):6578–6583

Chapter 21

Robust and Continuous Lipase-Catalyzed Reactions in Deep Eutectic Solvents: Low Viscosity and Double CLEA-LentiKats Immobilization (CLEA-LK-Lipases)

Nadia Guajardo and Pablo Domínguez de María

Abstract

Deep Eutectic Solvents (DES) are used as reaction media for lipase-catalyzed esterifications in continuous devices. In particular, DES may be useful for lipophilization-like reactions involving substrates with unpaired solubilities. Aspects to be considered are the viscosity of the solvent, as well as the stability of the enzyme in the non-conventional media. The viscosity can be decreased by adding buffer as cosolvent (up to 20% v/v) and keeping the non-conventional nature. Lipases can be stabilized by following a double immobilization pattern, comprising CLEA formation and entrapment in LentiKats®. The low viscosity and high stability of the CLEA-LK-lipase enable the use of DES under flow conditions.

Key words CLEA, LentiKats®, Immobilization, Lipases, Stability, Deep Eutectic Solvents, Flow Chemistry

1 Introduction

Biocatalysis comprises an ample number of applications that can be implemented at industrial scale to deliver sustainable solutions [1]. In the quest for environment-benign solvents, Deep Eutectic Solvents (DES) have gained importance in chemistry over the last decade. Particularly, in biocatalysis a number of enzymes and systems have been successfully reported [2]. Of note is the possibility of DES to dissolve substrates with unpaired polarity, which confer them a solid position to be applied for practical applications. Herein, key-to-success is the addition of buffer as cosolvent (up to 20%, v/v) to decrease the viscosity of the reaction media [3]. This enables the set-up of continuous biocatalytic processes, leading to powerful synergies for sustainable chemistry [4].

To confer robustness, enzymes can be immobilized, and the so-called cross-linking aggregates (CLEA) is a useful method to

Haralambos Stamatis (ed.), *Multienzymatic Assemblies: Methods and Protocols*,
Methods in Molecular Biology, vol. 2487, https://doi.org/10.1007/978-1-0716-2269-8_21,
© The Author(s), under exclusive license to Springer Science+Business Media, LLC, part of Springer Nature 2022

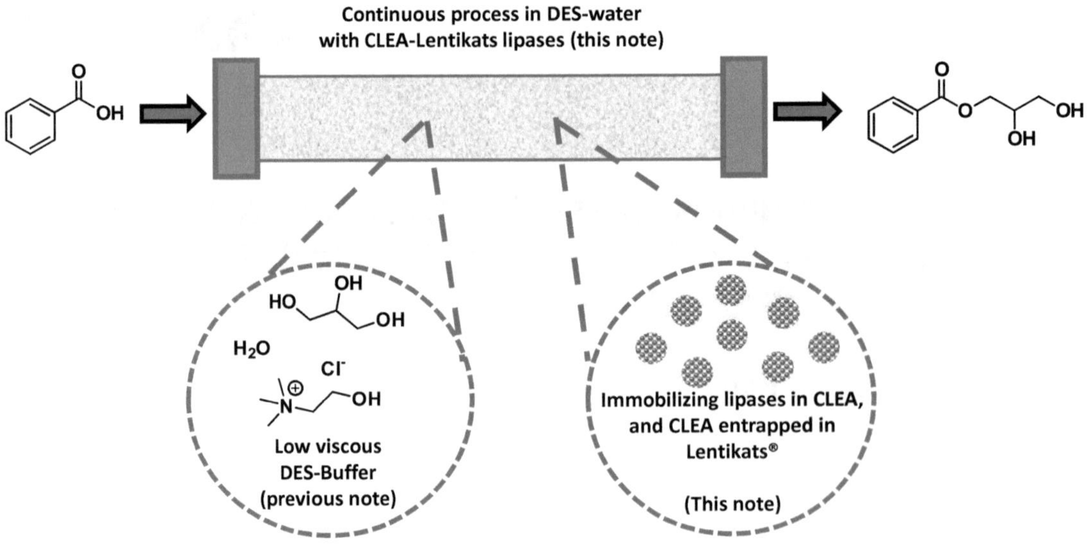

Fig. 1 Schematic process for lipase-catalyzed esterifications in DES-buffer media in continuous, using robust CLEA-LentiKats® immobilized enzymes [3, 7–9]

perform reactions in organic media [5, 6]. Particularly, CLEA immobilized derivatives display high stability and activity in DES, as reported elsewhere for Choline Chloride—Glycerol DES and *Candida antarctica* lipase B, leading to derivatives stable for more than 2 weeks [7]. Furthermore, to facilitate the recovery of enzyme derivatives, the subsequent entrapment of CLEAs in LentiKats® leads to largely stable catalysts with straightforward recyclability [8]. Overall, the combination of low viscous DES-buffer media with *doubly* immobilized enzymes (CLEA-LK) enables the option of setting continuous processes for esterification reactions [9].

In a previous note [10], we provided a protocol for the set-up of non-viscous DES-buffer media for lipase-catalyzed esterifications. In this note, we extend that approach by delivering doubly immobilized enzymes (CLEA-LK) with continuous designs. As model reaction, the esterification of benzoic acid and glycerol is performed, as it represents a useful example of substrates with unpaired solubility (Fig. 1).

2 Materials

2.1 Reagents

1. 4-nitrophenyl butyrate (*p*NPB).
2. 4-Nitrophenol (*p*NP).
3. Benzoic acid.
4. Glycerol.

5. Choline chloride.
6. Ultrapure water.
7. Methanol.
8. HPLC mobile phase: Methanol (45)/water (55)/H_3PO_4 (0.1) (v/v).
9. Phosphate Buffer 100 mM, pH 7: Acid solution: Prepare a 0.1 M monobasic potassium phosphate solution (molar mass is 136.08 g/mol). Basic solution: Prepare a 0.1 M disodium phosphate solution (molar mass is 141.96 g/mol). Mix the volumes of said solutions until reaching the desired pH (in this case 7).
10. Bovine serum albumin (BSA).
11. Glutaraldehyde solution 25%.
12. Ammonium sulfate.
13. Bradford reagent.
14. LentiKats®.

2.2 Enzymes

1. *Candida antarctica* lipase B (Commercially available, soluble, 8.2 $mg_{Protein}$/mL).
2. Other lipases (e.g., *Pseudomonas stutzeri* lipase).

2.3 Analytical Equipment

1. Karl Fischer titrator.
2. UHPLC (Altus A-30, PerkinElmer with a diode array detector): with a Kromasil Eternity-2.5-C18 UHPLC column (100 × 4.6 mm I.D).
3. UV-Visible Spectrophotometer.
4. Cole-Parmer peristaltic pump.
5. Shaker.

2.4 Equipment

1. Bioreactor: Cole-Parmer peristaltic pump, shaker (to control the temperature, hoses, glass column (0.6 cm of diameter × 7.5 cm of length)

3 Methods

3.1 DES Preparation and Lipase-Catalyzed Esterification (See also Ref. [10])

1. Gently mix choline chloride and glycerol (1:2 mol:mol) (*see* **Note 1**), and add phosphate buffer (100 mM, pH 7, 10–20% v/v). Pump the DES in the continuous reactor and add benzoic acid as substrate. Glycerol is part of the DES and substrate simultaneously. Add buffer, together with the immobilized enzyme, and reaction starts [10] (*see* **Note 2**).

3.2 Cross-Linking Aggregates Immobilization of Lipases (CLEA-Lipase)

1. Using CAL-B as model enzyme, CLEA-CALB was prepared by mixing the liquid commercial enzyme (2.5 mL) with 25 mM phosphate buffer pH 7 (2.5 mL) (1:1 v/v). Enrich the enzyme solution with BSA (mass ratio of Enzyme/BSA = 2.5) (*see* **Note 3**).
2. To this Enzyme/BSA solution, add dropwise 10 mL of a saturated solution of ammonium sulfate in a cold bath at 4 °C, and kept for 30 min. Then add dropwise glutaraldehyde (25% v/v) (molar ratio Glutaraldehyde/Enzyme 0.027; 2 mL) and left at 300 rpm for 1.5 h at 4 °C (*see* **Note 4**).
3. Centrifuge the resulting CLEA-containing suspension (5 °C, 10 min, 10,000 × *g*). Resuspend the obtained CLEA in 100 mM phosphate buffer pH 7, and centrifuge it again at 10,000 × *g*. Repeat this three times to wash the CLEA derivatives and store them at 4 °C until use.

3.3 Immobilization in LentiKats®

1. Dissolve 1 g of CLEA-CALB in 2 mL phosphate buffer, pH 7, 25 mM, dispersing the CLEA-CLAB granules. Melt the polyvinyl alcohol (LentiKats®) in a water bath at 40 °C and mix both solutions homogeneously (*see* **Note 5**).
2. Subsequently, seed the mixed solution in Petri dishes as small drops using a syringe or a device with several metal blades to accelerate printing. Record the initial weight, and dry the material avoiding that weight loss results more than 28% of initial weight. Then add the salt stabilizing solution (200 mL solution required for 5 mL of LentiKats®) (*see* **Note 6**).
3. Release the solution rested on the seeded plates and lentils with a rubber spatula. Place the lentils in a beaker with the stabilizing solution under gently stirring for 2 h. Discard the stabilizing solution and wash the lentils with buffer and stored in it until use (4 °C).

3.4 Set-up of a Continuous Process with CLEA-LK-lipases and Low Viscous DES-Buffer Media

1. Fill the packed-bed reactor (a glass column, diameter 0.6 cm; length 7.5 cm) with 1,7 g of biocatalyst (CLEA-LK-CALB). Carry out the reaction operation at a flow of 0.03 to 0.01 mL/min at 60 °C (*see* **Note 7**) and control it by the Core-Palmer peristaltic pump. Determine the residence time following the equation:

$$Conversion\ (\%) = \left[\frac{n_P - n_{Po}}{n_{So}} \times \frac{\nu_S}{\nu_P}\right] \times 100 \qquad (1)$$

where P = product; n_{Po} = initial product P amount (mol); n_{So} = initial amount of substrate S (mol); n_P = final amount of product P (mol); ν_S = stoichiometric factor for substrate S; and ν_P = stoichiometric factor for product P (*see* **Note 8**).

2. Quantify the reaction product (monoester of glycerol, Fig. 1) by UHPLC at 273 nm under the conditions indicated elsewhere [3].
3. Determine the enzymatic activity by the absorbance (348 nm) ($\varepsilon = 5.1[\mathrm{Abs}^*\mathrm{L}^*\mathrm{mmol}^{-1}]$) produced by release of *p*NP upon hydrolysis of 0.4 mM *p*NPB in 25 mM sodium phosphate buffer at pH 7.0 and 30 °C. An international unit of activity (IU) is defined as the amount of enzyme that hydrolyzes 1 μmol of *p*NPB per minute.

4 Notes

1. For the preparation of DES, choline chloride and glycerol must be mixed at 80 °C until a clear solution is obtained.
2. Other DES can be used, but if glycerol is not employed, add another alcohol as substrate for the esterification.
3. When mixing BSA, with the CAL-B solution, shake gently to avoid foam formation.
4. The recommended ratio provided by the supplier is, for every 80 mL of PVA solution, 20 mL of CLEA-CALB solution.
5. The cross-linking step must be carried out at a low temperature to avoid inactivation of the enzyme.
6. Salt stabilizing solution: 3 g of salt stabilizing powder in 200 mL of distilled water.
7. The reaction media must be kept at 60 °C before and during its pumping.
8. To determine the residence time (τ), it is necessary to determine the void fraction (ϵ), which corresponds to the free volume of biocatalyst within the column:

$$\tau = \frac{\varepsilon\, V_\mathrm{t}}{q}$$

where ϵ denotes the void fraction, V_t is the total bed volume, and q is the substrates flow rate.

Acknowledgements

NG acknowledges the financial support from Fondecyt N° 11150215 (Chile).

References

1. De Gonzalo G, Domínguez de María P (eds) (2017) Biocatalysis, an industrial perspective. RCS Publishing, Cambridge
2. Guajardo N, Müller CR, Schrebler R, Carlesi C, Domínguez de María P (2016) Deep eutectic solvents for organocatalysis, biotransformations, and multistep organocatalyst/enzyme combinations. ChemCatChem 8: 1020–1027
3. Guajardo N, Domínguez de María P, Ahumada K, Schrebler RA, Ramírez-Tagle R, Crespo FA, Carlesi C (2017) Water as cosolvent: non-viscous deep eutectic solvents for efficient lipase catalyzed esterifications. ChemCatChem 9:1393–1396
4. Guajardo N, Domínguez de María P (2019) Continuous biocatalysis in environmentally-friendly media: a triple synergy for future sustainable processes. ChemCatChem 11:3128–3137
5. Sheldon RA (2011) Characteristic features and biotechnological applications of cross-linked enzyme aggregates (CLEAs). Appl Microbiol Biotechnol 92:467–477
6. Wilson L, Fernández-Lorente G, Fernández-Lafuente R, Illanes A, Guisán JM, Palomo JM (2006) CLEAs of lipases and poly-ionic polymers: a simple way of preparing stable biocatalysts with improved properties. Enzyme Microbiol Technol 39:750–755
7. Guajardo N, Ahumada K, Domínguez de María P, Schrebler RA (2019) Remarkable stability of *Candida antarctica* lipase B immobilized via cross-linking aggregates (CLEA) in deep eutectic solvents. Biocatal Biotransformation 37:106–114
8. Guajardo N, Ahumada K, Domínguez de María P (2020) Immobilized lipase-CLEA aggregates encapsulated in lentikats® as robust biocatalysts for continuous processes in deep eutectic solvents. J Biotechnol 310:97–102
9. Guajardo N, Schrebler RA, Domínguez de María P (2019) From fed-batch to continuous packed-bed reactors: lipase-catalyzed esterifications in low viscous deep-eutectic-solvents with buffer as cosolvent. Bioresour Technol 273: 320–325
10. Guajardo N, Domínguez de María P (2018) Lipases in green chemistry: deep eutectic solvents (DES) as new green solvents. In: Sandoval G (ed) Lipases and phospholipases, Methods in molecular biology, vol 1835. Springer, Cham, pp 351–357

Chapter 22

Enzymatic Photometric Assays for the Selective Detection of Halides

Qingyun Tang, Askin S. Aslan-Üzel, Eva D. Schuiten, Christoffel P. S. Badenhorst, Ioannis V. Pavlidis, and Uwe T. Bornscheuer

Abstract

Halides are substrates and products of a number of biotechnologically important enzymes like dehalogenases, halide methyltransferases, and halogenases. Therefore, the determination of halide concentrations in samples is important. The classical methods based on mercuric thiocyanate are very dangerous, produce hazardous waste, and do not discriminate between chloride, bromide, and iodide. In this chapter, we describe a detailed protocol for the determination of halide concentrations based on the haloperoxidase-catalyzed oxidation of halides. The resulting hypohalous acids are detected using commercially available colorimetric or fluorimetric probes. The biocatalytic nature of the assays allows them to be implemented in one-pot cascade reactions with halide-generating enzymes. Since chloride is ubiquitous in biological systems, we also describe convenient photometric assays for the selective detection of bromide and iodide in the presence of chloride, obviating the need for laborious dialyses to obtain halide-free enzymes and reagents.

Key words Chloride, Bromide, Iodide, HOX assay, Hypohalous acid, Haloalkane, Halide methyltransferase, Haloperoxidase, Aminophenyl fluorescein, 3,3′,5,5′-Tetramethylbenzidine

1 Introduction

Many important molecules that we use as pharmaceuticals, agrochemicals, detergents, pigments, and polymers in modern life are halogenated compounds [1]. The carbon–halogen bonds (C–F, C–Cl, C–Br, and C–I) in these molecules are important for their potency and efficiency. However, halogenated compounds are also used as solvents, degreasing agents, pesticides, and intermediates in industrial chemical syntheses [2, 3]. In the last few decades, the production of these halogenated compounds has increased drastically, leading to increased pollution with these environmentally persistent toxins. Thus, both the incorporation and removal of halide ions are important reactions in organic chemistry and

Haralambos Stamatis (ed.), *Multienzymatic Assemblies: Methods and Protocols*,
Methods in Molecular Biology, vol. 2487, https://doi.org/10.1007/978-1-0716-2269-8_22,
© The Author(s), under exclusive license to Springer Science+Business Media, LLC, part of Springer Nature 2022

biocatalysis. Halogenases, including *S*-adenosyl-L-methionine (SAM)-dependent fluorinases [4] and chlorinases [5], flavin-dependent brominases [6] and iodinases [7], non-heme iron halogenases, and Fe/α-ketoglutarate-dependent halogenases [8, 9], introduce carbon–halogen bonds into molecules. In contrast, dehalogenases [3] such as the haloalkane dehalogenase LinB from *Sphingomonas paucimobilis* [10] and the halide methyltransferases recently reported for SAM and SAM analogue synthesis [11] release halide ions as products. More enzymes that use halides as substrates or generate halides as products according to the BRENDA database were summarized by Aslan-Üzel et al. [12, 13]. Therefore, the detection and quantification of halides are useful for the discovery and engineering of enzymes that consume or produce halide ions.

One of the most common assays for halide ions is known as the Iwasaki assay, which can detect low halide concentrations as long as large samples are available [14]. However, in the commonly employed microtiter plate format, the detection limit of this assay ranges from about 30 μM to about 150 μM. The assay also employs mercuric thiocyanate, which is extremely toxic and fatal when accidentally inhaled or ingested. Furthermore, the waste must be carefully disposed of since it is very harmful to aquatic ecosystems and can contribute to the bioaccumulation of mercury. For these reasons, we developed the halide oxidation (HOX) assay for the detection of chloride, bromide, and iodide under mild and environmentally friendly conditions [13]. The detection limits of the HOX assay range from 1 μM for chloride and iodide to 20 nM for bromide, making the assay up to three orders of magnitude more sensitive than other halide detection assays currently available. Reaction volumes can be as low as 40 μL or less, so very little sample is required. One of our studies demonstrated that the HOX assay could be used to detect enzymatic dehalogenation and it was recently used to find promiscuous dehalogenase activity in an epoxide hydrolase [15]. The HOX assay was used to quantify dehalogenase activity with different haloalkane substrates and validated using a GC-MS method. The assay can therefore be used to identify and characterize novel dehalogenases and may be of interest to those studying other halide-producing enzymes.

In the original HOX assay, halides are first oxidized by the *Curvularia inaequalis* vanadium-dependent chloroperoxidase (*Ci*VCPO) to form hypohalous acids (Fig. 1a). The *Ci*VCPO was selected due to its high stability in the presence of hydrogen peroxide, compared to heme-dependent haloperoxidases. The resulting hypohalous acids HOCl, HOBr, and HOI are bleaches that react with probes that generate absorbance, fluorescence, or even luminescence [16]. For the HOX assay, aminophenyl fluorescein (APF) was chosen for the detection of hypohalous acids since it is stable in the presence of hydrogen peroxide and produces fluorescein, a

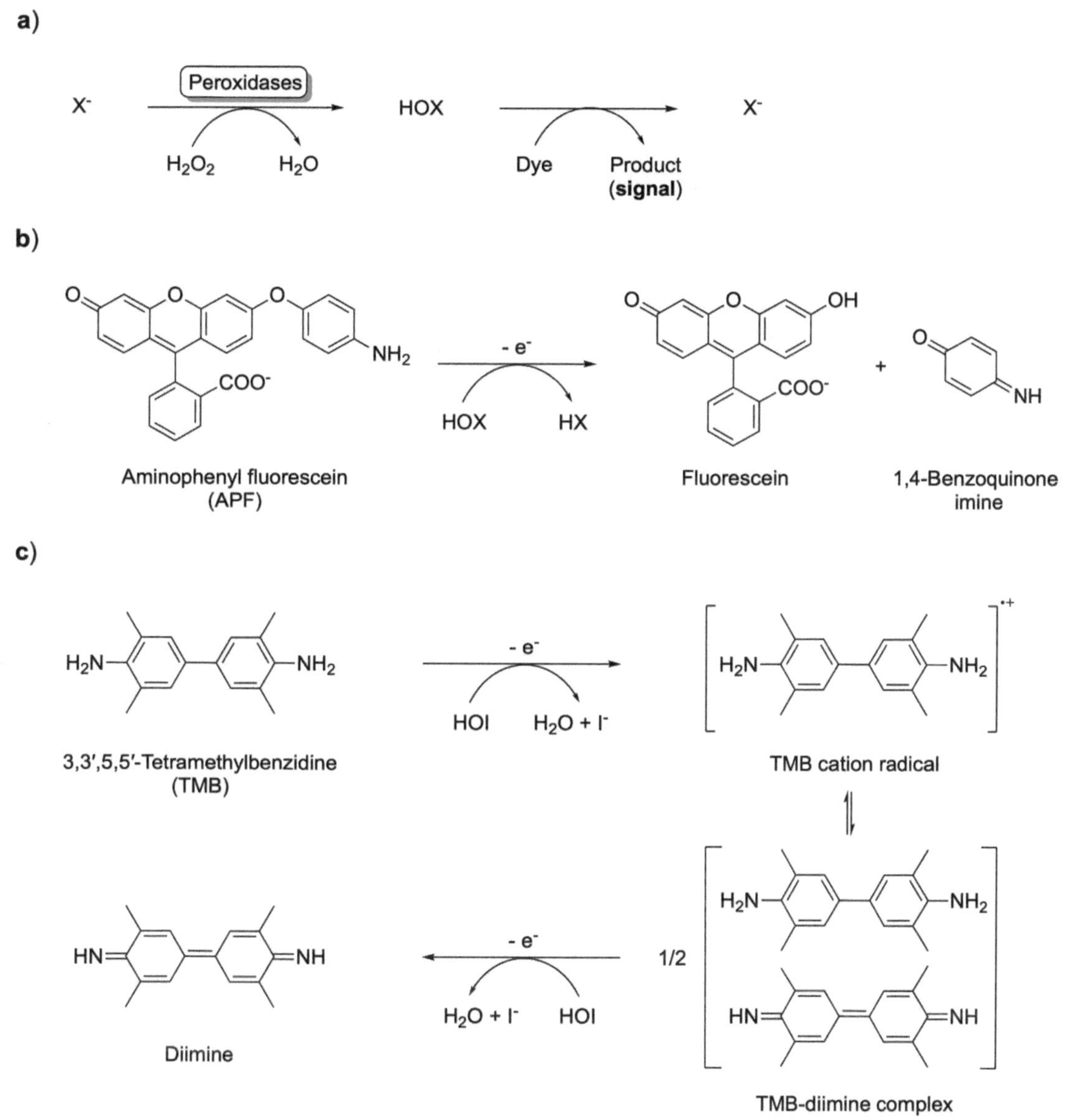

Fig. 1 Principles of the halide detection assays. (**a**) Halide ions are oxidized to hypohalous acids (HOX) using hydrogen peroxide (H_2O_2), catalyzed by a haloperoxidase. The hypohalous acids react with a probe to form compounds detectable by their colorimetric or fluorescent properties. (**b**) The principle behind the detection of halides by the original HOX assay. Oxidation of aminophenyl fluorescein (APF) by hypohalous acids results in the formation of fluorescein, a bright fluorescent dye that can be detected at nanomolar concentrations. APF can be used to detect chloride, bromide, and iodide ions. (**c**) The principle behind the selective detection of iodide using 3,3′,5,5′-tetramethylbenzidine (TMB). TMB is first converted by hypoiodous acid (HOI) to the one-electron oxidized cation radical which is in equilibrium with the TMB-diimine complex. The TMB-diimine complex is detectable at 570 nm and colors the reaction blue. The TMB-diimine is then converted by another molecule of hypoiodous acid (HOI) to the orange two-electron oxidized diamine product, which is detectable at 460 nm

bright fluorescent dye (excitation at 488 nm, emission at 525 nm) (Fig. 1b) [17]. The hypohalous acids oxidize the aminophenyl group of APF, which then releases 1,4-benzoquinone imine to form fluorescein. This assay is much less toxic than the Iwasaki assay. The most hazardous chemicals employed by the HOX assay are orthovanadate, used as a cofactor by vanadium-dependent haloperoxidases, and hydrogen peroxide, a common household bleach.

1.1 Selective Assays

We found that contamination of reagents and buffers with chloride is a common source of background signal, complicating the accurate quantification of halides and requiring enzymes and buffers to be dialyzed extensively. Therefore, we also developed variations of the HOX assay that could selectively detect bromide or iodide in the presence of chloride. This allows the enzymatic production of bromide and iodide to be detected even in the presence of high chloride concentrations, facilitating applications like screening recombinant enzyme variants in crude bacterial lysates without the need for desalting or dialysis.

The higher selectivity of the vanadium-dependent bromoperoxidase from *Corallina officinalis* (*Co*VBPO) for bromide enables selective bromide detection. Amines interfere with the original HOX assay since HOCl and HOBr react with amines to form haloamines. Bromamines are more reactive than chloramines toward HOX-sensors. Thus, amines like taurine can be used to suppress the signal from chloride, further enhancing the selectivity for bromide [17–19].

The selective iodide assay uses a different dye than the HOX assay for the specific detection of iodide in the presence of chloride. The hypohalous acid oxidizes the chromogen 3,3′,5,5′-tetramethylbenzidine (TMB) and generates visible colors, depending on the amount of bleach produced [20]. TMB is converted to the one-electron oxidized cation radical which is in equilibrium with the TMB-diimine complex. The TMB-diimine complex is detectable at 570 nm and colors the reaction blue. Eventually, the orange two-electron diamine product is formed, which is detectable at 460 nm [20]. This means that for lower concentrations of iodide, the assay turns blue and for higher concentrations the mixture turns purple, orange, then bronze, and finally black and the maximum absorption shifts from 570 nm to 460 nm [11]. In the process, the bleach HOI is reduced to iodide. Therefore, iodide is essentially used as a catalyst in the whole process, making the assay very sensitive. We measure production of the TMB-diimine complex at 570 nm and the rate of change ($v = A_{570\ \mathrm{nm}}/\mathrm{s}$) shows a linear relationship to iodide concentration from 5 μM to 400 μM (although nanomolar concentrations can be detected under optimized conditions). The linearity at 570 nm is lost once the diimine product of TMB is produced. Although TMB also reacts with

HOCl, the assay detects micromolar quantities of iodide but is relatively insensitive to even millimolar concentrations of chloride [11, 21]. Therefore, this assay can be used for detecting iodide in the presence of chloride-containing crude cell lysates, in vitro translation reactions, and whole-cell assays, and even for high-throughput screening. Apart from iodide, TMB can also be applied for the selective detection of bromide [21, 22].

2 Materials

Prepare all solutions using analytical grade reagents, analytical grade solvents, and Milli-Q water with a resistance of 18.2 MΩ·cm at 25 °C (or equivalent). Due to the high sensitivity of the general HOX assay towards chloride and the abundance of chloride in a biochemistry lab, extra care should be taken when preparing purified CiVCPO, buffers, and stock solutions for the HOX assay. Prepare and store all reagents at room temperature (unless indicated otherwise). Diligently follow all waste disposal regulations when disposing waste materials, especially when using hazardous substrates such as haloalkanes. All solutions with halogenated compounds must be discarded with halogenated organic waste according to local regulations. All buffers and solutions with orthovanadate should be disposed in the heavy metal waste. For any experiments involving bleach standards, extra care should be taken not to bleach surfaces and clothing and it is important to use appropriate personal protective equipment (lab coats, gloves, and safety glasses).

2.1 Peroxidases

The wild-type *Ci*VCPO (GenBank: CAA59686.1) is used for the general halide assay. A commercial *Co*VBPO from Sigma-Aldrich (Catalogue No.: B2170) is used for the selective bromide assay. This product contains 10 U of partially purified lyophilized powder of *Co*VBPO, containing MES buffer salts. According to the supplier, the chloroperoxidase activity of this bromoperoxidase is <1%, which enables selective detection of bromide in the presence of chloride. The *Ci*VCPO mutant P395D/L241V/T343A (referred to as the *Ci*VCPO triple mutant) was used for iodide assays [23]. However, the wild-type *Ci*VCPO or *Co*VBPO could also be used. *E. coli* TOP10 with pBAD vectors encoding *Ci*VCPO or the *Ci*VCPO triple mutant P395D/L241V/T343A were used for the expression of the two chloroperoxidase variants [23].

2.2 Cultivation and Purification of *Ci*VCPO and the Triple Mutant

1. Lysogeny Broth (LB medium): Dissolve 10 g tryptone, 5 g yeast extract, and 10 g sodium chloride (NaCl) in about 900 mL of distilled or deionized water and adjust the volume to 1 L. Sterilize by autoclaving for 20 min at 15 psi (1.05 kg/ cm^2) on liquid cycle.

2. Ampicillin stock solution (1000×, 100 mg/mL): Dissolve 1 g ampicillin in about 9 mL Milli-Q water and adjust volume to 10 mL. Filter the solution through a 0.22 μm sterile filter into sterile tubes and store at −20 °C.
3. L-Arabinose stock solution (500×, 10% w/w): Dissolve 10 g L-arabinose in about 90 mL Milli-Q water and adjust the volume to 100 mL. Filter the solution through a 0.22 μm sterile filter into sterile tubes and store at −20 °C.
4. Loading buffer for ion exchange chromatography (50 mM Tris sulfate, pH 8.1): Dissolve 6.06 g tris(hydroxymethyl)-aminomethane (Tris) in about 900 mL Milli-Q water. Adjust the pH with concentrated sulfuric acid (H_2SO_4) to pH 8.1 and adjust the volume to 1 L. Filter buffer through a 0.45 μm Corning filter and degas using vacuum or a sonicator.
5. Lysis buffer: Pour 50 mL loading buffer (*see* Subheading 2.2, **item 4**) into a 50 mL tube and add 100 mg lysozyme, 50 mg DNaseI and one EDTA-free protease inhibitor cocktail tablet (cOmplete™, Roche, Mannheim, Germany) to prepare a lysis buffer containing 2 mg/mL lysozyme, 1 mg/mL DNaseI and protease inhibitor cocktail. Precool lysis buffer at 4 °C or on ice before purification.
6. Elution buffer for ion exchange chromatography (50 mM Tris sulfate, 1 M NaCl, pH 8.1): Dissolve 6.06 g Tris and 58.44 g NaCl in about 900 mL Milli-Q water. Adjust the pH with concentrated sulfuric acid (H_2SO_4) to pH 8.1 and adjust the volume to 1 L. Filter the buffer through a 0.45 μm Corning filter and degas using vacuum or a sonicator.
7. Washing buffer for ion exchange chromatography (50 mM Tris sulfate, 100 mM NaCl, pH 8.1): For purification using a peristaltic pump, add 45 mL loading buffer (*see* Subheading 2.2, **item 4**) and 5 mL elution buffer (*see* Subheading 2.2, **item 6**) to a 50 mL tube and mix thoroughly. For purification on a fast protein liquid chromatography (FPLC) system, add 450 mL loading buffer and 50 mL elution buffer to a glass bottle and mix thoroughly. Filter buffer through a 0.45 μm Corning filter and degas using vacuum or a sonicator.
8. Dialysis buffer (50 mM sodium phosphate, 100 μM Na_3VO_4, pH 8.0; chloride-free) (*see* **Note 1**): Weigh 82.76 mg Na_3VO_4 (0.45 mmol) and put into a clean 5 L graduated beaker. Add 4.5 L 50 mM chloride-free sodium phosphate (pH 8.0) directly into the beaker according to the graduation scale or use a clean measuring cylinder. Put a clean magnetic stirring bar in the buffer and mix all components on a magnetic stirrer. Leave the magnet in the beaker for the dialysis step. Pour out a small amount of buffer (about 20 mL) into a 50 mL tube to confirm

buffer pH using a pH meter and adjust if necessary (*see* **Note 2**). Precool the dialysis buffer at 4 °C.

2.3 Halide Assay Reagents

1. 40 mM hydrogen peroxide solution: Add 4.0 μL 30% (w/w) hydrogen peroxide solution (1.110 g/mL, 9.79 M) to 996 μL Milli-Q water and mix thoroughly (*see* **Note 3**).
2. 5 mM Aminophenyl fluorescein (APF) solution (for the general HOX assay) (*see* **Note 4**): APF solutions are commercially available. The 5 mM solution supplied in dimethylformamide (DMF) (Sigma, Catalogue No.: A4108) is ready to use. For the solution supplied in methyl acetate (Cayman Chemicals, Item No.: 10157), the solvent needs to be removed before use. Completely evaporate methyl acetate from the bottle under a gentle stream of nitrogen. Before any APF is taken from the original bottle, add 472.4 μL DMF directly to the 1 mg APF and mix thoroughly to form a 5 mM APF stock solution. Store the APF solution at 4 °C. Alternatively, APF can also be synthesized from fluorescein and 4-fluoronitrobenzene as described in the literature [13]. Dissolve 4.23 mg APF in 2 mL DMF and mix thoroughly to form a 5 mM APF stock solution.
3. 500 mM Taurine solution (for the selective bromide assay): Dissolve 62.6 mg taurine in 1 mL Milli-Q water.
4. 3,3′,5,5′-tetramethylbenzidine (TMB) stock solution (for the selective iodide assay): A premixed solution of TMB and hydrogen peroxide in a mildly acidic buffer is commercially available (Liquid Substrate System for Membranes; Sigma-Aldrich; Catalogue No.: T0565). Although concentrations of components are not known, the solution can be used directly for the assay. Alternatively, prepare a 25 mM TMB stock solution by dissolving 6 mg TMB in 1 mL DMSO.
5. 25 mM Monochlorodimedone (MCD) solution (for determination of the activity of *Ci*VCPO and the triple mutant): Dissolve 44 mg MCD in 10 mL ethanol.
6. Reaction buffers (20 mM sodium phosphate, 1 mM Na_3VO_4; chloride-free) (*see* **Notes 1** and **4**): Weigh 183.9 mg Na_3VO_4 (1 mmol) into a 1 L graduated glass bottle. Add 1 L of 20 mM sodium phosphate buffer (pH 6.0). Seal the glass bottle with the lid and shake until all components are completely dissolved and mixed properly. Pour out a small amount of buffer (about 20 mL) into a 50 mL tube to confirm buffer pH using a pH meter (*see* **Note 2**). For the selective bromide assay, use 20 mM sodium phosphate, pH 6.5, for preparation of this solution.

Table 1
Preparation of 10 mL halide standard solutions from 10 mM stock solutions

Final concentration (μM)	0	20	50	100	200	300	400	500
10 mM standard solution (μL)	0	20	50	100	200	300	400	500
Milli-Q water (mL)	10.00	9.98	9.95	9.90	9.80	9.70	9.60	9.50

Table 2
Preparation of 50 mL bromide standard solutions from a 10 mM bromide stock solution

Final concentration (μM)	0	2	5	10	15	20
10 mM standard solution (μL)	0.00	10	25	50	75	100
Milli-Q water (mL)	50	49.99	49.98	49.95	49.93	49.90

2.4 Halide Standard Solutions

Halide standard solutions (10 mM): Weigh 29.2 mg NaCl or 37.3 mg KCl, 51.4 mg NaBr or 59.5 mg KBr, or 74.9 mg NaI or 83.0 mg KI directly in clean 50 mL tubes. Fill to 50 mL with Milli-Q water (*see* **Note 5**). Dilute stock solutions to give 10 mL standard solutions of 0–500 μM in 15 mL tubes and mix thoroughly (Table 1).

For the bromide assay, also dilute the 10 mM bromide stock solution to standard solutions of 0–20 μM in 50 mL tubes and mix thoroughly (Table 2).

2.5 Reaction Mixtures for the Assays

1. Reaction mixture for haloperoxidase activity: In 451 μL of the pH 6.0 reaction buffer (*see* Subheading 2.3), add 1 μL of 25 mM MCD solution, 132 μL of 40 mM hydrogen peroxide solution, 6 μL of 10 mM bromide standard solution, and mix thoroughly.
2. Reaction mixture for general halide assay: Add 50 μL 40 mM hydrogen peroxide solution, 5 μL of 5 mM APF, and 50 μL of 50 U/mL *Ci*VCPO to 895 μL of reaction buffer (pH 6.0) in a centrifuge tube and mix thoroughly (*see* **Notes 6** and 7).
3. Bromide assay solution: Add 50 μL 40 mM hydrogen peroxide solution, 5 μL 5 mM APF, and 50 μL 6 U/mL *Co*VBPO to 845 μL of reaction buffer (pH 6.5) in a centrifuge tube (*see* **Note** 7). Add 50 μL of 500 mM taurine to the assay solution in order to suppress the chloride signal and mix all components thoroughly (*see* **Note 8**).
4. Iodide assay solution: Mix 10 μL *Ci*VCPO triple mutant (137 U/mL) with 990 μL commercial TMB/hydrogen peroxide mixture solution (*see* **Notes** 7 and **9**). If the commercial TMB solution is not available, add 50 μL 40 mM hydrogen peroxide solution, 10 μL 25 mM TMB solution, and 10 μL

*Ci*VCPO triple mutant (137 U/mL) to 930 μL of reaction buffer (pH 6.0).

2.6 Consumables and Measuring Devices

Black 384-well microtiter plates are used to perform the fluorescence-based halide assays and transparent 384-well polystyrene microtiter plates are used for the iodide assays. Standard reaction volumes are 40 μL and 50 μL, respectively, but may be increased or decreased as desired. Black 96-well microtiter plates and transparent 96-well polystyrene microtiter plates can be used for higher reaction volumes, for example, 100 or 200 μL. The volume of each component should be adjusted accordingly. Fluorescence measurements are performed with plate readers that excite at 488 nm and measure emission at 525 nm. Plate readers that can measure absorbance at 570 nm are used for the iodide assay. Spectrophotometers that accept a 1 cm-pathlength quartz cuvette and measure at 290 nm are used for the MCD assay for determination of the activities of *Ci*VCPO and the triple mutant.

3 Methods

3.1 Cultivation and Purification of *Ci*VCPO and the Triple Mutant

1. Add 10 μL ampicillin stock solution (1000×) to 10 mL LB medium, for a final concentration of 100 μg/mL ampicillin, in a sterile 50 mL tube. Inoculate with a single colony of *E. coli* TOP10 transformed with the pBAD-*Ci*VCPO or the pBAD-*Ci*VCPO triple mutant plasmid. Incubate the 50 mL tube overnight at 37 °C, shaking at 180 rpm.
2. Add 1 mL ampicillin stock solution (1000×) to 1 L LB medium, for a final concentration of 100 μg/mL ampicillin. Inoculate the LB medium with the overnight culture. Incubate at 37 °C, shaking at 180 rpm. Monitor the OD_{600} of the culture until it reaches about 0.5. Then add 2 mL L-arabinose stock solution for a final concentration of 0.02% to induce expression of *Ci*VCPO. Incubate the culture at 25 °C, shaking at 180 rpm, for 24 h.
3. Harvest cells by centrifugation at 4,500 × *g* for 30 min in a precooled (4 °C) centrifuge. Resuspend cells in precooled lysis buffer (1 mL per gram of cells). Lyse cells on ice by sonication (three cycles of 5 min, 50% pulse, and 50% power) using a SONOPULS HD 2070 (BANDELIN electronic GmbH & Co. KG, Berlin, Germany) or an equivalent device. Centrifuge the crude cell lysate at 10,000 × *g* at 4 °C for 1 h for clarification. Transfer the clarified lysate to a 200 mL beaker.
4. Measure the volume of the clarified lysate. Add an equal volume of isopropanol, mix thoroughly and incubate on ice for 20 min. Centrifuge the lysate-isopropanol mixture at 10,000 × *g* at 4 °C for 30 min to remove precipitated proteins. Transfer the supernatant to clean tubes.

5. Equilibrate the 5 mL DEAE Sephacel column with 25 mL loading buffer (5 column volumes). Load the cell lysate on the DEAE Sephacel column. Wash the column with 25 mL loading buffer and then with 25 mL washing buffer to remove weakly bound proteins. Elute *Ci*VCPO from the column using 30 mL elution buffer and collect the eluate in fractions of 1–1.5 mL. Confirm presence and purity of the haloperoxidase by SDS-PAGE (*see* **Note 10**) and combine the fractions containing the haloperoxidase.
6. In order to obtain the chloride-free, active vanadium-bound *Ci*VCPO, dialyze the *Ci*VCPO eluate three times against the chloride-free, orthovanadate-containing dialysis buffer. After moistening the dialysis membrane (10 kDa MWCO) with some dialysis buffer, fold one end of the membrane twice and clamp it with clips. Transfer all *Ci*VCPO eluate carefully into the dialysis membrane, fold the open end twice and clamp (*see* **Note 11**). Dialyze the *Ci*VCPO eluate against 4.5 L precooled dialysis buffer on a magnetic stirrer in a cold room for 4 h. Replace the dialysis buffer and dialyze the eluate for another 4 h. Afterwards, replace the dialysis buffer and dialyze the eluate overnight (*see* **Note 12**). Aliquot the chloride-free, active vanadium-bound *Ci*VCPO into centrifuge tubes and store at 4 °C (*see* **Notes 13** and **14**). Measure the absorbance of *Ci*VCPO at 280 nm using a NanoDrop™ (Thermo Fisher, Hennigsdorf, Germany) or an equivalent device. Calculate the protein concentration of *Ci*VCPO (*see* **Note 15**).

3.2 Preparation of CoVBPO

The commercial *Co*VBPO described in Subheading 2.1 was used for selective bromide detection in the presence of chloride. Dissolve the entire vial of *Co*VBPO by adding 400 μL dialysis buffer. Transfer the solution to a dialysis device fit for small volumes like the Thermo Scientific Slide-A-Lyzer MINI Dialysis Devices (10 kDa MWCO, Catalogue No.: 69570). Dialyze it against 800 mL dialysis buffer for 1 h for three times to remove MES buffer and other salts and obtain the active vanadium-bound *Co*VBPO. Transfer the *Co*VBPO to a 2 mL tube and adjust the volume to 1.67 mL with the dialysis buffer to obtain a 6 U/mL *Co*VBPO stock solution. Store at 4 °C.

3.3 Determination of Haloperoxidase Activities

Specific activities of *Ci*VCPO and the triple mutant are determined using the MCD assay. Bromination of MCD results in a decrease in absorbance at 290 nm ($\Delta\varepsilon = 19{,}900\ M^{-1}cm^{-1}$). Add the haloperoxidase activity reaction mixture directly into a 1 cm-pathlength quartz cuvette and mix all components by pipetting. Final concentrations of MCD, hydrogen peroxide, and bromide in the reaction mixture are 42 μM, 8.8 mM, and 100 μM, respectively. Measurement of the brominating activity of *Ci*VCPO is initiated by adding

10 μL purified *Ci*VCPO or the triple mutant. The decrease in absorbance at 290 nm ($\Delta A_{290\ \mathrm{nm}}$) is monitored in a spectrophotometer. One unit of *Ci*VCPO or the triple mutant produces 1 μmol of brominated MCD per minute (*see* **Note 16**).

3.4 General Halide Assay (HOX Assay)

1. Add 38 μL assay solution per well of a black 384-well microtiter plate, then add 2 μL sample or the halide standard solutions (*see* **Notes 17** and **18**). Final concentrations of hydrogen peroxide, APF, and *Ci*VCPO in the reaction mixture are 2 mM, 25 μM, and 2.5 U/mL, respectively. Mix reactions thoroughly by pipetting or shaking on a microplate shaker (2.2 mm amplitude) at 600–800 rpm for 10 s. Cover the plate with a microtiter plate sealing film to avoid contamination and evaporation. Incubate the microtiter plate at room temperature for 60 min in the dark.
2. After incubation, measure the fluorescence of the reactions at 525 nm with excitation at 488 nm (*see* **Note 19**). For determination of the halide concentrations in samples, first obtain the standard curve of the corresponding halide by plotting the fluorescence signal against the concentrations of halide standards (*see* **Note 20**). Then calculate the halide concentrations in samples by comparing to the corresponding halide standard curve (*see* **Note 21**).

3.5 Selective Bromide Detection in the Presence of Chloride

This assay is performed at pH 6.5 due to the different pH optimum of the *Co*VBPO [24]. Add 38 μL bromide assay solution per well of a black 384-well microtiter plate, then add 2 μL sample or the bromide standard solutions. Final concentrations of hydrogen peroxide, APF, and *Co*VBPO in the reaction mixture are 2 mM, 25 μM, and 0.3 U/mL, respectively. Then follow the mixing, incubation and measurements in **steps 1–2** described in Subheading 3.4 (*see* **Note 22**).

3.6 Iodide Detection

1. Add 50 μL iodide assay solution to the wells of a transparent 384-well polystyrene microtiter plate. Final concentrations of hydrogen peroxide, TMB, and *Ci*VCPO in the reaction mixture are 2 mM, 250 μM, and 1.37 U/mL, respectively. Then add 1 μL sample or the iodide standards to the assay solution. Mix reactions thoroughly by pipetting or shaking on a microplate shaker (2.2 mm amplitude) at 800 rpm for 10 s.
2. Immediately measure sample absorbance at 570 nm every 10 s for at least 2 min. Calculate the initial rate of increase in absorbance at 570 nm ($v = A_{570\ \mathrm{nm}}/s$) (*see* **Note 23**). For determination of the iodide concentrations in samples, first obtain the standard curve of iodide by plotting the $A_{570\ \mathrm{nm}}/s$ against the concentrations of iodide standards (*see* **Note 24**).

Then calculate the iodide concentrations in samples by comparison to the iodide standard curve (*see* **Note 19**).

3. For high-throughput screening, repeat **step 1** in Subheading 3.6 and incubate reactions at room temperature for at least 30 min. By comparing the colors of sample reactions to the iodide standard reactions, iodide concentrations of samples can be estimated (*see* **Note 25**).

4 Notes

1. Try to use fresh buffer components and keep them only for the halide assays to avoid contamination with chloride. Rinse all glassware (glass bottles, measuring cylinders) and reusable utensils (spatulas and magnetic stirring bars) at least three times with Milli-Q water in addition to normal cleaning before use.
2. Avoid putting pH probes directly into the chloride-free buffers. Since pH probes are stored in 3 M KCl, chloride contamination from the electrode is highly likely. Thus, check the pH of chloride-free buffers using an aliquot that will be discarded after verification of the pH.
3. Prepare fresh 40 mM hydrogen peroxide solution prior to the assay.
4. Aliquot assay components after preparation. This prevents contamination of the reagent stocks.
5. Iodide standard solutions can be kept at room temperature for a few months. Discard when they turn yellow.
6. To avoid accidental contamination by chloride, use pre-packed (filter) pipette tips for preparing assay solutions and during the assay.
7. Prepare fresh assay master mix prior to the assay. We recommend using 250 M of TMB rather than 25 mM as described by Tang et al [11].
8. Taurine is added as a scavenger of HOCl. It forms taurine chloramine when reacting with HOCl. Chloramines are less reactive than bromamines, allowing HOBr to be detected selectively.
9. Prepare assay solutions in brown tubes and keep out of direct sunlight.
10. Expected size of *Ci*VCPO and the triple mutant is 67.5 kDa.
11. Wear clean gloves when handling the dialysis membrane to avoid introduction of chloride from sweat on the hands.

12. It is advisable to dialyze the enzyme after purification instead of using PD10 desalting columns. At 98% desalting capacity, PD10 columns do not remove enough of the salts used in purification. In contrast, three rounds of dialysis (100× sample volume, >4 h per round) will be sufficient.
13. Make sure to dialyze the purified *Ci*VCPO and the *Ci*VCPO triple mutant eluate three times to completely remove chloride and Tris since the HOX assay is highly sensitive to residues of chloride, and amines can inhibit the HOX assay.
14. *Ci*VCPO and the triple mutant are highly stable and can be stored in the fridge for extended periods.
15.

$$\text{Concentration of } Ci\text{VCPO or mutant (mg/mL)} = \frac{\text{Absorbance at 280 nm}}{\text{Extinction coefficient} \times \text{Pathlength}} \times \text{Molecular weight} = \frac{\text{Absorbance at 280 nm}}{91915\ \text{M}^{-1}\text{cm}^{-1} \times 1\ \text{cm}} \times 67546\ \text{g/mol};$$

16.

$$\text{Specific activity of } Ci\text{VCPO or mutant } (\mu\text{mol}/(\text{mg min})) = \frac{\Delta A_{290\ \text{nm}} \times \text{Reaction volume}}{\text{Extinction coefficient of MCD} \times \text{Pathlength} \times \text{Enzyme volume} \times \text{Enzyme concentration}} = \frac{\Delta A_{290\ \text{nm}}(\text{min}^{-1}) \times 600\ \mu\text{L}}{19{,}900\ \text{M}^{-1}\text{cm}^{-1} \times 1\ \text{cm} \times 0.01\ \text{mL} \times \text{Enzyme concentration (mg/mL)}}$$

17. After preparation of the assay master mix, first fill the sample volume by Milli-Q water and measure the background signal as a negative control. If the background signal is too high, check all components for chloride contamination and discard any contaminated aliquots.
18. Space out samples on the assay plate to avoid cross-contamination by, for example, accidentally pipetting into the wrong well (especially with 384-well plates).
19. Perform each assay in triplicate to increase accuracy of the assay. The amount of *Ci*VCPO and triple mutant used in the assays can be adjusted according to different requirements. However, make sure to use the same master mix for the standard curves and samples.
20. Fluorescence signal of the HOX assay is linear to halide ions in the following ranges: 0–25 μM for chloride, 0–1000 nM for bromide, and 0–10 μM for iodide [13].

21. Avoid using buffer reagents with amino groups in the samples. Amines can inhibit the HOX assay.
22. Buffer reagents with amino groups can be used in the samples in this assay.
23. Calculate the initial rate of change in absorbance at 570 nm ($v = A_{570\ \mathrm{nm}}/\mathrm{s}$) using the first few measured absorbance values that are linearly proportional to the measurement time. The linearity will be lost when the diimine product of TMB starts forming.
24. Initial rate of change in absorbance at 570 nm ($v = A_{570\ \mathrm{nm}}/\mathrm{s}$) of the selective iodide assay is linear to iodide standards ranging from 0 μM to 400 μM [11]. When using self-prepared TMB and hydrogen peroxide mixture in reaction buffer, the range of iodide detection can be different.
25. Reaction mixtures will turn blue, purple, bronze, orange, and eventually black with increasing iodide concentration and incubation time [11].

Acknowledgement

Q. T. thanks the China Scholarship Council for financial support of her PhD thesis project (File No.: 201606150073). A. S. A.-Ü. and E. D. S. thank the European Union (722610 ES-CAT).

References

1. Menon BRK, Richmond D, Menon N (2020) Halogenases for biosynthetic pathway engineering: toward new routes to naturals and non-naturals. Catal Rev:1–59. https://doi.org/10.1080/01614940.2020.1823788
2. Janssen DB, Stucki G (2020) Perspectives of genetically engineered microbes for groundwater bioremediation. Environ Sci Processes Impacts 22(3):487–499. https://doi.org/10.1039/c9em00601j
3. Koudelakova T, Bidmanova S, Dvorak P, Pavelka A, Chaloupkova R, Prokop Z, Damborsky J (2013) Haloalkane dehalogenases: biotechnological applications. Biotechnol J 8(1):32–45. https://doi.org/10.1002/biot.201100486
4. Dong CJ, Huang FL, Deng H, Schaffrath C, Spencer JB, O'Hagan D, Naismith JH (2004) Crystal structure and mechanism of a bacterial fluorinating enzyme. Nature 427(6974): 561–565. https://doi.org/10.1038/nature02280
5. Eustaquio AS, Pojer F, Noel JP, Moore BS (2008) Discovery and characterization of a marine bacterial SAM-dependent chlorinase. Nat Chem Biol 4(1):69–74. https://doi.org/10.1038/nchembio.2007.56
6. Neubauer PR, Widmann C, Wibberg D, Schroder L, Frese M, Kottke T, Kalinowski J, Niemann HH, Sewald N (2018) A flavin-dependent halogenase from metagenomic analysis prefers bromination over chlorination. PLoS One 13(5). https://doi.org/10.1371/journal.pone.0196797
7. Gkotsi DS, Ludewig H, Sharma SV, Connolly JA, Dhaliwal J, Wang YP, Unsworth WP, Taylor RJK, McLachlan MMW, Shanahan S, Naismith JH, Goss RJM (2019) A marine viral halogenase that iodinates diverse substrates. Nat Chem 11(12):1091–1097. https://doi.org/10.1038/s41557-019-0349-z
8. Hayashi T, Ligibel M, Sager E, Voss M, Hunziker J, Schroer K, Snajdrova R, Buller R (2019) Evolved aliphatic halogenases enable regiocomplementary C-H functionalization of a pharmaceutically relevant compound. Angew

Chem Int Ed 58(51):18535–18539. https://doi.org/10.1002/anie.201907245

9. Voss M, Malca SH, Buller R (2020) Exploring the biocatalytic potential of Fe/alpha-ketoglutarate-dependent halogenases. Chem Eur J 26(33):7336–7345. https://doi.org/10.1002/chem.201905752
10. Marek J, Vevodova J, Smatanova IK, Nagata Y, Svensson LA, Newman J, Takagi M, Damborsky J (2000) Crystal structure of the haloalkane dehalogenase from *Sphingomonas paucimobilis* UT26. Biochemistry 39(46):14082–14086. https://doi.org/10.1021/bi001539c
11. Tang Q, Grathwol CW, Aslan-Üzel AS, Wu S, Link A, Pavlidis IV, Badenhorst CPS, Bornscheuer UT (2021) Directed evolution of a halide methyltransferase enables biocatalytic synthesis of diverse SAM analogs. Angew Chem Int Ed 60(3):1524–1527. https://doi.org/10.1002/anie.202013871
12. Jeske L, Placzek S, Schomburg I, Chang A, Schomburg D (2019) BRENDA in 2019: a European ELIXIR core data resource. Nucleic Acids Res 47(D1):D542–D549. https://doi.org/10.1093/nar/gky1048
13. Aslan-Üzel AS, Beier A, Kovar D, Cziegler C, Padhi SK, Schuiten ED, Dorr M, Bottcher D, Hollmann F, Rudroff F, Mihovilovic MD, Buryska T, Damborsky J, Prokop Z, Badenhorst CPS, Bornscheuer UT (2020) An ultrasensitive fluorescence assay for the detection of halides and enzymatic dehalogenation. ChemCatChem 12(7):2032–2039. https://doi.org/10.1002/cctc.201901891
14. Iwasaki I, Utsumi S, Ozawa T (1952) New colorimetric determination of chloride using mercuric thiocyanate and ferric ion. Bull Chem Soc Jpn 25(3):226–226. https://doi.org/10.1246/bcsj.25.226
15. Schuiten ED, Badenhorst CPS, Palm GJ, Berndt L, Lammers M, Mican J, Bednar D, Damborsky J, Bornscheuer UT (2021) Promiscuous dehalogenase activity of the epoxide hydrolase CorEH from *Corynebacterium sp.* C12. ACS Catal 11(10):6113–6120. https://doi.org/10.1021/acscatal.1c00851
16. Brestel EP (1985) Co-oxidation of luminol by hypochlorite and hydrogen peroxide implications for neutrophil chemiluminescence. Biochem Biophys Res Commun 126(1):482–488. https://doi.org/10.1016/0006-291X(85)90631-X
17. Flemmig J, Remmler J, Zschaler J, Arnhold J (2015) Detection of the halogenating activity of heme peroxidases in leukocytes by aminophenyl fluorescein. Free Radic Res 49(6):768–776. https://doi.org/10.3109/10715762.2014.999676
18. Chen XQ, Lee KA, Ren XT, Ryu JC, Kim G, Ryu JH, Lee WJ, Yoon J (2016) Synthesis of a highly HOCl-selective fluorescent probe and its use for imaging HOCl in cells and organisms. Nat Protoc 11(7):1219–1228. https://doi.org/10.1038/nprot.2016.062
19. Sokolov AV, Kostevich VA, Kozlov SO, Donskyi IS, Vlasova II, Rudenko AO, Zakharova ET, Vasilyev VB, Panasenko OM (2015) Kinetic method for assaying the halogenating activity of myeloperoxidase based on reaction of celestine blue B with taurine halogenamines. Free Radic Res 49(6):777–789. https://doi.org/10.3109/10715762.2015.1017478
20. Josephy PD, Eling T, Mason RP (1982) The horseradish peroxidase-catalyzed oxidation of 3, 5, 3', 5'-tetramethylbenzidine. Free radical and charge-transfer complex intermediates. J Biol Chem 257(7):3669–3675. https://doi.org/10.1016/S0021-9258(18)34832-4
21. Bozeman PM, Learn DB, Thomas EL (1990) Assay of the human leukocyte enzymes myeloperoxidase and eosinophil peroxidase. J Immunol Methods 126(1):125–133. https://doi.org/10.1016/0022-1759(90)90020-v
22. Senthilmohan R, Kettle AJ (2006) Bromination and chlorination reactions of myeloperoxidase at physiological concentrations of bromide and chloride. Arch Biochem Biophys 445(2):235–244. https://doi.org/10.1016/j.abb.2005.07.005
23. Hasan Z, Renirie R, Kerkman R, Ruijssenaars HJ, Hartog AF, Wever R (2006) Laboratory-evolved vanadium chloroperoxidase exhibits 100-fold higher halogenating activity at alkaline pH: catalytic effects from first and second coordination sphere mutations. J Biol Chem 281(14):9738–9744. https://doi.org/10.1074/jbc.M512166200
24. Carter JN, Beatty KE, Simpson MT, Butler A (2002) Reactivity of recombinant and mutant vanadium bromoperoxidase from the red alga *Corallina officinalis*. J Inorg Biochem 91(1):59–69. https://doi.org/10.1016/S0162-0134(02)00400-2

Index

Haralambos Stamatis (ed.), *Multienzymatic Assemblies: Methods and Protocols*,
Methods in Molecular Biology, vol. 2487, https://doi.org/10.1007/978-1-0716-2269-8,

© The Editor(s) (if applicable) and The Author(s), under exclusive license to Springer Science+Business Media, LLC, part of Springer Nature 2022

FSC
www.fsc.org
MIX
Papier aus verantwortungsvollen Quellen
Paper from responsible sources
FSC® C105338

If you have any concerns about our products,
you can contact us on
ProductSafety@springernature.com

In case Publisher is established outside the EU,
the EU authorized representative is:
Springer Nature Customer Service Center GmbH
Europaplatz 3, 69115 Heidelberg, Germany

Printed by Libri Plureos GmbH
in Hamburg, Germany